Lecture Notes of the Institute for Computer Sciences, Social Informatics and Telecommunications Engineering   126

Kristin Glass   Richard Colbaugh
Paul Ormerod   Jeffrey Tsao (Eds.)

Complex Sciences

Second International Conference
COMPLEX 2012
Santa Fe, NM, USA, December 5-7, 2012
Revised Selected Papers

 Springer

Volume Editors

Kristin Glass
Sandia National Laboratories
Santa Fe, NM, USA
E-mail: klglass@sandia.gov

Richard Colbaugh
Sandia National Laboratories
Santa Fe, NM, USA
E-mail: colbaugh@comcast.net

Paul Ormerod
Volterra Partners, London, UK
E-mail: pormerod@volterra.co.uk

Jeffrey Tsao
Sandia National Laboratories
Albuquerque, NM, USA
E-mail: jytsao@sandia.gov

ISSN 1867-8211   e-ISSN 1867-822X
ISBN 978-3-319-03472-0   e-ISBN 978-3-319-03473-7
DOI 10.1007/978-3-319-03473-7
Springer Cham Heidelberg New York Dordrecht London

Library of Congress Control Number: 2013951876

CR Subject Classification (1998): I.6, I.2, K.4, F.1.3, F.2, G.2, H.3

Typesetting: Camera-ready by author, data conversion by Scientific Publishing Services, Chennai, India

Printed on acid-free paper

Springer is part of Springer Science+Business Media (www.springer.com)

Preface

Welcome to Santa Fe for the 2nd International Conference on Complex Sciences: Theory and Applications. The EAI COMPLEX conference series started in 2009 in Shanghai, People's Republic of China, and the conference proceedings have been published by the Springer Lecture Notes in Computer Science, Social Informatics, and Telecommunications Engineering (LNICST) series. We are delighted to have your participation in this, the second installment of the series, and hope your time in "The Land of Enchantment" will be stimulating and rewarding.

Much has been learned about the structure and function of complex systems during the past three decades. This research has informed numerous individual scientific fields in important ways, and has led to maturing the understanding of the extent to which complex systems from diverse domains share common underlying properties and characteristics. The recent availability of innovative measurement techniques (e.g., in biology) and new modes of social interaction (e.g., via social media) are enabling vast data sets to be compiled, and with the emergence of these data comes the promise that progress in complex systems research will further accelerate. COMPLEX 2012 aims to provide a stimulating and trans-disciplinary forum that permits researchers and practitioners in all areas of complex systems science to exchange ideas and present their latest results. This year's conference theme has drawn much interest and we are very grateful to all who submitted papers and to our keynote speakers Geoffrey West and Paul Ormerod for delivering fascinating and thought-provoking lectures.

The conference would not have been a success without the help of many people, and we would like to acknowledge their contributions. First, we would like to thank all the authors for their excellent submissions to COMPLEX 2012. We also express our most sincere appreciation to our Technical Program Committee co-chairs Kristin Glass, Paul Ormerod, and Jeffrey Tsao, and Technical Program Committee members Nancy Brodsky, Michael Gabbay, Gil Gallegos, Curtis Johnson, Marshall Kuypers, Hilton Root, and Antonio Sanfilippo. Special thanks to Erica Polini from the European Alliance for Innovation for her invaluable support with venue arrangements, finances, and registration. Many thanks to our local arrangements chair Kristin Glass, and our student supporters from the New Mexico Highlands University and New Mexico Institute of Mining and Technology. We are grateful to the European Alliance for Innovation, the Institute for Computer Sciences, Social Informatics, and Telecommunications

Engineering, CREATE-NET, and Sandia National Laboratories for sponsorship and financial support. Finally, we wish to express our sincere gratitude to Kristin Glass, the COMPLEX 2012 publication chair, for making this proceedings a reality.

September 2012   Richard Colbaugh
Tobias Preis
Gene Stanley

Organization

General Co-chairs

Richard Colbaugh   Sandia National Laboratories, USA
Tobias Preis   Warwick Business School, UK
Gene Stanley   Boston University, USA

Steering Committee

Francesco De Pellegrini   CREATE-NET, Italy
Gaoxi Xiao   Nanyang Technological University, Singapore

Technical Program Committee Co-chairs

Richard Colbaugh   Sandia National Laboratories, USA
Kristin Glass   Sandia National Laboratories, USA
Paul Ormerod   Volterra Partners and University of Durham, UK
Jeffrey Tsao   Sandia National Laboratories, USA

Technical Program Committee

Nancy Brodsky   Sandia National Laboratories, USA
Michael Gabbay   University of Washington, USA
Gil Gallegos   New Mexico Highlands University, USA
Curtis Johnson   Sandia National Laboratories, USA
Marshall Kuypers   Sandia National Laboratories, USA
Hilton Root   George Mason University, USA
Antonio Sanfilippo   Pacific Northwest National Laboratory, USA

Publication Chair

Kristin Glass   Sandia National Laboratories, USA

Conference Organizer

Erica Polini   European Alliance for Innovation, Italy

Conference Sponsors

European Alliance for Innovation, Italy
ICST: Institute for Computer Sciences, Social Informatics and
    Telecommunications Engineering
CREATE-NET, Italy
Sandia National Laboratories, USA

Local Arrangements Chair

Kristin Glass   Sandia National Laboratories, USA

Table of Contents

Regular Papers

The Crossover Point:
Comparing Policies to Mitigate Disruptions

Matthew Antognoli[1, 2], Marshall A. Kuypers[1, 3], Z. Rowan Copley[4],
Walter E. Beyeler[1], Michael D. Mitchell[1], and Robert J. Glass[1]

[1] Complex Adaptive System of Systems (CASoS) Engineering
Sandia National Laboratories, Albuquerque, New Mexico, USA
{mantogn,mkuyper,webeyel,micmitc,rjglass}@sandia.gov
http://www.sandia.gov/CasosEngineering/
[2] School of Engineering, University of New Mexico
Albuquerque, New Mexico, USA
[3] Management Science and Engineering Department, Stanford University
Stanford, California, USA
[4] St. Mary's College of Maryland
St. Mary's City, Maryland

Abstract. Companies, industries, and nations often consume resources supplied
by unstable producers. Perturbations that affect the supplier propagate
downstream to create volatility in resource prices. Consumers can invest to
reduce this insecurity in two ways; invest in and impose security on the
suppliers, or can invest in self-sufficiency so that shocks no longer present
devastating consequences. We use an agent-based model of a complex adaptive
system to examine this tradeoff between projecting security and investing in
self-sufficiency. This study finds that the significance of tradeoffs correlates
with the dependence of the consumer on the supplier.

Keywords: agent-based modeling, complex systems, projecting security, self-
sufficiency, trade.

1 Introduction

Unstable nations often control critical resources. Iraq and Nigeria are large oil
exporters and the Democratic Republic of the Congo contains reserves of cobalt, gold
and copper [1]. Other unstable regions contain deposits of heavy metals used in
computer chip manufacturing. Trade agreements between an economically-stable
nation and an unstable nation are risky. Instability in producing countries can cause
disruptions in supply leading to volatility in resource prices. Stable countries have an
interest in encouraging the stability of their trading partners. The United States has
historically maintained a military presence in the Middle East to secure the oil fields
of its trading partners. Recently, nations are giving more consideration to another
option for securing resources. Consumer countries are investing in technologies to
develop more efficient native production of critical resources, reducing the need to
import from unstable countries. The U.S. is investing in alternative sources of energy,

K. Glass et al. (Eds.): COMPLEX 2012, LNICST 126, pp. 1–10, 2013.

and photovoltaic manufacturers are studying new materials that could replace rare heavy metals. Analyzing the costs and benefits associated with protecting unstable suppliers and those with investing in self-sufficiency can aid in designing effective policies that lower resource cost and increase resource security.

We present an agent-based model that represents resource exchanges among nations. We initialize three nations. One nation produces a surplus of oil and the other two have deficits of oil. The oil supplier is subject to perturbations which prevent the consumer nations from obtaining sufficient oil for their needs. One of the consumer nations has the ability to provide security for the supplier for a certain cost. Alternatively, the nation can invest in technology which slowly lowers their dependence on oil. We vary the amount this 'Policy Maker' chooses to invest in technological and/or military measures in a variety of resource distributions.

2   Model Overview

The Exchange Model (ExM) is an agent-based model developed at Sandia National Laboratories that represents interacting nations which exchange resources [2]. Each nation is comprised of sectors and markets. A sector is comprised of a collection of agents (entities) that consume particular resources to produce other unique resources. Each resource a sector produces is consumed by entities belonging to other sectors. Each entity's viability is based directly on consumption of resources and indirectly on production of resources. In this way, a hierarchical co-dependence among interacting entities, sectors, and nations is formed.

We use a metric, 'health,' which is a measurement of an entity's ability to survive. An entity's health has a nominal level that, if maintained, allows continued homeostasis. An entity's health fluctuates based on its ability to consume particular resources at a specific rate. If the entity is not able to consume resources at this specific rate, its health will deteriorate. Alternately, if the entity exceeds this rate of consumption it will become healthier than its nominal level.

Entities exchange resources through markets, which may be international (meaning entities from all nations can use the market to trade resources) or domestic (which prohibits entities from trading unless they are members of the nation in which the market resides). Entities establish an offer and bid price for the resources being traded based on their health, money, and resource levels. Markets allow resources to be matched between entities that consume and produce the same resource using the double auction algorithm.

Perturbations can be introduced into an ExM model simulation to represent resource shortages. Perturbations are imposed on a particular sector within a nation, causing all entities in that sector to lose their store of output resources. This scenario affects both the perturbed entities which now have nothing to sell and the entities who require the lost resource for their own production. The resulting shock ripples throughout the system as entities respond to the disruption.

Some nations are configured with a specialized sector called the government. For the purposes of this model, the government taxes domestic trade, using the money to buy labor. The government uses labor to create a military resource (that buffers perturbations) or a technology resource (that improves the production rate of a resource).

Each national sector consumes the resources of every other sector to produce a single unique resource. In this configuration, the modeled system's network structure is characterized as fully connected with the exception of the government sector which is asymmetric. The effect of network structure on the model's dynamics has been studied by Kuypers et al. [3].

3   Implementing the Experiment

To study the effectiveness of investment in the protection of a critical resource imported from an unstable region versus investment in self-sufficiency, we initialized simulations made up of interactions among three nations; Supplier, Policy Maker, and Drone. Each nation is comprised of five sectors except Policy Maker which has an additional sector called government. The five sectors used here are named to represent relevant commodities: labor, goods, farming, mining, and oil. At the start of every simulation, each nation's production and consumption rates are set for each commodity sector.

The Supplier nation's economy is balanced for all resources except oil. For the labor, goods, farming, and mining sectors production and consumption levels are balanced, the nation produces as much as it consumes. The oil sector produces more oil than the other sectors consume, resulting in a surplus of oil.

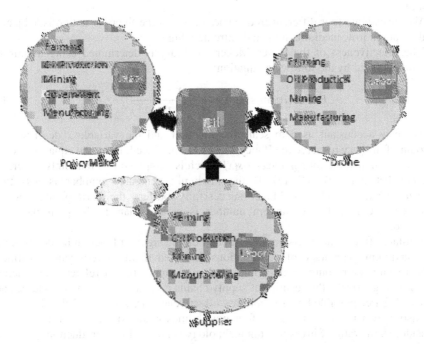

Fig. 1. Diagram of modeled entities (nations and sectors), disrupted production sector and international market for oil showing commodity flow

The Policy Maker's national economy is balanced except for the oil sector as well. However, instead of a surplus, the Policy Maker has an oil deficit. If the Policy Maker cannot trade in the international market to obtain more oil, the health of its sectors falls very low. The Policy Maker's government sector consumes a small fraction of its produced labor resource to produce units of military and/or technology resource. The military can be used to deflect perturbations, the technology can improve domestic oil production.

Similar to the Policy Maker, the Drone nation has a balanced economy except for an oil production deficit and the absence of a government sector to either defend against perturbations or invest in technology.

The Supplier, Policy Maker, and Drone nations have complementary economies. The Supplier has a surplus of oil while the Policy Maker and Drone have a deficit of oil. All resources are traded internationally except for labor.

3.1   Parameter Sweeps

We use a 3-dimensional parameter sweep in which the frequency of perturbations within the Supplier nation, the effectiveness of the Policy Maker's military at buffering those perturbations and the ratio of investment Policy Maker applies towards producing technology versus military are varied to examine the tradeoffs between projecting security and investing in self-sufficiency to buffer resource scarcity.

We investigate four different frequencies at which the Supplier nation is subjected to disruption: low, medium, high, and extremely high.

The effectiveness of the Policy Maker's military is determined using a sigmoid function defined by the following equation:

$$Effectiveness(x) = \frac{x * p}{(1 + p * x^2)} \tag{1}$$

where p is a constant used to scale the function and the variable x describes the amount of military resource the Policy Maker has produced. From the equation we are able to obtain a value for $Effectiveness(x)$ which is compared to a randomly generated number between 0.0 and 1.0. If the randomly generated number is less than $Effectiveness(x)$ the Policy Maker successfully buffers the perturbation and no resources are lost. If the random number is greater than 1, the perturbation is executed.

Military effectiveness sets an upper bound on the sigmoid function by determining the maximum percentage of perturbations deflected if all Government investment went to military resource production. Setting an upper bound reflects the inherent inability to guarantee the security of supply chains no matter how many resources are invested. Three possible levels of military effectiveness are shown in Table 1.

Also shown in Table 1 are the four possibilities modeled for the third parameter considered, the ratio of investment in technology versus military production.

Table 1. Military effectiveness and tradeoff options. The military effectiveness defines the maximum perturbation deflection probability. The tradeoff data show four ratio options for government investment in military and technology.

Military Effectiveness	Perturbation Deflection Probability
Strong	91%
Average	56%
Weak	20%

Tradeoff	Military Investment	Technology Investment
Option A	60%	40%
Option B	40%	60%
Option C	20%	80%
Option D	0%	100%

We also consider dependence and abundance of oil. Dependency is a measure of oil production rates relative to the total amount of oil production in the system. We vary this parameter over ten different scenarios. Abundance is a measure of the total production and consumption of oil in the system. If there is a global oil surplus then the system has high abundance, while a global deficit of oil means low abundance. For example, a scenario in which global oil production is greater than global oil consumption and, as we have defined for this study, the Supplier nation produces much more oil than either the Policy Maker or Drone nations, a condition of high abundance and high dependence would exist.

4   Results

The multidimensional parameter sweep generates large amounts of data. The results we present here were chosen to highlight the understanding we have gained regarding the general characteristics of the model.

4.1   Low Dependence, High Abundance

First, we consider the case of low dependence and high abundance of oil. The Policy Maker is able to produce enough oil initially to supply its internal demand without importing and thus has a low dependence on the Supplier.

We vary the Policy Maker's investment allocation between technology and military resources and take the average health value of the oil sector of each set of simulations. We find the Supplier's health is highly dependent upon the Policy Maker providing protection from perturbations. The health of the Policy Maker grows as it increases its investment in technology (for more efficient native oil production) over military production (Figure 2).

The price of oil fluctuates widely in each of the 48 individual simulations. The frequency of perturbations to Supplier oil production and the low dependence of the Drone and the Policy Maker results in a weaker demand for oil. When Supplier oil production is perturbed, the other nations are in less need of the oil it supplies, so the Supplier has a harder time recovering. When dependence is high the Supplier recovers from disruption more quickly as the other nations immediately need oil and are willing to pay a premium, resulting in increased money flow to the Supplier.

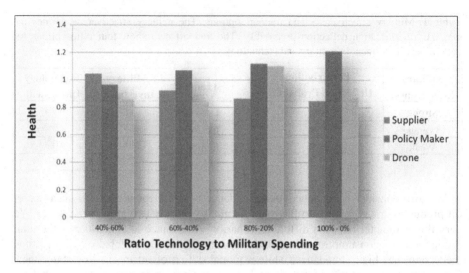

Fig. 2. In a disruption scenario under low dependence and high abundance conditions, the greater the Policy Maker investment in technology, the lower Supplier's health falls and the higher Policy Maker's health rises

4.2   Medium Dependence, Medium Abundance

In the medium dependence and medium abundance disruption scenario, the Policy Maker no longer produces enough oil to supply its demand, thus needing to import oil from the Supplier.

In this case, we find the Supplier's average health is slightly dependent upon the Policy Maker providing protection from perturbations but has a greater than nominal health regardless of the Policy Maker's investment decisions. As seen in Figure 3, since the Policy Maker is dependent on the Supplier for a small fraction of its oil consumption needs, its health increases as it chooses to invest in technology over military production. In this scenario, Policy Maker reaches its nominal health level when the government invests 100% in technology.

Looking at the individual runs, we see a decrease in the magnitude of the price spikes. In this scenario there is a greater dependence on oil from the Supplier requiring the Policy Maker and Drone nations to buy from the Supplier (potentially at a premium), making the consequences of perturbation less devastating to the Supplier.

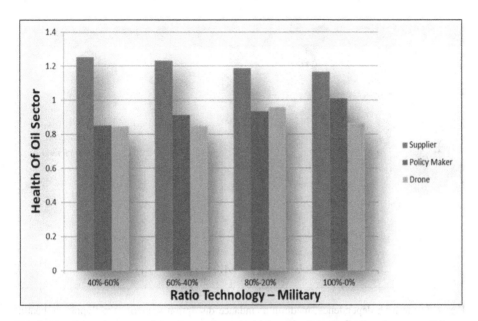

Fig. 3. In a disruption scenario under medium dependence and medium abundance conditions, the Policy Maker's health rises as it invests more in technology. Although this investment policy diminishes the Supplier's health, its level remains above nominal.

4.3 High Dependence, Medium Abundance

When high dependence and medium abundance conditions exist, the model imposes strict constraints on the nations in the simulations. The Policy Maker is highly dependent on importing oil from the Supplier nation to fulfill its consumption demand.

The Supplier's average health is no longer dependent upon Policy Maker protection from the perturbations it experiences. As illustrated in Figure 4, the Supplier also exhibits a greater than nominal health independent of the Policy Maker's investment decisions. The average health of the Policy Maker's oil production sector rises as the nation increases technology investment over military spending up to a 4-to-1 spread. Without Policy Maker providing any protection to the Supplier's oil production sector and despite a 100% investment in technology, all three nations see a decrease in health.

In the individual simulations, the cost of oil can be seen to rise very little. The demand is so great that in all cases, the Supplier is able to withstand the effects of disruptions to its oil production sector.

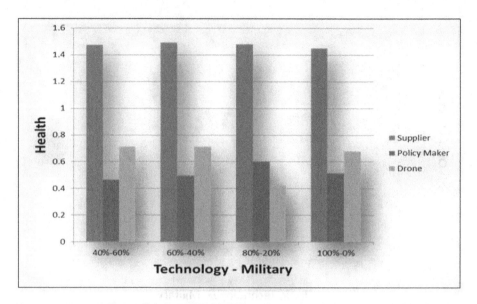

Fig. 4. In the high dependence, medium abundance disruption scenario, the Supplier's health remains very high regardless of the Policy Maker's investment strategy. The health of both the Policy Maker and the Drone fall well below nominal.

4.4   Effects of Dependency

We measure the effect of dependency by taking the difference between the Supplier and the Policy Maker's initial oil production rates and varying them inversely by small increments. In this manner, the world's total production rate of oil never changes, only the dependence of the Policy Maker on foreign oil.

As the Policy Maker's dependence on foreign oil increases, the average health of the Supplier grows. As the Policy Maker becomes more dependent on the Supplier there is less advantage to the health of either the Supplier or the Policy Maker from manipulating military and technology investment strategies. When there is low dependence, different allocation strategies can affect the health levels of each nation's sectors more than under high dependence conditions.

Figure 5 shows the health of the Supplier's oil production sector with respect to the level of the Policy Maker's dependence. Each curve represents a ratio of Policy Maker contributions to technology versus military resources. Figure 6 illustrates the Policy Maker's oil sector health considering the same parameters. To obtain this data from the simulations, we found the average health of the oil sectors having varied military effectiveness, frequency of perturbation and percentage of technology versus military production.

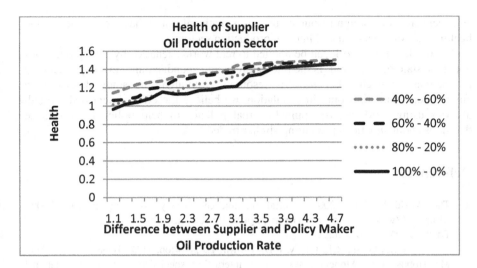

Fig. 5. Health of Supplier's oil production sector. The health values of different allocation strategies converge as the Supplier and Policy Maker's oil production levels diverge.

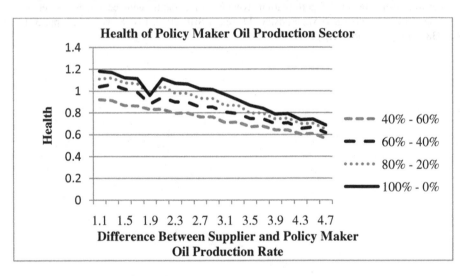

Fig. 6. Health of Policy Maker's oil production sector. The health values associated with different investment ratios converge as the oil production levels diverge.

5   Conclusion

By varying the frequency of perturbations on the Supplier, the Policy Maker's military effectiveness, the oil dependence of the Policy Maker, and global abundance of oil we have been able to comprehensively study the effects of Policy Maker's military and technology resource investment. Counter to our intuition, we found that as dependence on foreign resources increases, national policy has less impact on the health levels of the nation and internal production sectors.

The results of this work can be applied to real world scenarios by asking questions about the state of the world. Is the relevant resource abundant, and how dependent are the nations on each other for this resource? Although this study describes an environment that exists only in simulation, we believe the effects identified through analysis using ExM can be applied to real policies to help policy makers make decisions about situations presenting similar tradeoffs.

References

1. The World Factbook 2009. Central Intelligence Agency, Washington, DC (2009), https://www.cia.gov/library/publications/the-world-factbook/index.html
2. Beyeler, W.E., Glass, R.J., Finley, P.D., Brown, T.J., Norton, M.D., Bauer, M., Mitchell, M., Hobbs, J.A.: Modeling systems of interacting specialists. In: 8th International Conference on Complex Systems (2011)
3. Kuypers, M.A., Beyeler, W.E., Glass, R.J., Antognoli, M., Mitchell, M.D.: The impact of network structure on the perturbation dynamics of a multi-agent economic model. In: Yang, S.J., Greenberg, A.M., Endsley, M. (eds.) SBP 2012. LNCS, vol. 7227, pp. 331–338. Springer, Heidelberg (2012)

Progress Curves and the Prediction
of Significant Market Events

Sofia Apreleva[1,*], Neil Johnson[2], and Tsai-Ching Lu[1]

[1] HRL Laboratories, LLC, Malibu, CA
[2] Department of Physics, University of Miami, Miami, FL
svapreleva@hrl.com

Abstract. Progress curves have been used to model the evolution of a wide range of human activities -- from manufacturing to cancer surgery. In each case, the time to complete a given challenging task is found to decrease with successive repetitions, and follows an approximate power law. Recently, it was also employed in connection with the prediction of the escalation of fatal attacks by insurgent groups, with the insurgency "progressing" by continually adapting, while the opposing force tried to counter-adapt. In the present work, we provide the first application of progress curves to financial market events, in order to gain insight into the dynamics underlying significant changes in economic markets, such as stock indices and the currency exchange rate and also examine their use for eventual prediction of such extreme market events.

Keywords: Progress Curve fitting, stock indexes, currency exchange rates, prediction.

1   Introduction

Progress curves have been used to model the evolution of a wide range of human activities -- from manufacturing to cancer surgery. In each case, the time to complete a given challenging task is found to decrease with successive repetitions, and follows an approximate power law. Recently, it was also employed in connection with the prediction of the escalation of fatal attacks by insurgent groups, with the insurgency 'progressing' by continually adapting, while the opposing force tried to counter-adapt. In the present work, we provide the first application or progress curves to the temporal evolution of financial market events, in order to gain insight into the dynamics underlying significant changes in economic indicators, such as stock indices and the currency exchange rate – and also examine their use for eventual prediction of such extreme market events.

Our use of a progress curve function to analyze how specific features develop within a financial time series, is reasonable given that daily changes of stock markets are driven by news from various domains of human activity. News about a rise in the unemployment rate or inflation leads to a drop in stock market prices reflecting the state of the economy. News about civil unrest, riots or insurgencies are also likely to have a negative impact with market reaction expressing fear of impending instability.

K. Glass et al. (Eds.): COMPLEX 2012, LNICST 126, pp. 11–28, 2013.

One example of the influence of the political situation on stock markets is given in Figure 1. Here we show daily changes (%) of IBVC stock index from the Caracas Stock Exchange (Venezuela) in 2012. The stock index rose 99% in the period from January to April, with the main driver for this growth being the absence of reliable information about President Hugo Chavez's health.

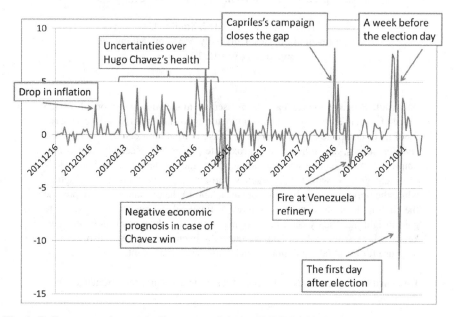

Fig. 1. Daily percent changes in Caracas stock index IBVC for 2012. Text boxes show events related to significant changes.

The arrival of news can stimulate a range of responses among market participants. After they have each weighed the news as being more or less significant for any particular financial instrument such as a stock, their aggregate action then gets recorded as a subsequent market movement. This suggests that the process generating market movements can be viewed as a series of 'blocks', where each block contains a sequence of events related to a particular story. As we can see in Figure 1, the current year 2012 is an election year in Venezuela and the majority of significant events are related to it. It is therefore natural to divide the entire time span into distinct periods with a progress-curve escalation (or de-escalation) dynamics within each. In this paper we show that it is indeed possible to divide the time span into intervals employing the notion of stationary/nonstationary processes. This approach not only avoids assuming that the market obeys stationary dynamics, it represents a conscious effort to capture and quantify the short-term ebbs and flows of trading behavior at the collective level.

We model individual financial time series as a sequence of periods (epochs) with the goal of capturing the successive buildup of behaviors prior to significant market changes. Each of these periods is a combination of escalation processes with switching processes. We then obtain time-series of the Z-score which shows standardized changes in the original financial time series. These Z-scores are a convenient tool for comparing the behavior of markets in different countries. We focus on Z-score values calculated over certain periods of the original time series (e.g. 30- and 90- day periods). This enables us to partition the Z-score time-series into periods where a stable region is followed by an escalation, before calming down in a final stage. Using progress curves to model the escalation within these periods, we find that our method can predict significant events defined as absolute Z-score value > = 4 for daily time differences with 30 day period, and Z-score value > = 3 with a 90 day period.

In the paper we present explicitly the three steps which comprise the approach:

1. A *time series partitioning* method that derives the periods within which there are distinctive escalation processes. We call these derived periods E-periods.

2. An *escalation parameter fitting method* that trains on historical data within each E-period, in order to derive escalation parameters for the progress curve model;

3. A *progress curve prediction* method that predicts significant events (e.g. absolute Z-score value > = 4 daily time differences with 30 day period, and Z-score value > = 3 with 90 day period) based on parameters obtained at the training stage.

All current methods for modeling and forecasting financial time series can be divided into four categories: 1). statistical modeling of time series; 2). machine learning methods; 3) analysis of influence of social behavior on economic markets (behavioral finance); 4) empirical findings of patterns in time series behavior (most of them obey exponential laws). The most popular data for financial prediction are time series of returns, stock indexes (closing/opening prices), volumes of transactions, interest rates. Here we briefly present the most prominent papers discussing these topics.

Classical econometric approaches for financial markets study and prediction rely on statistical models of time series to reveal the trends, i.e. the connections between consecutive time points. This relation serves as a basis for the prediction of time series values during the next time period. Auto-regression models, hidden Markov models, random walk theory, efficient market hypothesis (EMH) and other models have found large application in the world of finance [4, 9, 10, and 12]. For example, Wang, Wang, Zhang and Guo in Ref. [9] employed a hybrid approach combining three methods for modeling and forecasting the stock market price index. Also used are the exponential smoothing model (ESM), autoregressive integrated moving average model (ARIMA), and the back propagation neural network (BPNN). The weight of the proposed hybrid model (PHM) is determined by a genetic algorithm (GA). The closing of the Shenzhen Integrated Index (SZII) and the opening of the Dow Jones Industrial Average Index (DJIAI) are used as illustrative examples to evaluate the performances of the PHM and to compare with traditional methods including ESM, ARIMA, BPNN, the equal weight hybrid model (EWH), and the random walk model (RWM).

A self-excited multifractal statistical model describing changes of particular time series rather than the time series themselves, was proposed in [14]. There the authors defined the model such that the amplitudes of the increments of the process are expressed as exponentials of a long memory of past increments. The principal feature of the model lies in the self-excitation mechanism combined with exponential nonlinearity, i.e. the explicit dependence of future values of the process on past ones. Distributions of daily changes of stock markets share the same features as distributions of values of the Z score: turbulent flows, seismicity or financial markets, multifractality, heavy tailed probability density functions.

Machine learning models, on the other hand, gained their popularity by incorporating patterns and features obtained from historical data [2, 5]. The authors of Ref. [2] compared three machine learning techniques for forecasting - multilayer perceptron, support vector machine, and hierarchical model. The hierarchical model is made up of a self-organizing map and a support vector machine, with the latter on top of the former. The models are trained and assessed on a time series of a Brazilian stock market fund. The results from the experiments show that the performance of the hierarchical model is better than that of the support vector machine, and much better than that of the multilayer perceptron.

Around 1970, behavioral finance developed into a mature science, focusing the explanation of financial time series variations on the collective behavior of individuals involved into the market [3,7,11]. Recently this vision of how society influences economic indicators has changed to incorporate a broad range of social, political and demographic processes [1,8]. Most of these works has an empirical character in order to overcome the limitations of existing classical approaches.

The work of Bollen, Mao and Zeng [1] explores the influence of society mood states on economic markets, in contrast to behavioral finance which focuses on the collective psychology of individuals involved in sale processes. In particular, Bollen et al. investigate whether measurements of collective mood states derived from large-scale Twitter feeds are correlated to the value of the Dow Jones Industrial Average (DJIA) over time. The text content of daily Twitter feeds is analyzed. A Granger causality analysis and a Self-Organizing Fuzzy Neural Network are used to investigate the hypothesis that public mood states are predictive of changes in DJIA closing values. Their results indicate that the accuracy of DJIA predictions can be significantly improved by the inclusion of specific public mood dimensions, but not others.

Zantedeschi, Damien, and Polson [13] employed dynamic partition models to predict movements in the term structure of interest rates. This allowed the authors to investigate large historic cycles in interest rates and to offer policy makers guidance regarding future expectations on their evolution. They used particle learning to learn about the unobserved state variables in a new class of dynamic product partition models that relate macro-variables to term structures. The empirical results, using data from 1970 to 2000, clearly identify some of the key shocks to the economy, such as recessions. Time series of Bayes factors serve as a leading indicator of economic activity, validated via a Granger causality test.

Polson and Scott [10] propose a model of financial contagion that accounts for explosive, mutually exciting shocks to market volatility. The authors fit the model using country-level data during the European sovereign debt crisis, which has its roots in the period 2008-2010 but was continuing to affect global markets as of October 2011.

Analysis presented in the paper shows that existing volatility models are unable to explain two key stylized features of global markets during presumptive contagion periods: shocks to aggregate market volatility can be sudden and explosive, and they are associated with specific directional biases in the cross-section of country-level returns. Their proposed model rectified this deficit by assuming that the random shocks to volatility are heavy-tailed and correlated cross-sectionally, both with each other and with returns.

The authors of Ref. [8] present a novel approach resulting from studying patterns in transaction volumes, in which fluctuations are characterized by abrupt switching creating upward and downward trends. They have found scale-free behavior of the transaction volume after each switching; the universality of results has been tested by performing a parallel analysis of fluctuations in time intervals between transactions. The authors believe that their findings can be interpreted as being consistent with the time-dependent collective behavior of financial market participants. Taking into account that fluctuations in financial markets can vary from hundreds of days to a few minutes, the authors raise the question as to whether these ubiquitous switching processes have quantifiable features independent of the time horizon studied. Moreover they suggest that the well-known catastrophic bubbles that occur on large time scales—such as the most recent financial crisis—may not be outliers but single dramatic representatives caused by the formation of increasing and decreasing trends on time scales varying over nine orders of magnitude, from very large down to very small.

In contrast to these previous works, the novelty of the approach presented in this paper lies in the unique characterization of system behaviors into escalation and de-escalation periods, and the detection of their switching points. We demonstrate the results of fitting a progress curve to the time series of stock indexes and currency exchange rates in chosen Latin America countries, and make an estimate of the prediction of significant events using this model. The paper presents preliminary results which suggest an extension of the approach in order to take into account the current situation in a country and incorporate filtering/weighting of different events included in the escalation process.

2   Methods and Results

2.1   Overview of the Approach

Figure 2 provides a systematic view of our progress curve prediction system. The time series are partitioned into periods of escalation/de-escalation (E-periods) in order to define time frames for PCM fitting: a stable region is followed by an escalation, before calming down in a final stage. Within each period, sequences of inter-event time intervals and intensities are constructed and the regression parameters Θ_h are found. Prediction of future significant events for the current period can be performed based on regression coefficients obtained from the historical data, or directly from the current E-period if the number of inter-event sequences is $>=4$.

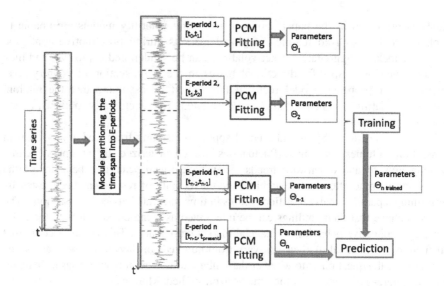

Fig. 2. Schematic overview of significant event prediction using progress curve model

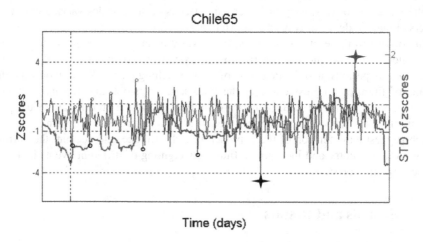

Fig. 3. An event with a Z-score=4.2 for Chile65 stock index, can be predicted using the progress curve formula approximately 220 days in advance. A negative event for the same escalation period with a sigma value of -4.3, can be predicted approximately 65 days in advance. Dashed line shows the beginning of new E- period. Red and black circles mark points forming time series corresponding to positive and negative events.

To define distinct periods of Z-score time series, we use the standard deviation (STD) of the Z-score as the measure of market stability. We observe (see Figure 4 and Section 2.2) that a roughly periodical structure emerges when the STD of the Z-score is calculated over different time intervals: the longer the interval, the more evident is the separation into periods of low and high rate-of-change in the Z-score values. We build two time series τ_n and v_n of points with increasing absolute sigma values $>= 1$ for inter-event time and inter-event intensity, and then fit progress curve models on a log-log scale. We consider positive and negative significant events separately by fitting separate escalation parameters. We have found that dependences of the regression coefficient on intercepts for each market indicator obey a simple linear relationship.

To make a prediction as to when a significant event with high sigma value will happen in the future, we use simple extrapolation and regression parameters obtained from historical data. The algorithm described above was tested on 16 market indicators (9 stock indices and 7 exchange rates). Below in Figure 3, we give an example of the predictions made for two significant events for Chile65 stock index marked on the figure with a red star (Z-score=4.2) and a black mark (Z-score = -4.3).

2.2 Partition of Z-Score Time Series into E-Periods

In order to characterize significant changes in economic indicators, the raw time series has been transformed to Z-score time series, with time windows of 30 and 90 days. In Figure 4 (top panel) we show an example of Z-score time series for BVPSBVPS index for the time period of July 2007 to July of 2012. It can be seen that the series can be viewed as a sequence of 'waves of burstiness' with a large amplitude of oscillation. We aim to separate these 'waves' and consider them as separate objects of interest representing escalation process followed by the period of calming down. It is an open question if each of these burst periods has to be considered separately, or sometimes their sequence with successfully magnifying amplitude reflects the building up of one large escalation process with unique underlying dynamics. We can assume that each E-period is manifested by a significant event with absolute value of Z-score $>= 4$ calculated over 30-days' time window and $|Z - score| >= 3$ of 90-days' time window. This definition implies the unification of several (two or three) burst periods into one E-period.

The following steps are performed to define E-periods:

1. *Derive Z-score time series* from the daily difference of original time series (e.g. daily closing price of BVPSBVPS) with chosen moving time windows (30 or 90 days).

2. *Derive the standard deviation time series* of Z-score time series (called SD time series) with respect to different time windows (e.g., 10, 30, 90 day windows).

3. *Apply standard min/max identification* algorithms to identify local minima and maxima in the identified SD time series from Step 2.

4. *Output E-periods as the duration* of the sequences of local minima and maxima identified in Step 3.

In Figure 4, we show the SD time series (the bottom three panels – 10, 30, 90 days) for the Z-score time series (the top panel) of BVPSBVPS. Step 3 results in the identification of local minima and maxima (dashed lines) that lead to partition and the output of corresponding E-periods.

Intuitively, the daily difference of closing prices represents the first derivative of the original time series. The Z-score time series of this first derivative represents the second derivative - volatility. The SD time series, the standard derivation of Z-score time series, represents the third derivative – momentum of volatility. The identification of local minima and maxima of SD time series naturally give rise to the identification of switching of momentum of volatility – E-periods. This novel partition process enables us to identify periods where the escalation or de-escalation process operates within the identified boundary.

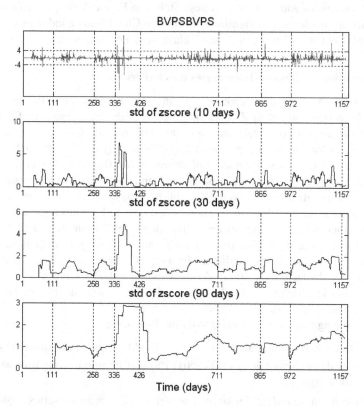

Fig. 4. Illustration of E-periods identification: The top panel shows the BVPSBVPS z-score of 30 day time window. The following three panels show standard deviation of z-score with different time windows: 10, 30, and 90 days. The dashed line partitioned time series into different (de)escalation periods based on minima and maxima of standard deviation of z-score (30 days).

2.3   Progress Curve Modeling

Following Johnson's Progress Curve modeling approach for escalation process [6], we use the following formula:

$$T_n = T_1 n^{-b} \tag{1}$$

where T_n is a time interval between *(n-1)-th* and *n-th* days, *n* is the number of days and *b* is an escalation rate. The challenge in Progress Curve modeling is that behavior process may switch from escalation to de-escalation with different scales and intensity. In the previous section, we presented a novel approach to identify escalation periods, inter-event time and intensity. In this section, we show the steps for escalation parameter fitting modules, which are described as follows:

1. *Extract inter-event time and intensity for escalation trends.* For each period, we identify 'dark' events with absolute vales >= 1 separately for positive and negative sequences. For each 4-sigma (-4-sigma) event, we identify prior 'dark' events to build up the escalation (de-escalation) event trends and then derive inter-event time and intensity. As a result we work with series of τ_n for inter-event times and v_n for inter-event intensities. Figure5 shows an example of points picked to identify escalation process within an E-period with a following Progress Curve fit for each of constructed sequences.

2. *Fit Progress Curve models to derive escalation parameters.* Given identified inter-event times and intensities for each period, we fit progressive curve model to identify parameters: slope and intercepts (we use the Progress Curve formula in log-log scale). We then fit regression line on (de)escalation parameters from all periods. This gives us a basis for identifying escalation parameter (slop) given intern-event time and intensity (intercepts). We perform the same steps for inter-event intensity. Figure 6 demonstrates the general scheme of algorithm.

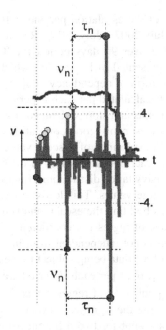

Fig. 5. Illustration of building of inter-event time series τ_n and v_n. Red circles refer to the significant events, yellow circle - to 'dark' positive events, purple circles denote 'dark' negative eve.

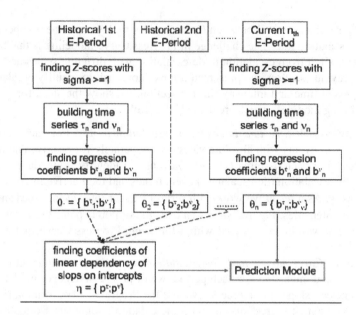

Fig. 6. Scheme of the fitting with PCM. Parameters used in the prediction are obtained from current E-period ($\theta_n = \{ b^\tau_n; b^v_n \}$ or from the historical data sets ($\eta = \{p^\tau ; p^v \}$). Fitting performed separately for time series with positive and negative values.

To illustrate the working of the escalation parameter fitting module, we show an example for IGBVL index (July 2007 – July 2012). Results of PCM fitting module of Z-score time series calculated over 90 days period for IGBVL index starting from July 2007 to July 2012 are presented in Figure 7b., which includes pooled together regression coefficients in the form of dependency of slope on intercept, and distribution of correlation coefficients calculated per each E-period.

One can observe that intercepts and slopes obtained from all E-periods obey a simple linear relationship with a reasonably good correlation (see Figure 7). This probably can be explained by the limited range of Z-score values and low frequency of the events to happen. It is known fact that time series of differences of stock market indexes belong to the class of heavy tailed distribution (Figure 7a) with rare occurrences of events with significantly large values. This leads to the limited range of numbers drawn to construct time series building the escalation/excitation process, which leads as a result to the limited range of regression coefficients. We have to note here also that variance of the intercepts is not independent from the one of the slopes: they are related through the variance of the data being subject to the regression.

Once parameters $\theta_n = \{ b^\tau_n; b^v_n \}$ per each E-period are calculated and parameters $\eta = \{p^\tau ; p^v \}$ characterizing pooled set of regression coefficients is obtained, we can proceed to the module predicting the significant event for the current E-period. Two sets of parameters serve as an input for two different paths used for prediction. The prediction module is described in more detail in the next section.

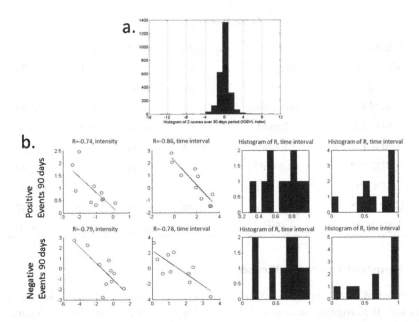

Fig. 7. Figure 7 – Results for PCM fit for IGBVL index: a). basic statistics for inter-event time series τ_n and v_n. X axis of the first two columns of plots is an intercept of the fit, Y axis is a slop of the fit. Here R is a Pearson correlation coefficient; b). typical heavily tailed distribution of Z-score values.

PCM fitting has been performed on 16 financial time series for the same time period (07/2007 – 07/2012) (stock indices and currency exchange rate) given in the Table 1. Table includes the precision criteria as a measure of the accuracy of the prediction, so this metrics will be discussed further in the text.

Table 1. Stock indexes and currency exchange rate used for PCM fitting

Stock Index/ Currency Exchange Rate	Country	Precision			
		P30	N30	P90	N90
BVPSBVPS	Panama	0.88	0.71	0.42	0.57
CHILE65	Chile	1.00	0.67	0.78	0.73
COLCAP	Colombia	0.75	1.00	0.90	0.89
CRSMBCT	Costa Rica	0.10	0.07	0.10	0.20
IBOV	Brazil	na	0.35	0.42	0.13
IBVC	Venezuela	0.69	0.50	0.64	0.23

Table 1. (*continued*)

IGBVL	Peru	1.00	0.50	0.85	0.91
MERVAL	Argentina	0.60	0.50	0.70	0.75
MEXBOL	Mexico	1.00	0.50	0.67	0.67
USDARS	Argentina	0.67	0.50	0.67	0.75
USDBRL	Brazil	0.50	0.00	0.75	0.00
USDCLP	Chile	0.50	0.00	0.80	0.80
USDCOP	Colombia	1.00	1.00	0.80	0.80
USDCRC	Costa Rica	0.50	0.50	0.80	0.40
USDMXN	Mexico	0.80	0.00	0.75	0.67
USDPEN	Peru	0.57	0.00	0.71	0.50

2.4   Statistical Test of Significance

We performed three significance tests to confirm the validity of fitting financial time series with the progress curve model. We employ time series reshuffling techniques at the level of the entire time span of interest with fixed E-periods (method I); reshuffling of all Z-score values within E-period (method II). Method III works with re-shuffled inter-event time and intensities sequences also within particular E-periods.

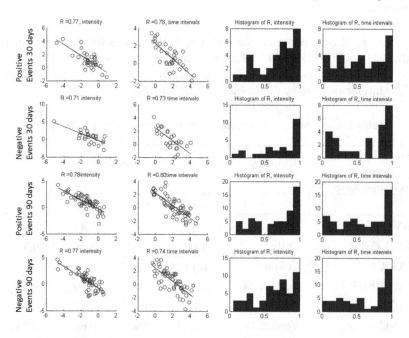

Fig. 8. Results for PCM fit for 16 time series. X axis of the first two columns of plots is an intercept of the fit, Y axis is a slop of the fit.

To carry out the statistical tests, we had to pool parameters $\theta_n = \{b^\tau_n; b^\nu_n\}$ per each E-period from all indexes and currency rates to reach a statistical power allowing us to make statistical tests. Pooling together was possible because Z-scores values are standardized metrics as oppose to daily changes of market Z-score, thus they can be used to reveal and investigate behavior and features common for time series representing different indexes. Regression coefficients for all 16 economic indicators and distribution of all corresponding correlation coefficients are shown in the Figure 8. We find here that pooled set of regression parameters reveal properties similar to those observed for individual fits for each indicator. Distributions shown on this figure are used to test for significance and compare against those obtained from randomizing procedures.

Table 2. Results of tests for significance for all types of escalation time series (negative and positive sequences for 30 days Z score time series, negative and positive for 90 days time series). R stands for Pearson correlation coefficients, ν_n for inter-event intensities and τ_n for inter-event time intervals.

Pos. 30D

Method	Intercepts ν_n	Slops ν_n	Intercepts τ_n	Slops τ_n	R ν_n	R τ_n
I	0.19481	0.03011	0.00082	0.00193	0.00725	0.00834
II	0.97556	0.38436	0.71327	0.69671	0.49038	0.21443
III	0.00173	0.00011	0.04379	0.04379	0.00008	0.01428

Neg 30D

Method	Intercepts ν_n	Slops ν_n	Intercepts τ_n	Slops τ_n	R ν_n	R τ_n
I	0.02745	0.08955	0.14402	0.72344	0.02222	0.87667
II	0.60838	0.22055	0.8285	0.13333	0.42567	0.01061
III	0.00385	0.01535	0.24464	0.21098	0.00482	0.65643

Pos. 90D

Method	Intercepts ν_n	Slops ν_n	Intercepts τ_n	Slops τ_n	R ν_n	R τ_n
I	0.00001	0.00029	0.22755	0.05098	0.00002	0.67475
II	0.49325	0.57649	0.39936	0.12133	0.40901	0.04833
III	0.00007	0.00001	0.11758	0.04093	0.00001	0.14779

Neg.90D

Method	Intercepts ν_n	Slops ν_n	Intercepts τ_n	Slops τ_n	R ν_n	R τ_n
I	0.00713	0.01248	0.00965	0.10357	0.01189	0.13235
II	0.45183	0.67793	0.13995	0.72016	0.89812	0.79491
III	0.11394	0.02779	0.14126	0.10413	0.14228	0.02081

Table 2 demonstrates p-values for T test (distributions of regression coefficients) and Wicoxon non parametric rank sum test for distributions of correlation coefficients. Tests for regression coefficients were made for intercepts and slopes separately.

Method II (reshuffling of Z score values within E-period) does not show statistical significance. As we can see from the table, the majority of p-values are < 0.05 for method I and method III, which proves that the observed patterns and predictions based on PCM fit does not happen at random.

2.5   Prediction Using PCM

In the previous section we have demonstrated that the progress curve approach can be used to analyze the behavior of financial time series of stock indexes and currency exchange rates: increasing of absolute values of Z-score within one escalation period can be described using progress curve formula to a satisfactory accuracy (see distribution of correlation coefficients for inter-event intensities and time intervals presented in Figure 7b). We also have shown that parameters $\theta_n = \{b^\tau{}_n; b^v{}_n\}$ per each E-period can be pooled together and dependence of slopes on intercepts $\eta = \{p^\tau ; p^v \}$ can be obtained. To predict a date of the significant event for the current E-period, two approaches are proposed and described in this section. The first set θ_n is used in an extrapolation algorithm and the second set η enters the second block utilizing the historical data ('trained') linear dependency.

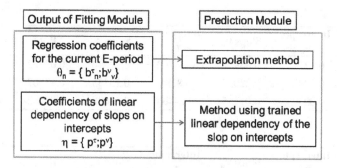

Fig. 9. Scheme of the fitting/prediction module. Parameters used in the prediction are obtained from current E-period ($\theta_n = \{ b^\tau{}_n; b^v{}_n\}$ or from the historical data sets ($\eta = \{p^\tau ; p^v \}$). Fitting performed separately for time series with positive and negative values.

The simple relationship between inter-event time and intensity provide a tool for prediction of future significant events. As soon as several points building escalation process are observed, we can predict at what moment in the future the significant event exceeding 4 (-4) sigma for 30 days Z – score time series and 3 (-3) for 90 days Z-score time series will happen. To predict this event, a simple formula for integration over inter-event intensities is used:

$$n = \exp\left(\frac{(I - Z_0)(-b_v + 1)}{v_1}\right)\bigg/(-b_v + 1) \qquad (2)$$

where I is a value of significant Z-score; Z_0 is a Z-score value at the time of the first sigma event; b_v is a slope of the PCM fit for the intensities sequence; v_1 is an intercept of the fit. Knowing the number of inter-event intervals n and parameters of PCM fit for time interval, one can calculate the final date of the significant event. This formula represents the heart of the extrapolation algorithm. Results of this simple procedure are shown in the Figure 10. Green dots on the picture representing predicted dates are connected with the real dates of events (blue circles).

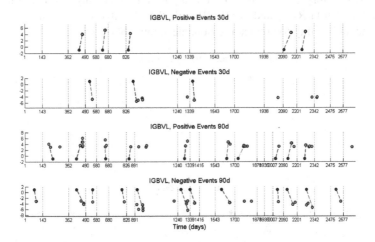

Fig. 10. Results for prediction using extrapolation method: Green circle represent significant events, blue circle – predicted date of event, red line connects predicted and actual events, dashed line separates E-periods

As we mentioned above, the slopes and intercepts of PCM fits for both inter-event time intervals and intensities are in reasonable agreement and regression parameters for this linear relationship can be used for prediction. The advantage of this approach as opposed to an extrapolation procedure, is that one needs only the first time interval between 'dark' events comprising the escalation process. So the final algorithm uses the same formula (2) as the extrapolation method, but regression parameters for both sequences are taken from historical data. To test the prediction using the distribution of regression coefficients obtained from historical data, we have used a longer time span which has been available for stock indexes given in the Table 1 – from the beginning of 2002 to July 2012. This helped us to increase the statistical power since, as we mentioned before, significant events are relatively rare events. Prediction results for IGBVL index using regression coefficients obtained from historical data are displayed on Figure 11.

As can be seen from Figures 10 and 11, the majority of significant events for IGBVL for all 4 types of events can be predicted reasonably well. The following metrics have been calculated to characterize prediction accuracy – precision, D-time and L-time. The two latter metrics serve to estimate the time interval between predicted dates and actual (D-time) and between the time points when prediction is made and the actual time of the event (L-time). Precision is defined as a ratio of events predicted with an absolute value of |D-time|<=50 days. Although at this stage we cannot predict each of multiple significant events within an E-period, this can be done in the future as an extension and refinement of the method. The present approach gives only a rough estimate of the escalation processes. D-times range from 3 to 30 days and L-time usually depends on the duration and constitutes 50 days on average.

In Table 1, the precision metric is displayed. Precision varies from 0 to 1 demonstrating satisfactory performance of the algorithm. It is only for two time series CRSMBCT (Costa Rica) and IBOV (Brazil) that the algorithm shows poor performance and an inability to predict a reasonable amount of events.

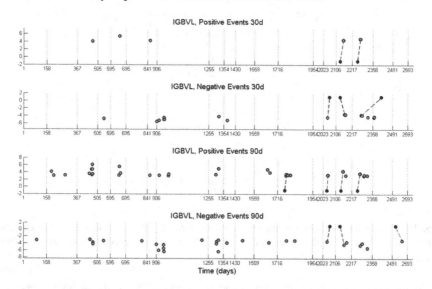

Fig. 11. Results for prediction using training method. Green circles represent significant events, blue circle – predicted date of event, red line connects predicted and actual events, and dashed lines separate E-periods. Two third of all E-periods have been used to obtain distribution of regression coefficients. Prediction is made for the rest of the time span.

3   Conclusions and Future Work

The present work shows the approach to fit financial time series, such as stock indexes and the currency exchange rates, with a progress curve applied within a period with a distinctive build-up of the escalation processes. The idea of this approach has been justified by the modeling of a broad range of human activities using this class of functions, and the main goal was to find progress curve patterns in financial markets.

To enable the fitting we have proposed the algorithm which consists of two steps: 1) partitioning of the entire time series into periods of stationary/nonstationary behavior using standard deviation time series of Z-score values; 2) fitting block which comprises the construction of inter-event time intervals and intensities and fitting itself. Regression coefficients found in the fitting module can be further used for prediction of significant events.

Results of the application of the entire scheme to the financial times series for Latin America countries (9 stock indexes and 6 currency exchange rates) demonstrate the validity of the approach. The results also demonstrate the predictive power of the approach for rough estimation.

Future work can be directed to refine the approach in order to improve the definition of the E-periods, concentrating and incorporating the current situation of the country. This would help to unite or separate several small 'burst' periods depending on their underlying context. Also the studying of news which had the most significant impact, could help to filter non-relevant 'dark' events and improve the predictability of the method. Another possible direction is to build a statistical model which models the escalation processes with a better accuracy.

Acknowledgement. Supported by the Intelligence Advanced Research Projects Activity (IARPA) via Department of Interior National Business Center (DoI / NBC) contract number D12PC00285. The U.S. Government is authorized to reproduce and distribute reprints for Governmental purposes notwithstanding any copyright annotation thereon. The views and conclusions contained herein are those of the authors and should not be interpreted as necessarily representing the official policies or endorsements, either expressed or implied, of IARPA, DoI/NBE, or the U.S. Government.

References

1. Bollen, J., Mao, H., Zeng, X.: Twitter mood predicts the stock market. Journal of Computational Science 2(1), 1–8 (2011)
2. Carpinteiro, O.A.S., Leite, J.P.R.R., Pinheiro, C.A.M., Lima, I.: Forecasting models for prediction in time series. Artificial Intelligence Review 38(2), 163–171 (2012)
3. Fama, E.F.: Market efficiency, long-term returns, and behavioral finance. Journal of Financial Economics 49(3), 283–306 (1998)
4. Fox, E., Sudderth, E.B., Jordan, M.I., Willsky, A.S.: Bayesian non-parametric inference of switching dynamic linear models. IEEE Transactions on Signal Processing 59(4), 1569–1585 (2011)
5. Huang, W., Nakamori, Y., Wang, S.: Forecasting stock market move-ment direction with support vector machine. Computers and Operations Research 32(10), 2513–2522 (2005)
6. Johnson, N., et al.: Pattern in escalations in insurgent and terrorist activity. Science 333(6038), 81–84 (2011)
7. Lu, B., Song, X., Li, X.: Bayesian analysis of multi-group nonlinear structural equation models with application to behavioral finance. Quantitative Finance 12(3), 477–488 (2012)
8. Preis, T., Schneider, J.J., Stanley, H.E.: Switching processes in financial markets. Proceedings of the National Academy of Sciences of the United States of America 108(19), 7674–7678 (2011)

9. Wang, J., Wang, J., Zhang, Z., Guo, S.: Stock index forecasting based on a hybrid model. Omega 40(6), 758–766 (2012)
10. Polson, N.G., Scott, J.C.: Predictive macro-finance with dynamic partition models (2011), http://faculty.chicagobooth.edu/nicholas.polson/research/papers.html
11. Shiller, R.J.: From efficient markets theory to behavioral finance. Journal of Economic Perspectives 17(1), 83–104 (2003)
12. Watanabe, K., Takayasu, H., Takayasu, M.: A mathematical definition of the financial bubbles and crashes. Physica A: Statistical Mechanics and its Applications 383(1 spec. iss.), 120–124 (2007)
13. Zantedeschi, D., Damien, P., Polson, N.G.: Predictive macro-finance with dynamic partition models. Journal of the American Statistical Association 106(494), 427–439 (2011)
14. Filimonov, V., Sornette, D.: Self-excited multifractal dynamics. EPL (Europhysics Letters) 94(4), 46003 (2011)

Bifurcation as the Source of Polymorphism

Ernest Barany

New Mexico State University, Las Cruces, NM 88001, USA
ebarany@nmsu.edu

Abstract. In this paper we present a symmetry breaking bifurcation-based analysis of a Lotka-Volterra model of competing populations. We describe conditions under which equilibria of the population model can be uninvadable by other phenotypes, which is a necessary condition for the solution to be evolutionarily relevant. We focus on the first branching process that occurs when a monomorphic population loses uninvadability and ask whether a symmetric dimorphic population can take its place, as standard symmetry-breaking scenarios suggest. We use Gaussian competition functions and consider two cases of carrying capacity functions: Gaussian and quadratic. It is shown that uninvadable dimorphic coalitions do branch from monomorphic solutions when carrying capacity is quadratic, but not when it is Gaussian.

Keywords: Symmetry breaking bifurcation, population model, evolutionary stability.

1 Introduction

Polymorphic population models, that is models in which organisms with more than one value of a specified physical trait can coexist asymptotically in time, can arise from models that allow only organisms with a single specified trait value via steady state bifurcations. The states in these bifurcations are the physical traits and the bifurcation parameters are environmental variables. This kind of phenomena is typical of systems that exhibit *spontaneous symmetry breaking*. Since it is known that spontaneous symmetry breaking is generic for systems with symmetry [1] (as many common population models such as the Levene model or the Lotka-Volterra models do), the possibility presents itself that speciation processes may be driven by symmetry breaking bifurcations. The process by which a population can split is referred to as *disruptive selection* by biologists (see the review by [2]) who often study the evolutionary mechanisms of branching using complicated stochastic "individual-based" simulations. The work in this paper shows that many features of population equilibrium models are fixed by the stability properties of the deterministic dynamical model.

In this paper we present a bifurcation-theoretic analysis of a Lotka-Volterra model to show how the possibility of stable branching from a monomorphic population to dimorphic depends on the details of the competition and carrying capacity functions. In particular, we consider Gaussian competition functions as

K. Glass et al. (Eds.): COMPLEX 2012, LNICST 126, pp. 29–39, 2013.

is commonly done, and take two cases for carrying capacity functions: Gaussian and a quadratic function. An interesting feature is that though the two carrying capacity models are essentially identical at quadratic order, so that the local bifurcation behavior near the origin of phenotype space might be expected to be similar. In fact, the two cases give quite different results.

The basic model under consideration can be written

$$\frac{dn_j}{dt} = r_j[K(x_j) - \sum_{k=1}^{M} C(x_j, x_k)n_k]n_j, \tag{1}$$

where $j = 1, 2, \ldots, M$ indexes the competing phenotypes, and n_j is the population of organisms that have phenotype x_j, which we assume to be a real number. The functions $C(x_j, x_k)$ describe competition between the jth and k^{th} phenotypes and the function $K(x_j)$ gives the carrying capacity for the jth phenotype; these functions will be discussed below. The function r_j is the growth rate, which has been scaled by $K(x_j)$ relative to a normal logistic model in order to simplify the analysis. Since the basic growth rates and the carrying capacity are all positive, it follows that the r_j's have no effect on the stability issues which are the main topic of this analysis.

We are interested in describing the simplest splitting event: monomorphism to dimorphism. Therefore it is sufficient to consider a model with three competing phenotypes (i.e., $M = 3$), though the results presented for this case generalize to arbitrary M. More competitors can lead to different long term results, and can alter the rate at which the morphic types of the population change, but if the phenomenon of an uninvadable dimporhism cannot occur for three competitors, it will not be able to occur no matter how many competitors exist, so the present analysis can be thought of as a necessary conditition for branching that any more realistic model must be consistent with.

The remainder of this section will be organized as follows. First, general expressions will be derived for the eigenvalues of the jacobian of the system for monomorphic and dimorphic populations in terms of the as yet unspecified competition and carrying capacity functions. Then the idea of invadability of asymptotically stable equilibria will be discussed. Finally, detailed analyses will be presented for specific competition and carrrying capacity functions to show that the possibility of stable branching depends on the specifics of these functions.

2   Stability and Invadability of Equilibria

Before proceeding, we note a property of (1), namely *symmetry* [1]. This term refers to a specific mathematical property that has profound effects, but which will be used only cursorily here. The basic operation of symmetry in this context amounts to the relabeling of phenotypes, which can be thought of as a permutation of the M indices. Mathematically, the set of such rearrangements comprises the group S_M of permutations on M objects. The key observation is that the set of M equations (1) is invariant under the operations of S_n. That is, the particular equations are permuted among themselves in the same way that the indices are permuted.

The usefulness of this observation in the current case is that it means that all equilibrium solutions with a fixed number of nonzero phenotypes, (for example, all monomorphic solutions) are equivalent, in that each such solution is the image of all other such solutions under the action of some element of S_n (just pair-wise interchanges in the monomorphic situation). Therefore, only one such solution need be analyzed, say the solution with $n_1 = K(x_1)$ and all other $n_j = 0$. Similarly only one dimorphic solution need be studied, say the one with n_1 and n_2 nonzero.

2.1 Monomorphism

The local stability of the monomorphic solution $n_j^{Mon} = K(x_j)\delta_{j1}$ is determined by the eigenvalues of the jacobian matrix evaluated at the monomorphic equilibrium:

$$J^{Mon} = \begin{pmatrix} -2r_1K(x_1) & -r_1C(x_1,x_2)K(x_1) & -r_1C(x_1,x_3)K(x_1) \\ 0 & r_2(K(x_2) - C(x_2,x_1)K(x_1)) & 0 \\ 0 & 0 & r_3(K(x_3) - C(x_3,x_1)K(x_1)) \end{pmatrix} \tag{2}$$

which are just the diagonal elements of the matrix, since it is triangular.

The $(1,1)$ element of J^{Mon} is clearly negative, and amounts to the fact that if an organism with any phenotype is introduced into the environment with no other organisms present, the population will grow. The $(2,2)$ and $(3,3)$ elements contain the same information as each other, which refers to whether an invading phenotype x_2 or x_3 can out-compete a resident population with phenotype x_1. That is, the condition that a resident with phenotype x will be stable against invasion by a phenotype y can be written

$$\Lambda^{Mon}(x,y) = K(y) - C(y,x)K(x) < 0. \tag{3}$$

Note that this result is the same as would be obtained for any number of competitors M.

A property of particular interest is whether the monomorphic equilibrium is *uninvadable*, see, e.g., [3], which will be so for some phenotype $x = x^*$ if the condition (3) holds for all $y \neq x^*$. Note that $\Lambda^{Mon}(x,x) = 0$ for all x, which is intuitively obvious since it amounts to the statement that "invasion" by an identical phenotype can always occur, so a necessary condition for phenotype x^* to be uninvadable is that the function $\Lambda^{Mon}(x^*,y)$ have a critical point as a function of y for $y = x^*$, that is,

$$\frac{\partial \Lambda}{\partial y}(x^*,y)|_{y=x^*} = 0. \tag{4}$$

Moreover, the critical point x^* will, in fact, be univadable if the local extremum is a maximum, that is, if

$$\frac{\partial^2 \Lambda}{\partial y^2}(x^*, y)|_{y=x^*} < 0, \tag{5}$$

and will be invadable by all nearby phenotypes if the opposite sign condition holds. Examples for specific cases will be shown below, but a useful property that holds for a wide class of competition functions can be seen as follows. We assume that $C(y, x) \le 1$ for all phenotypes x, y and that equality hold only for $y = x$. This is not a strong assumption and can be thought of a a normalization condition since species with the same phenotype will be maximally competitive. Under this assumption, $\frac{\partial C}{\partial y}(y, x)|_{y=x} = 0$, and a necessary condition for any such competition function (such as a Gaussian function) to allow the existence of an uninvadable monomorphism x^*, is that x^* must be a critical point of the carrying capacity function,

$$K'(x^*) = 0. \tag{6}$$

and a sufficient condition for uninvadability is that

$$\frac{\partial^2 \Lambda^{Mon}}{\partial y^2}(x^*, y)|_{y=x^*} = K''(x^*) - \frac{\partial^2 C}{\partial y^2}(y, x^*)|_{y=x^*} K(x^*) < 0. \tag{7}$$

The opposite sign condition indicates that a resident population with phenotype x^* can be invaded by all nearby phenotypes.

2.2   Dimorphism

We next present an analysis that parallels the previous case for the case of dimorphic solutions. As above, we assume that only phenotypes x_1 and x_2 have nonzero populations, the analysis for other pairs of nonzero populations is equivalent. The solution can easily be found to be

$$n_1^{Di} = \frac{K(x_1) - C(x_1, x_2)K(x_2)}{1 - C(x_1, x_2)C(x_2, x_1)} \tag{8}$$

$$n_2^{Di} = \frac{K(x_2) - C(x_2, x_1)K(x_1)}{1 - C(x_1, x_2)C(x_2, x_1)} \tag{9}$$

$$n_3^{Di} = 0. \tag{10}$$

Note that the numerators of the nonzero populations are identical to the stability eigenvalue function Λ^{Mon}, above, so that the dimorphic solution becomes feasible (positive) when the monomorphic solution loses stability. This is the signal of a transcritical bifurcation.

After some simplification, the jacobian matrix at the dimorphic solution can be written

$$J^{Di} = \begin{pmatrix} -r_1 n_1^{Di} & -r_1 C(x_1, x_2)n_1^{Di} & -r_1 C(x_1, x_3)n_3^{Di} \\ -r_2 C(x_2, x_1)n_2^{Di} & -r_2 n_2^{Di} & -r_2 C(x_2, x_3)n_2^{Di} \\ 0 & 0 & r_3(K(x_3) - C(x_3, x_1)n_1^{Di} - C(x_3, x_2)n_2^{Di}) \end{pmatrix} \tag{11}$$

Two of the stability eigenvalues of the dimorphic equilibrium are the eigenvalues of the 2×2 matrix

$$
\begin{pmatrix}
-r_1 n_1^{Di} & -r_1 C(x_1, x_2) n_1^{Di} \\
-r_2 C(x_2, x_1) n_2^{Di} & -r_2 n_2^{Di}
\end{pmatrix}
\tag{12}
$$

which are easily seen to be positive as long as n_1^{Di} and n_2^{Di} are positive and $C(x_1, x_2) < 1$ as long as $x_1 \neq x_2$. The third eigenvalue measures the ability of a third phenotype (say z) to invade an existing dimorphic population (with phenotypes x and y), and its sign is determined by the function

$$
\Lambda^{Di}(x, y, z) = K(z) - \frac{C(z,x) - C(z,y)C(y,x)}{1 - C(x,y)C(y,x)} K(x) - \frac{C(z,y) - C(z,x)C(x,y)}{1 - C(x,y)C(y,x)} K(y).
\tag{13}
$$

As before, note that for fixed x and y, $\Lambda^{Di}(x, y, x) = \Lambda^{Di}(x, y, y) = 0$, so again, in order for a dimorphic population with phenotypes x^* and y^* to be uninvadable by a third phenotype z, it must be true that $\Lambda(x^*, y^*, z)$ viewed as a function of z must have critical points for $z = x^*$ and $z = y^*$, and they must be local maxima. Before considering specific examples, we consider some issues of a general character relative to the issue of critical behavior for dimorphic solutions which is a necessary condition for uninvadability. We assume that the competition functions $C(x, y)$ obey the conditions $C(x, y) = C(y, x)$ and $\frac{\partial C}{\partial x}(x, y) = -\frac{\partial C}{\partial y}(x, y)$, which are true for many important special cases, such as any function that depends on x and y through the quantity $(x - y)^2$, i.e., $C(x, y) = C((x - y)^2)$, of which a Gaussian function is a familiar special case. Under these assumptions, there is always a class of symmetric dimorphic solutions that satisfies $y^* = -x^*$, and for definiteness and simplicity, we will focus on these solutions in this paper. For the symmetric branching solution, he stability eigenvalue becomes

$$
\Lambda^{Di*}(x, z) = K(z) - \frac{C(z, x) + C(z, -x)}{1 + C(x, -x)} K(x),
\tag{14}
$$

and in this case, the two criticality conditions become identical. Specifically, if $C^* = C(-x^*, x^*)$ and $C^{*'} = \frac{\partial C}{\partial y}(y, x^*)|_{y=-x^*}$, then the critical value x^* must satisfy

$$
K'(x^*) - \frac{C^{*'}}{1 + C^*} K(x^*) = 0.
\tag{15}
$$

Further, if $C^{*''} = \frac{\partial^2 C}{\partial y^2}(y, x^*)|_{y=-x^*}$ and $C^{0''} = \frac{\partial^2 C}{\partial y^2}(x^*, x^*)$, a sufficient condition that ensures that the critical points defined by (2.2) describe an uninvadable dimorphism is

$$
K''(x^*) - \frac{C^{0''} + C^{*''}}{1 + C^*} K(x^*) < 0,
\tag{16}
$$

and if the opposite sign condition holds then all nearby phenotypes can invade the dimorphism.

As a final comment in this section, we mention that including more competitors in the system (1) does not change the previous conclusions. Additional eigenvalues occurring in such cases are equivalent to Λ^{Di}.

3   Gaussian Competition Functions

In the previous section, we derived conditions for univadable monomorphism and uninvadable dimorphism in an ecological models of competing populations with particular values of some phenotype. So far, no assumptions whatsoever have been made about the carrying capacity function $K(x)$, and only minimal assumptions about the competition functions $C(x, y)$, specifically that C is a function of the square of the distance in phenotype space $C(x, y) = C((x - y)^2)$, and that $C(x, x) = 1$ for all phenotypes x. To proceed, we must be more specific. In this paper we will take the approach of specifying the competition function once and for all, and then considering various subcases of carrying capacity. In particular, we will assume the competition function to be Gaussian $C(x, y) = e^{\frac{-(x-y)^2}{2\sigma_C^2}}$, as is commonly done in the biological literature [4,5]. In this case, the conditions for existence of uninvadable mono- and dimorphism are given below. Recall from above that for each class of polymorphism, there are two conditions: a criticality condition in the form of an equation that a phenotype corresponding to an uninvadable population must satisfy which in practice is what must be solved for x^*, and an inequality that states that the curvature of the eigenvalue at the critical point is such that the critical phenotype is a local maximum of the eigenvalue.

3.1   Univadable Monomorphism

$$\text{Criticality: } K'(x^*) = 0 \tag{17}$$

Recall that this condition for uninvadability depends only on the requirement that $C(x, x) = 1$ for all x.

$$\text{Curvature: } K''(x^*) + \frac{1}{\sigma_C^2} K(x^*) < 0 \tag{18}$$

3.2   Uninvadable Dimorphism

$$\text{Criticality: } K'(x^*) - \frac{x^*}{\sigma_C^2} \left(\tanh(\frac{x^{*\,2}}{\sigma_C^2}) - 1 \right) K(x^*) = 0 \tag{19}$$

$$\text{Curvature: } K''(x^*) + \frac{2x^*}{\sigma_C^2} K'(x^*) + \frac{1}{\sigma_C^2} K(x^*) < 0 \tag{20}$$

4 Gaussian Carrying Capacity: Competitive Exclusion Principle

We now specialize to Gaussian carrying capacity functions. That is, we will consider the two carrying capacity functions

$$K_G^{(1)}(x) = K_0 e^{\frac{-x^2}{2\sigma_K^2}} \tag{21}$$

which describes a single Gaussian distribution centered about the location $x = 0$ (chosen without loss of generality). There is a great deal of work on the model above. This is a classic example where the competitive exclusion principle would be expected to hold, wherein two species competing for a single resource must result in one species outcompeting the other. We will see that monomorphic uninvadability can happen in this context, but this univadable state can also lose stability as ecological parameters are varied, and, as we will see, no uninvadable dimorphism exists emerges to take its place.

From (17), the only possible uninvadable phenotype is $x = 0$, and the curvature condition there is

$$\frac{1}{\sigma_C^2} - \frac{1}{\sigma_K^2}, < 0 \tag{22}$$

that is, the state that maximizes the carrying capacity is uninvadable as long as the variance of the carrying capacity is less than the variance of the competition function. This is well known in the biology literature, [6,4]. If we define the ecological parameter whose variation we study to be $\rho = (\frac{\sigma_K}{\sigma_C})^2$, the condition for an uninvadable monomorphism at $x = 0$ to exist can be written $\rho < 1$. This is a well known result.

Next we consider dimorphic solutions. Recall that dimorphic solutions become asymptotically stable in the subspace associated with the dimorphic phenotypes when the monomorphic solution loses asymptotic stability. As mentioned previously, this occurs by virtue of a transcritical bifurcation as ρ crosses one. However, as we will see, none of the symmetric dimorphisms can possibly be uninvadable, meaning that when the monomorphic solutions become unstable due to variation of a parameter, no stable, univadable dimorphism emerges from a branching event as we will see does occur in the two resource case.

To see all this, note that the condition (19) that determines possible critical symmetric dimorphisms x^* becomes

$$\frac{K'(x^*)}{K(x^*)} = -\frac{x^*}{\sigma_K^2} = \frac{1}{\sigma_C^2}[\tanh(\frac{x^{*\,2}}{\sigma_C^2}) - 1] \tag{23}$$

which can be solved to obtain

$$x^* = \frac{\sigma_C}{\sqrt{2}} \sqrt{\ln(\frac{2\sigma_K^2 - \sigma_C^2}{\sigma_C^2})} = \frac{\sigma_C}{\sqrt{2}} \sqrt{\ln(2\rho - 1)} \tag{24}$$

from which it is seen that the critical symmetric dimorphic solution exists only if $\rho > 1$, which is the same as the condition for the monomorphism with $x = 0$ to become unstable. The curvature condition for uninvadability of the symmetric dimorphism at x^* can be written

$$\frac{K(x^*)}{\sigma_K^2}[\rho - 1 + \ln(2\rho - 1)\frac{1 - 2\rho}{2\rho}] < 0 \qquad (25)$$

or

$$k(\rho) \doteq [\rho - 1 + \ln(2\rho - 1)\frac{1 - 2\rho}{2\rho}] < 0 \qquad (26)$$

and it is seen in Figure (1) that the condition is never satisfied for $\rho > 1$, meaning that the dimorphic solution can be invaded by all nearby phenotypes. This result is also known by biologists who suggest that the end result of this property might be a population with a continuum of phenotypes [5,6,7,8].

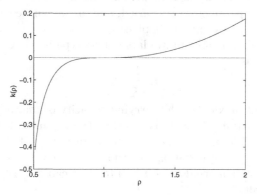

Fig. 1. Curvature of critical branching dimorphism as a function of $\rho = \left(\frac{\sigma_K}{\sigma_C}\right)^2$. Since $k(\rho) > 0$ for all $\rho > 1$, the dimorphism is always invadable.

This result implies that no evolutionarily stable coalitions with more than one phenotype can exist.

5   Quadratic Carrying Capacity: Stable Coexistence

To conclude, we show that if instead of Gaussian, the carrying capacity is a quadratic function

$$K(x) = \begin{cases} K_0(1 - (\frac{x}{a})^2), & |x| \leq a \\ 0, & |x| > a \end{cases} \qquad (27)$$

where a is a measure of the region of parameter space where the capacity is nonzero, that there can be a bifurcation in which an uninvadable monomorphism loses stability as an uninvadable dimorphism branches from the monomorphic critical point. Since the quadratic function is identical to a Gaussian at quadratic order near the critical point when $\frac{a}{\sqrt{2}}$ is identified with σ_K, it is interesting that the details of the functions away from the bifurcation point can result in such different evolutionary behavior.

For the quadratic carrying capacity, the critical point is again $x^* = 0$, and the curvature condition 17 becomes

$$\sigma_C > \frac{a}{\sqrt{2}} \tag{28}$$

so we again obtain the result that the critical monomorphism is uninvadable for sufficiently large variance of the competition function σ_C.

Turning to the symmetric dimorphism, the criticality condition for the phenotype of the dimorphism 19 becomes

$$\frac{2\frac{\sigma_C^2}{a^2}}{1 - \frac{x^2}{a}} = 1 - \tanh\left(\frac{x}{\sigma_C}\right)^2. \tag{29}$$

Since the left side of 29 for $x < a$ has a minimum at $x = 0$, while the right hand side has a maximum at $x = 0$, it is obvious that solutions to this equation for $x < a$ must bifurcate off the origin as σ_C decreases. The curves achieve tangency at $x = 0$ when $\sigma_C = \frac{a}{\sqrt{(2)}}$, so that a critical symmetric dimorphism begins to exist exactly when the critical monomorphism at $x^* = 0$ becomes invadable, just as in the case of Gaussian carrying capacity. See figure 2

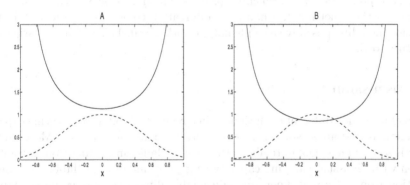

Fig. 2. The left hand side (*solid curve*) and right hand side (*dashed curve*) of 29 for $a = 1$. Figure A is for $\sigma_C = .65$ and Figure B is for $\sigma_C = .75$.

The curvature condition for uninvadability of the critical dimorphism 20 becomes

$$1 - \frac{2\sigma_C^2}{a^2} - \frac{5x^2}{a^2} < 0, \tag{30}$$

and unlike the case of Gaussian carrying capacity, the condition 30 does hold on the critical symmetric dimorphism for values of σ_C less than the critical value $\frac{a}{\sqrt{2}}$, so that in this case, the population does exhibit an uninvadable dimorphism. See Figure 20

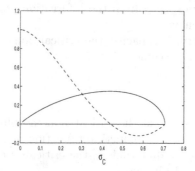

Fig. 3. The value of x^* (*solid curve*) and the curvature condition 30 (*dashed curve* plotted against σ_C

An interesting feature of Figure 3 is that the critical dimorphism becomes invadable around $\sigma_C = 0.4410$ which also is where the critical nonzero phenotype x^* achieves its maximum value. This suggests that the branching process will continue as σ_C continues to decrease with uninvadable coalitions with more and more phenotypes present as the competition between nearby phenotypes becomes weaker. The analysis becomes somewhat more complicated since there is not obvious what kind of coalition might replace the dimorphism when it becomes invadable. For example, the critical dimorphism could split into a four-phenotype population if a branching process similar to that shown in Figure 2 occurs. On the other hand, a non-local bifurcation could result in a trimorphic coalition including possibly the a phenotype with $x = 0$. This will be investigated in future work.

6   Discussion

The fact that such different evolutionary behavior is exhibited by carrying capacity functions that are so similar near the critical point is unexpected. Most likely, there is something degenerate about the case of Gaussian competition in combination with Gaussian carrying capacity that prevents an univadable dimorphic coalition from emerging when an uninvadable dimorphism loses uninvadability. Therefore, it is possible that the speculation that a real population would approach a continuous distribution of phenotypes may not be accurate. The mathematically generic behavior of branching processes suggest that symmetry breaking cascades may be the mechanism by which multi-phenotype populations arise in real populations.

References

1. Golubitsky, M., Stewart, I., Schaeffer, D.: Singularities and groups in bifurcation theory: Volume 2. Applead Mathematical Sciences, vol. 69. Springer-Verlag, New York, Inc., New York (1988)

2. Rueffler, C., Van Dooren, T.J.M., Leimar, O., Abrama, P.A.: Disruptive selection and then what? Trends in Ecology and Evolution 21, 238–245 (2006)
3. Otto, S.P., Day, T.: A biologists guide to mathematical modeling in ecology and evolution. Princeton University Press, Princeton (2007)
4. Dieckmann, U., Doebeli, M.: On the origin of species by sympatric speciation. Nature 400, 354–357 (1999)
5. Bolnick, D.I.: Mulit-species outcomes in a common model of sympatric speciation. J. Theoretical Biology 241, 734–744 (2006)
6. Roughgarden, J.: Evolution of niche width. American Naturalist 106, 683–718 (1972)
7. Barton, N.H., Polechova, J.: The limitations of adaptive dynamics as a model of evolution. J. of Evolutionary Biology 18, 1186–1190 (2005)
8. Polechove, J., Barton, N.H.: Speciation through competition: A critical review. Evolution 59, 1194–1210 (2005)

Large-Scale Conflicts
in Massively Multiplayer Online Games

Rogelio E. Cardona-Rivera[1,2], Kiran Lakkaraju[1],
Jonathan H. Whetzel[1], and Jeremy R. Bernstein[1,3]

[1] Sandia National Laboratories
P.O. Box 5800
Albuquerque, NM, 87185
[2] Digital Games Research Center
North Carolina State University, Campus Box 8206
Raleigh, NC 27695
[3] University of New Mexico
1924 Las Lomas Rd.
Albuquerque, NM, 87131
{recardo,klakkar,jhwhetz,jrberns}@sandia.gov

Abstract. Complex systems are of interest to the scientific community
due to their ubiquity and diversity in daily life. Popularity notwithstanding, the analysis of complex systems remains a difficult task, due to the
problems in capturing high-volume data. Massively Multiplayer Online
Games (MMOGs) have recently emerged as a tractable way to analyze
complex system interactions, because these virtual environments are able
to capture a great amount of data and at high-fidelity, often tracking
the actions of many individuals at a time resolution of seconds. MMOGs
have been used to study phenomena such as social networks and financial
systems; our focus is to identify behaviors related to Large-Scale Conflict (LSC). In this paper, we review how one particular MMOG allows
large-scale complex behavior to emerge and we draw parallels between
virtual-world LSCs and real-world LSCs. The LSC-related behavior that
we are interested in identifying deals with the conditions that lead a
participant in the virtual environment (a game player) to engage and
actively participate in a LSC, with the goal of informing an agent-based
model that predicts when any one player is likely to engage in conflict.
We identify virtual world behavioral analogues to real-world behavior of
interest (i.e. insurgent behavior), and link the virtual behavior to a (previously derived) theoretical framework that analyzes the determinants of
participation in the civil war of Sierra Leone. This framework identifies
three general theories that collectively predict participation in civil war
(a type of LSC); we operationalize one of the theories (Theory of Social
Sanctions), and look at how insurgent behavior can occur as a function
of community networks, which are assumed to impose social sanctions
for non-participation in an LSC.

K. Glass et al. (Eds.): COMPLEX 2012, LNICST 126, pp. 40–51, 2013.
© Institute for Computer Sciences, Social Informatics and Telecommunications Engineering 2013

1   Introduction

Understanding and anticipating changes in complex social systems, such as those relating to economies, financial institutions, and conflict is a problem of importance for national security. The complexity of these systems, such as the large number of factors and the human element, makes gathering data and running controlled experiments difficult. Promising methods such as modeling and simulation have made headway, however they are also subject to additional complexity issues and may face limited applicability.

Recently, a new source of data has emerged that can help in understanding complex social systems – data from Massively Multiplayer Online Games [1]. MMOGs afford and promote complex social interactions amongst hundreds to thousands of players in online fictional worlds, attracting players from a wide variety of backgrounds, age groups, and genders. MMOGs serve as a tractable way of analyzing complex social interactions, due to two important features. Firstly, they serve as environments with a high-degree of expressivity, i.e., they allow the participants (also known as "players") to pursue a wide variety of complex social actions, in broad categories such as peer-to-peer and group communication, economic trading, and congregating with other players. Secondly, due to the virtual nature of the environments, MMOG's are able to capture a great amount of data and at high-fidelity, often simultaneously tracking the actions of all individuals in near-real time. MMOG's have been used to study phenomena such as education [2,3], social networks [4,5], and financial systems [6,7]; our focus is to identify behaviors related to Large-Scale Conflict (LSC).

In this paper, we review how one particular MMOG allows large-scale complex behavior to emerge and we draw parallels between virtual-world LSC's and real-world LSC's. Our data set covers the actions of players in this MMOG for a period greater than one year, which we analyzed to identify the following LSC behavior of interest: under what conditions does a game player participate in a LSC? We analyzed this information with the goal of informing an agent-based model that predicts when any one person is likely to engage in conflict. Our methodology involves identifying virtual world behavioral analogues to real-world behavior of interest (i.e. insurgent behavior), and analyzing the virtual behavior with real-world predictive models of participation in LSC. To that effect, we employ a (previously derived) theoretical framework that analyzes the determinants of participation in the civil war of Sierra Leone. This framework identifies three general theories that collectively predict participation in civil war (a type of LSC); we operationalize one of the theories (the *Theory of Social Sanctions*), and look at how virtual insurgent behavior can occur as a function of community networks, which are assumed to impose social sanctions for non-participation in a LSC. These communities are defined by communication patterns, as well as virtual group co-memberships. Generally, our hypotheses predict that the more members of a players community are involved in a LSC, the more likely the player will engage and be active in the LSC. Our results apply to a virtual setting, and we discuss how they might generalize to real-world setting, which is of primary concern.

2   Details of Our MMOG Environment

Our data was obtained from a Massively Multiplayer Online Game, which is confidential in nature. To protect the identity of the parties involved in the game's creation and management, we have anonymized some of the terminologies in the game's description. None of the game's interactions or mechanics have been changed.

2.1   Overview

Game X is an open-ended free-to-play Massively Multiplayer Online Game. Players can pursue a variety of different roles and interact with other players (real and artificial) in the virtual world. In Game X, every player commands one vehicle, with a set cargo capacity as well as defensive and offensive capabilities. Players use this vehicle to explore an open, persistent game world.

Game X is unique in that the game does not impose an explicit goal structure and a player cannot win the game. Instead, the game encourages players to make up their own goals, role-play and to acquire wealth, fame, and power, in an environment driven by several societal factors, such as friendship, cooperation, competition, and conflict.

Players have a limited number of "turns" per day. Nearly all actions cost some number of turns to execute. Turns are replenished per hour. The basic categories of actions players can undertake are: economic, social, and combat.

2.2   Economic Activities

Players can mine goods integral to improving their economic performance in game. To obtain other goods, they may engage in trade, which can be a form of barter or via the use of *marks* an in-game currency. With enough game experience and the proper amount of marks, a player can construct a factory outlet, which can manufacture sellable goods. In addition, a player can eventually earn enough resources to build and maintain a market center, which can facilitate bartering and selling with passerby players.

Players can also be strategic about the location of their factories and market centers, by exploring critical city areas that are ripe for developing profitable trade routes. By the same token, players must take care to avoid establishing business in areas that are targeted by pirates; other players can plunder factory outlets, market centers, as well as vehicles.

In addition to piracy, other "illegal" (in terms of society, not in terms of what is permissible in game) options for economic activity also exist: a player may elect to bootleg illegal goods. If caught, a player can be subject to social sanctions and be barred from bartering and (if it was a repeat offense) from being able to engage socially with other players. Other sanctions include: not being able to repair your vehicle, and not being able to approach/visit city areas.

2.3   Social Activities

Players in Game X can also socialize and associate with other players through a number of different ways. In fact, to be successful in the game world, it behooves players to forge social partnerships.

To socialize, players can post on public forums, send personal messages to other players as well as broadcast messages top specific groups of players. An important feature of communication methods in Game X is that they do not cost players any amount of available turns; communication is "free of charge."

To associate, players can join one of three pre-defined nations; players can only affiliate with one nation at a time and doing so yields certain benefits, such as the possibility to gain access to nation-specific technology. Players may also elect to not join a nation, which allows them to be exempt from nation-centric wars. In addition, players can create or join a guild, which are entities designed to combine groups of players and allow them to operate for (possibly) common aims. Guild membership is independent of nation membership. For both nation and guild memberships, if a player accumulates sufficient in-game experience, he or she can be promoted to a senior level, which commands a higher influence in the respective nation or guild. In fact, senior level members of nations command considerable political power in the decision to go to war (see Section 2.5).

Finally, players can designate other players as friends or hostiles, which facilitates or hinders communication and other game activities with those players. Friend/hostile tables are completely private, meaning that no one except the labeling and labeled players has information about ties between them (i.e. it is not possible to see second degree neighbors, such as friends of friends).

2.4   Combat Activities

Players can engage in combat with other players (real and artificial), as well as with factory outlets and market centers. Players can outfit their vehicles with a variety of different weapons and defensive armors that (alongside a player's skill) can be used to give certain advantages in battle.

Players have an array of skills they can improve based upon their successful in game battles. Higher skill values increase the probability of successful combat in the future.

2.5   Potential for Large Scale Conflict

Large Scale Conflicts (i.e. wars) are socially centric and very related to combat activities. Wars are only possible between the three pre-defined nations. Each nation can have one of the following diplomatic relations to all others: Benign, Neutral, Strained, or Hostile.

The senior members of a nation constitute the nation's governing body. Every day, each nation's governing body convenes and each of the senior members chooses a disposition with regards to diplomatic relations with the other nations. Non-senior members cannot vote, but can exert influence by lobbying senior members to vote a certain way. If enough members of a governing body select hostile diplomatic relations against another nation, a war is declared between the respective nations.

When a war has broken out, additional combat actions are available for the warring nations. In particular, war quests are available, which provide medals of valor to the players that wish to undertake and complete the quests. Any attack against the opposing nation (be it in the form of a war quest or not) results in accumulating a set number of war points. When the war ends, these war points determine the "winner" of the large-scale conflict. A war situation will (via the game's design) gravitate towards a state of peace. Each of the respective governing bodies must maintain a majority vote to continue the war effort. Over time, the amount of votes required to continue is increased by the game itself. Eventually, no amount of votes will suffice and the nations return to a state of peace.

2.6   A Player's Death

A player cannot permanently die in Game X. If an enemy destroys a player's vehicle, then the player loses a fixed amount of skill points, as well as all the cargo on his or her vehicle and in addition loses some available actions for the play session.

3   Validity of Our Work

Our primary interest is to identify under what conditions is a player likely to engage in conflict. As previously mentioned, the idea is to use our MMOG environment as a testbed for theories that predict participation in large-scale conflict; we posit that our virtual environment is a reasonable proxy for behavior that we could expect to see in the real world, an issue which we address in this section.

3.1   Mapping Virtual Worlds to Real World Phenomena

Clearly several differences exist between a virtual world and the real world; especially when looking at phenomena such as combat. Of most pertinent interest is the fact that players cannot die or experience physical harm in the virtual world. As mentioned before, Game X does not have permanent death.

Although these are important issues, we think they are negligible given that players are interested in their characters. Decisions made by the players may not be as emotionally driven (for instance, they may not care if their friends are being harmed), but as long as there is involvement in the characters by the players, their decisions may be driven by the need to preserve their characters' well-being. Since many of the players we have studied have played a significant amount of time, we argue that players are involved. While intensity of emotions will vary (from the real world), the underlying actions are still driven by a need to preserve the character well-being and social relations, similar to the real world.

One may also ask whether players are choosing different persona's or acting differently as they would in the real world. This question is out of the scope of this work but it is a question that is being intensively studied in this general area. Several studies indicate that player characteristics influence player behaviors in the virtual world; for instance, personality traits are correlated with behaviors in game ([8,9]).

4   Large Scale Conflicts in Our MMOG

4.1   Analogues to Real World Large Scale Conflict

Despite some of the issues highlighted in the previous section regarding mapping virtual world to real world phenomena, we noticed very clear analogues between virtual- and real-world LSC. In particular, during both war periods of Game X, we noticed a significant spike in both the number of combat attacks and the number of messages sent, as seen in Figure 1. This is consistent with real-world accounts of conflict, in which participants of armed conflict see an increase in mobilization and coordination for combat.

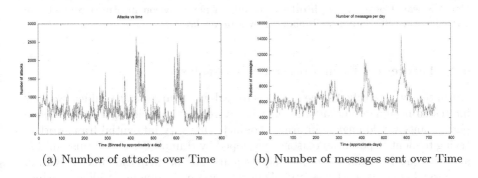

(a) Number of attacks over Time   (b) Number of messages sent over Time

Fig. 1. Combat and messaging patterns throughout the entire Game X data set. The number of combat attacks and the number of messages sent both spike during the periods of war, consistent with real world accounts of conflict.

In addition, our virtual LSC is consistent with real LSC as it relates to the number of participants that participate in combat actions. Specifically, our virtual world LSC exhibits a pattern over the number of attackers that pursue different levels of attacks, which is consistent with the power-law distribution [10] as exhibited in other accounts of participation in armed conflict [11]. This distribution is illustrated in Figure 2.

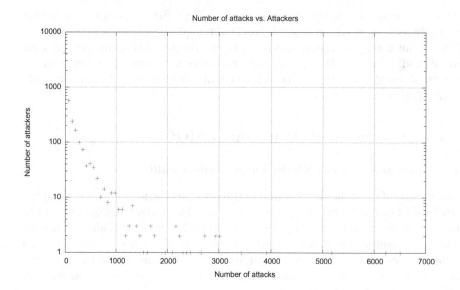

Fig. 2. Distribution of Number of Attacks v. Number of Attackers during the first Game X war. Consistent with other accounts of participation in armed conflict, our virtual world LSC data is well-modeled by a power law.

4.2   Theoretical Framework for LSC Analysis

Given that we identified some parallels to real-world large-scale conflict, we were interested in understanding *a priori* what were the theoretical potential reasons participants would engage in armed conflict. One very influential framework in trying to identify potential reasons, developed by Humphreys and Weinstein [12], analyzed the determinants of participation and non-participation in the civil war of Sierra Leone. Humphreys and Weinstein originally sought to identify which of three "competing" theories better explained insurgent and counter-insurgent participation and non-participation in Sierra Leone. Through their analysis, they found support for all three theories, suggesting that the theories should not be taken in contrast to each other, but rather as an ensemble, capable of identifying multiple influencing factors that affect participation in an armed conflict. One of the three theories was particularly interesting due to its applicability to our game environment: *The Theory of Social Sanctions*. This theory predicts that an individual's participation in large-scale conflict is a function of the community that the individual is a part of. If the community is *strong*, then it can bring to bear a social pressure that will prompt individuals to fight in the conflict on behalf of their respective community. A strong community is (for instance) defined by (1) shared core beliefs and values, (2) close and many-sided relationships between the community's constituents, and (3) activities of reciprocity between the community's constituents [13].

4.3 Behavior of Interest: Dimensions of Combat Behavior as a Function of Community Networks

We operationalized the Theory of Social Sanctions in the context of our game, and were interested in answering the following questions, solely on the basis of an individual's community:

- Will the person engage or not? (i.e. participation)
- Will the person be an active agent in the engagement or not? (i.e. activeness)
- How fast will the person engage? (i.e. time to first response)

Thus, we developed the following set of hypotheses, that are explored in the remainder of this paper.

◇ H1: The greater the amount of community participation for a player, the more likely the player will participate in conflict.
◇ H2: The greater the amount of community participation for a player, the more the player will participate in conflict.
◇ H3: The greater the amount of community participation for a player, the faster the player will participate in conflict.

5 Experimental Methodology

5.1 Operationalizing the Hypotheses

To make some of the hypotheses more precise, we introduced operational definitions for the terms "participation," "community," and "community participation" To be considered a participant of the virtual LSC, an individual had to commit at least one combat action during the war period under study. An individual's community could be defined in three ways; each of them represents a different dimension of interaction between the players of Game X and tried to capture the spirit of the definitions used by Humphreys and Weinstein [12].

Definition 1 (Friendship Community Definition). *Let p and pc be players. For any pc ≠ p, pc is in p's community if p and pc are bidirectional friends and pc is not in p's hostile table. A bidirectional friendship between two players p and pc exists when p is on pc's friend table and vice-versa.*

Definition 2 (Communication Community Definition). *Let p and pc be players. For any pc ≠ p, pc is in p's community if pc actively communicates with p. Specifically, p and pc are in each other's community if they send and receive at least 4 messages between them during the war period under study.*

The threshold of 4 messages is arbitrary, and was chosen on the basis of the trend of sent and received messages across all players during the war period under study. Specifically, during the first war period, approximately 50% of players had less than 4 messages sent and received.

Definition 3 (Guild Co-Membership Community Definition). *Let p and pc be players. For any $pc \neq p$, pc is in p's community if p and pc belong to the same guild for a majority of the war period under study.*

A natural inclination is to use the intersection of all three communities as a definitive measure of community. However, such a combination did not yield statistically significant results. In addition, we also felt it better to study different types of communities to see whether or not the expected behaviors appeared throughout. Therefore, each hypothesis has three variants, one for each definition of community. Finally, to define community participation, we chose to represent it as the proportion of players within a community that were active during the war period under study. This ensured that the numbers weren't too biased for large communities, by ensuring that all the community participation statistics varied within a common range ($[0.0 - 1.0]$).

5.2   Participants

Our data set includes data for over 50,000 players across >700 days. The data set is historical, beginning in 2007. All participant data has been anonymized and all players agreed as part of Game X's sign-up process to have their data collected for purposes of scientific research. Despite the magnitude of the data, only a small percent of players were actually considered as part of the analysis; several players did in fact sign up, but did not participate enough in Game X to consider their presence meaningful. A great majority of players did not sign-in to play for more than 10% of the entire data set time period. After excluding these players, our participant pool was reduced to 6,156 players. Approximately 13.56% of players were female and 86.44% were male.

Participants of Interest. Our focus was on the first Game X war. After having filtered the data once to remove inactive players, we filtered the data again, this time filtering by two measures: "cumulative actions taken prior to war period" and "log-in percentage". In our description of Game X in Section 2.1, we discussed how players had a limited number of turns per day; "cumulative actions taken prior to war period" is a measure of how many turns they have taken per day across all days prior to the start of the first Game X war. The threshold for consideration was 500,000 turns taken prior to the start of the first Game X war. The reason for filtering by actions taken was because we wanted to control for players who were signing-in and not doing anything, which does not represent the behavior we were interested in studying. The measure "log-in percentage" is defined similarly to how it was defined previously. However, the threshold for consideration was 80% as opposed to the original 10%. These combined filters reduced our participant pool from 6,156 players to 981 players. Of these, approximately 11.62% were female and 88.38% were male.

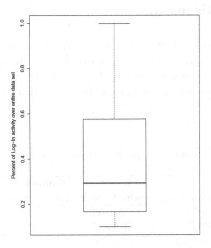

Fig. 3. Box plot of log-in percentage. Log-in percentage is determined by the number of days that players logged in over the total number of days in our data set. We filtered the data for all players whose log-in percentage was less than 10%.

5.3 Hypothesis Tests

Our hypothesis testing was restricted to the first Game X war. For all cases of the hypothesis tests, the variable `communityParticipation` is the proportion of the player's community that was active during the first Game X war. Given the operational definitions, and our hypothesis set, we tested each hypothesis as follows:

◇ H1: Likelihood → Logistic Regression of the variable `hadWarAction`, which took value 1 if the player had at least one combat action during the first Game X war and 0 otherwise, over the variable `communityParticipation`. H1 thus predicts that the greater the proportion of `communityParticipation`, the more likely the response variable `hadWarAction` will be 1.

◇ H2: Amount → Ordinary Least Squares Regression of the variable `numWarAction`, which is a count of the number of combat actions taken during the first Game X war, over the variable `communityParticipation`. H2 thus predicts that the greater the proportion of `communityParticipation`, the higher the variable `numWarAction` will be.

◇ H3: Time to First Attack → Survival Analysis of the variable `timeOfFirstAttack`, which is a number indicating the day the player's first combat action during the first Game X war was registered, over the variable `communityParticipation`. H3 thus predicts that the variable `communityParticipation` will lower the survival function of the variable `timeOfFirstAttack`; in other words, `communityParticipation` will predict how quickly a player commits his or her first combat action as measured by `timeOfFirstAction`;

6 Results and Discussion

The results of our hypothesis tests are shown in Table 1. Survival analysis for H3 did not yield statistical significance for any case, thus we omit the results from the table. We achieved different results under different community definitions, with the communication community not yielding statistical significance for either hypothesis.

Table 1. Hypothesis Test Results

Hypotheses	Community Definitions		
	Guild Community	Friend Community	Communication Community
H1: Likelihood	1.633	-4.703	0.404
(Logit)	[0.604]**	[0.826]***	[0.247]
H2: Amount	0.669	-9.748	-1.952
(OLS)	[1.697]	[2.394]***	[1.124]

Notes: Standard errors are in brackets. **Significant at 5%; ***Significant at 1%.

The results found indicate a surprising interplay – while friend community participation does *not* affect one's likelihood to participate, your guild community does. This seems to indicate the importance of guild membership and community over other relationships. The non-significant result for H2 with guild community is odd, since from H1 we would assume a greater number of attacks. We posit that a "free rider" effect may be occurring, where individuals in a community may participate, but leave the bulk of combat to others who are better suited for combat. In further work we are looking at the nature of guilds and whether they have a heterogeneous set of players (in terms of skills).

7 Conclusion

Massively Multiplayer Online Games (MMOGs) are a new source of data that help in understanding complex systems. MMOGs serve as a tractable way of analyzing complex social interactions, due to two important features. Firstly, they serve as environments with a high-degree of expressivity, i.e., they allow the participants (also known as "players") to pursue a wide variety of complex social actions, in broad categories such as peer-to-peer and group communication, economic trading, and congregating with other players. Secondly, due to the virtual nature of the environments, MMOG's are able to capture a great amount of data and at high-fidelity, often simultaneously tracking the actions of all individuals in near-real time.

In this report we have compared one particular type of complex behavior, large scale conflict (LSC), in a MMOG (Game X) and in the real world. A high level similarity was seen – a power law distribution of the number of attacks vs. the number of attackers – which corresponds to known patterns in real world conflict. Assessment of community influence on player participation was surprising, as individual friendships did not have a positive influence; however guild communities did.

This lack of correspondence is interesting, and there may be several causes for it. Firstly, the hypotheses we tested were developed (and evaluated) with data from a civil war. This type of conflict may not be as relevant for Game X conflicts. Secondly, there could be a strong "free rider" effect – the more my community is willing to assume the cost of war, the more likely I am going to abstain and "free-ride" on their participation. This could explain the negative interaction for the friend community case. Thirdly, there may be a division of labor within guilds; with some individual being more combat oriented and others more economically oriented. This could explain why there was a positive effect for H1+guild community, but a non-significant effect for H2+guild community – others may be taking on the combat roles.

Further work will focus on addressing these issues to identify the reasons for why players participate in LSCs.

References

1. Bainbridge, W.S.: The Scientific Research Potential of Virtual Worlds. Science 317(5837), 472–476 (2007)
2. Steinkuehler, C.A.: Learning in Massively Mulitplayer Online Games. In: Proceedings of the 6th International Conference on Learning Sciences, pp. 521–528 (2004)
3. Steinkuehler, C.A.: Cognition and Literacy in Massively Multiplayer Online Games. In: Handbook of Research on New Literacies, pp. 1–38. Erlbaum, Mahwah (2008)
4. Ducheneaut, N., Yee, N., Nickell, E., Moore, R.J.: Alone together?: Exploring the Social Dynamics of Massively Multiplayer Online Games. In: Proceedings of the SIGCHI Conference on Human Factors in Computing Systems, pp. 407–416 (2006)
5. Szell, M., Thurner, S.: Measuring Social Dynamics in a Massively Multiplayer Online Games. Social Networks 32(4), 313–329 (2010)
6. Alves, T., Roque, L.: Using Value Nets to Map Emerging Business Models in Massively Multiplayer Online Games. In: Proceedings of the Pacific Asia Conference on Information Systems (2005)
7. Papagiannidis, S., Bourlakis, M., Li, F.: Making Real Money in Virtual Worlds: MMORPGs and Emerging Business Opportunities, Challegens and Ethical Implications in Metaverses. Technological Forecasting and Social Change 75(5), 610–622 (2008)
8. Yee, N., Ducheneaut, N., Nelson, L., Likarish, P.: Introverted elves & conscientious gnomes: The expression of personality in world of warcraft. In: Proceedings of CHI 2011 (2011)
9. Yee, N., Harris, H., Jabon, M., Bailenson, J.N.: The expression of personality in virtual worlds. Social Psychological and Personality Science 2(1), 5–12 (2011)
10. Clauset, A., Shalizi, C.R., Newman, M.E.J.: Power-law Distributions in Empirical Data. Arxiv preprint arxiv:0706.1062 (2007)
11. Bohorquez, J.C., Gourley, S., Dixon, A.R., Spagat, M., Johnson, N.F.: Common Ecology Quantifies Human Insurgency. Nature 462(7275), 911–914 (2009)
12. Humphreys, M., Weinstein, J.M.: Who Fights? The Determinants of Participation in Civil War. American Journal of Political Science 52(2), 436–455 (2008)
13. Taylor, M.: Rationality and Revolution. Cambridge University Press (1988)

Impact of Global Edge-Removal
on the Average Path Length

Lock Yue Chew[1], Ning Ning Chung[2], Jie Zhou[2], and Choy Heng Lai[3,4]

[1] Nanyang Technological University, Division of Physics and Applied Physics,
21 Nanyang Link, Singapore 637371
lockyue@ntu.edu.sg
http://www.ntu.edu.sg/home/lockyue/
[2] National University of Singapore, Temasek Laboratories, Singapore 117508
[3] National University of Singapore, Beijing-Hong Kong-Singapore Joint Centre
for Nonlinear and Complex Systems (Singapore), Kent Ridge 119260, Singapore
[4] National University of Singapore, Department of Physics, Singapore 117542

Abstract. In this paper, we further explore into the impact of link removal from a global point of view. While diseases spread more efficiently through the best spreaders and removal of local links attached to them can have great impact, it is also important to have a method to estimate the cost of edge-removal from a global point of view since the removal of a link may also affect certain global properties of the network. We discuss global strategies on link removal and study their effectiveness in controlling the propagation of infectious diseases based on the spreading control characteristics (SCC). The SCC framework opens up a comprehensive way for researchers to assess and compare the efficacy of their strategies against the potential cost of their implementation from a global perspective.

Keywords: complex network, epidemic control, epidemic spreading, social and economic cost.

Finding an efficient way to slow down the propagation of infectious diseases within a society has always been an important subject in network sciences. Over the past decades, with the availability of large-scale, comprehensive data sets that trace the movement and interaction of the population, there have been increasing number of investigation on the control of the spreading of transmittable diseases [1,2,3,4,5,6,7,8,9,10,11,12,13]. Since the spreading of epidemics affects the society as a whole, the study of its control should not be separated from the associated economic and social dimensions. For this reason, we have recently examined into the efficacy of local control strategies on the propagation of infectious diseases by removing local connections attached to the best spreaders with the associated economic and social costs taken into account [14]. When local links attached to the best spreader are removed, we found an increase in the spreading time while the centrality betweenness of the best spreader decreases as a result of the reduced connectivity of the network topology. Nevertheless, our studies reveal that it is possible to trade minimal reduction in connectivity

K. Glass et al. (Eds.): COMPLEX 2012, LNICST 126, pp. 52–57, 2013.
© Institute for Computer Sciences, Social Informatics and Telecommunications Engineering 2013

of an important hub with efficiencies in epidemic control by removing the less busy connections. In other words, we have uncovered the surprising results that removing less busy connections can be far more cost-effective in hindering the spread of the disease than removing the more popular connections.

In this article, we study the impact of edge-removal on the global property of the network. Specifically, when a link is removed, instead of investigating the local cost paid by the best spreader, we are interested to know the impact of the removal on the average path length (APL) of the network. Most real-world networks have very short average path lengths such that communication within the network can be carried out efficiently. In the context of a small world, this means that any two nodes in the network are well connected through a very short path. As edges are removed, connectivity of a network decreases. Hence, if an edge-removal strategy is employed for epidemic control, a cost has to be paid by all of the nodes in the network, i.e. a decrease in the efficiency of information or mass transport in the network. Our aim is to look for an optimal edge-removal strategy that allows us to trade a minimal reduction in the connectivity of a network with efficiencies in epidemic control.

For this, we quantify the relative effectiveness (E_{ij}) of the removal of an edge ij in increasing the epidemic threshold by the decrease in the extreme eigenvalue ($\Delta\lambda_{ij}$) of the network adjacency matrix, normalized by the maximum eigenvalue of the original network, i.e.

$$E_{ij} = \frac{\Delta\lambda_{ij}}{\lambda_m}. \tag{1}$$

Note that E_{ij} captures how removal of link ij increases the difficulty for the epidemic outbreak to take place. On the other hand, the global cost (C_{ij}) of an edge removal is quantified by the decrease in the network average inverse path length (AIPL) [15], ΔL_{ij}, normalized by the average inverse path length of the original network, L_0:

$$C_{ij} = \frac{\Delta L_{ij}}{L_0}. \tag{2}$$

Here, AIPL is used instead of APL since edge-removal may fragment the network into disconnected components, causing APL of the network to diverge, but AIPL of the network remains finite.

Then, we simulate the global impact of the removal of a single edge in three real-world complex networks from different fields, i.e.: 1) the US air transportation network [16], 2) the collaboration network in computational geometry [17] and 3) the Gnutella peer-to-peer internet network [18,19]. As shown in Fig. 1, E_{ij} and C_{ij} relate differently to edge betweenness (B_E) [20] of the removed link. The most effective link-removal does not necessary comes with the largest cost to pay. This implies that it is possible to have a optimized global gain in the control of epidemic spreading if the link to be removed is picked properly. For global edge-removal, we define the gain in spreading control for removing edge ij as:

$$G_{ij} = \frac{\Delta\lambda_{ij}}{\lambda_m}\frac{L_0}{\Delta L_{ij}}. \tag{3}$$

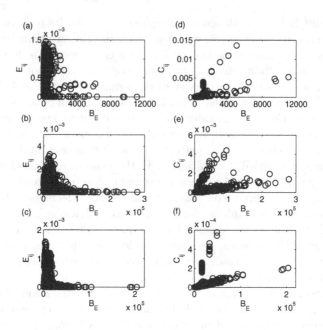

Fig. 1. Dependence of E_{ij} and C_{ij} on B_E of the removed link for (i) the US air transportation ((a) and (d)), (ii) the collaboration((b) and (e)) and (iii) the Gnutella ((c) and (f)) networks

Next, we look at four different global edge-removing strategies which involve removal of more than one link, namely, removing the edges following (1) the decreasing order of the edge betweenness, (2) the increasing order of the edge betweenness, (3) the decreasing order of the product of the k_s [21,22] values, and (4) the decreasing order of the gain. Note that the product of the k_s values of an edge ij is defined as the product of the k_s values of nodes i and j. For each strategies, edges are removed one by one according to their original ranking without further updating of the ranking during the process of removal. We measure the cost of removing q links (following a specified order according to the adopted strategy) on the network by the decrease in the network average inverse path length (AIPL), $\Delta_q L$, normalized by the average inverse path length of the original network, L_0:

$$C_q = \frac{\Delta_q L}{L_0}. \tag{4}$$

Meanwhile, the effectiveness in epidemic control of removing the q links are measured by

$$E_q = \frac{\Delta_q \lambda}{\lambda_m}, \tag{5}$$

where λ_m is the maximum eigenvalue before edge removal. By studying the global spreading control characteristics (GSCC) curves, i.e. the $E_q - C_q$ curve, one can analyze the difference between various global edge-removing strategies.

In Fig. 2, we show the GSCC curves for the three networks when edges are removed according to the four global edge removal strategies. Strategy 2 performs better than strategy 1 in both the US airline and the Gnutella networks. In these cases, removal of the less busy connections shows higher cost-efficiency than removal of the busy connections. However, in the collaboration network, strategy 2 performs better than strategy 1 only at the earlier stage of edge-removal. For strategy 4, we observe that the gradient near the end of the curves are larger than the earlier parts of the curves. This implies that the edge-removal strategy is not optimized here to yield the largest gain at each step. In particular, as shown in Fig. 2 (b) for the collaboration network, strategy 4 performs better only at the early stage of the edge-removal. After which, the curve is flat with almost no increase in E_q as C_q increases to 0.4. The gradient of the curve increases after that and drops quickly to zero again. In contrast, the performance of the third strategy is consistently better in all the three networks.

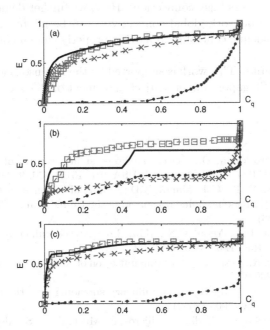

Fig. 2. Global spreading control characteristics for (a) the US air transportation, (b) the collaboration and (c) the Gnutella networks. Note that SCC are plotted as dashed curves with dots, dashed curves with crosses, dashed curves with squares and solid curves for the first, second, third and fourth strategies respectively.

Compared to a local edge-removal strategy, a global edge-removal strategy involves not only the edges attached to the best spreader but all the edges in the network. Hence, the number of removed edges can be a lot larger when a global edge-removal strategy is considered. After the removal of a large number of edges, the topological structure of the network can be very different and the ranking of the gain of the rest of the edges can change drastically. When this happens, one cannot obtain the optimal tradeoff between spreading control effectiveness and the cost to pay following the original order of the gain. Therefore, for global link-removal, removing edges following the highest gain does not consistently give optimal epidemic control efficiency as it does for local link removal. While one can update the ranking of the gain repeatedly each time after a link is removed to obtain an optimized result, the operation requires a huge amount of computational power. Specifically, it takes a time of $O(MN)$ to calculate ΔL of all the M edges in a network with N nodes. Therefore, we suggest to adopt strategy 3 if more than 50% of the effectiveness is required.

In conclusion, from both the local and global point of view, removal of the most popular connections is not the most cost-effective solution in slowing down the spreading of either viruses or rumors. In our earlier study on the local edge removal strategies, we show that it is possible to maximize the effectiveness in epidemic control with minimal reduction in the centrality of the network's hub by removing the less busy connections. Here, we further demonstrated that the decrease in the network global connectivity can be minimized with better performance in epidemic control by removing properly chosen edges.

Acknowledgements. This work is supported by the Defense Science and Technology Agency of Singapore under project agreement of POD0613356.

References

1. Hufnagel, L., Brockman, D., Geisel, T.: Forecast and control of epidemics in a globalized world. Proc. Natl. Acad. Sci. USA 101, 15124–15129 (2004)
2. Brownstein, J.S., Wolfe, C.J., Mandl, K.D.: Empirical evidence for the effect of airline travel on inter-regional influenza spread in the United States. Plos. Medicine 3, 1826–1835 (2006)
3. Holme, P., Kim, B.J., Yoon, C.N., Han, S.K.: Attack vulnerability of complex networks. Phys. Rev. E 65, 056109 (2002)
4. Pastor-Satorras, R., Vespignani, A.: Immunization of complex networks. Phys. Rev. E 65, 036104 (2002)
5. Marcelino, J., Kaiser, M.: Reducing influenza spreading over the airline network. PLOS Currents, RRN1005 (2009)
6. Kitsak, M., Gallos, L.K., Havlin, S., Liljeros, F., Muchnik, L., Stanley, H.E., Makse, H.A.: Identification of influential spreaders in complex networks. Nature Phys. 6, 888–893 (2010)
7. Zhou, J., Liu, Z.: Epidemic spreading in communities with mobile agents. Physica A 388, 1228–1236 (2009)
8. Chen, Y., Paul, G., Havlin, S., Liljeros, F., Stanley, H.E.: Finding a better immunization strategies. Phys. Rev. Lett. 101, 058701 (2008)
9. Colizza, V., Barrat, A., Barthélemy, M., Vespignani, A.: The role of the airline transportation network in the prediction and predictability of global epidemics. Proc. Natl. Acad. Sci. USA 103, 2015–2020 (2006)
10. Newman, M.E.J.: Spread of epidemic disease on networks. Phys. Rev. E 66, 016128 (2002)

11. Pastor-Satorras, R., Vespignani, A.: Epidemic spreading in scale-free networks. Phys. Rev. Lett. 86, 3200–3203 (2001)
12. Lessler, J., Kaufman, J.H., Ford, D.A., Douglas, J.V.: The cost of simplifying air travel when modeling disease spread. Plos One 4, e4403 (2009)
13. Crépey, P., Barthélemy, M.: Detecting robust patterns in the spread of epidemics: A case study of influenza in the United States and France. Am. J. Epidemiol. 166, 1244–1251 (2007)
14. Chung, N.N., Chew, L.Y., Zhou, J., Lai, C.H.: Impact of edge removal on the centrality of the best spreaders. Europhys. Lett. 98, 58004 (2012)
15. Beygelzimer, A., Grinstein, G., Linsker, R., Rish, I.: Improving network robustness by edge modification. Physica A 357, 593–612 (2005)
16. Colizza, V., Pastor-Satorras, R., Vespignani, A.: Reaction-diffusion processes and metapopulation models in heterogeneous networks. Nature Physics 3, 276–282 (2007)
17. Jones, B.: Computational Geometry Database (February 2002), http://compgeom.cs.edu/~jeffe/compgeom/biblios.html
18. Leskovec, J., Kleinberg, J., Faloutsos, C.: Graph evolution: densification and shrinking diameters. ACM Transactions on Knowledge Discovery from Data (ACM TKDD) 1(1) (2007)
19. Ripeanu, M., Foster, I., Iamnitchi, A.: Mapping the Gnutella Network: Properties of Large-Scale Peer-to-Peer Systems and Implications for System Design. IEEE Internet Computing Journal (2002)
20. Newman, M.E.J.: Scientific collaboration networks. II. Shortest paths, weighted networks, and centrality. Phys. Rev. E 64, 016132 (2001)
21. Seidman, S.B.: Network structure and minimum degree. Social Networks 5, 269–287 (1983)
22. Carmi, S., Havlin, S., Kirkpatrick, S., Shavitt, Y., Shir, E.A.: A model of internet topology using k-shell decomposition. Proc. Natl. Acad. Sci. USA 104, 11150–11154 (2007)

Assessing the Spatial Impact on an Agent-Based Modeling of Epidemic Control: Case of Schistosomiasis

Papa Alioune Cisse[1,2], Jean Marie Dembele[1,2],
Moussa Lo[1,2,3], and Christophe Cambier[2]

[1]LANI, UFR SAT, Université Gaston Berger, BP 234 Saint-Louis, Sénégal
[2]UMI 209 UMMISCO
[3]LIRIMA, M2EIPS
papaaliounecisse@yahoo.fr,
{jean-marie.dembele,moussa.lo}@ugb.edu.sn,
christophe.cambier@ird.fr

Abstract. Given that most mathematical models of schistosomiasis are based on ordinary differential equations (ODE) and therefore do not take into account the spatial dimension of the schistosomiasis spread, we use an agent-based modeling approach to assess environmental impact on the modeling of this phenomenon. We show that taking into account the environment in the modeling process somehow affects the control policies that must be established according to the environmental characteristics of each system that is meant to be studied.

Keywords: Complex systems, mathematical modeling, ODE, Agent-based modeling, Simulation, Schistosomiasis, Epidemic control.

1 Introduction

Epidemiological phenomena often involve a large number of entities - host, vector, pathogen, environment, etc. - that can interact and give rise to complex dynamics ranging over several spatiotemporal scales [9]. These dynamics can have serious health consequences like the spread over large geographical areas and the contamination of a large number of persons. Epidemiological phenomena, because of their evolution that results from the elements interactions, can be described as complex systems. To efficiently study them, it is necessary to go through a process of modeling and simulation [8] in order to produce prediction tools and define prevention and control policies. This is what this paper is meant to, with the specific case of the Schistosomiasis spread.

There are two main approaches to model and simulate epidemiological systems: a mathematical approach based on solving continuous equations and a computational approach resulting on individual-based models. For many infectious diseases including schistosomiasis, mathematical modeling has proved to be a valuable tool for predicting epidemic trends and design control strategies [2]. But, still, they suffer from some conceptual limitations.

K. Glass et al. (Eds.): COMPLEX 2012, LNICST 126, pp. 58–69, 2013.

In this paper we thus propose an agent-based model of Bilharzia spread. The approach consists in explicitly representing in the model the individuals, the environment, and their interactions. The methodology adopted is as follows. In a first step, we build a discrete and individual-based model not taking into account the spatial dimension. In that first model, humans, snails and other pathogen agents are randomly distributed and also move in a similar random way. With this model, we run a first simulation to make a sensitivity analysis of some control strategies on the epidemic spread. One can notice that, this proposed model is close to the mathematical description; our simulations indeed produce similar results with respect to a mathematical model of control chosen as reference [1]. In the following step, we extend our model by including the space. The goal is to evaluate and assess the impact of space heterogeneity on the basic individual-based model. Our simulations show that taking into account the spatial dimension influences control strategies. These ones must then be defined according to the study case and the environmental specificities of each system.

After presenting in Section 2 the basic elements and the underlying dynamics of schistosomiasis, we briefly expose in section 3 the mathematical modeling approach, showing how the system could be represented in a continuous way. Section 4 will therefore come back on the way to model with agents a phenomenon that is mathematically described. The so-called agents-based models are then presented in sections 5 and 6 and we terminate the paper in section 7 with the conclusion and some discussions.

2   Schistosomiasis and Its Propagation Dynamics

Schistosomiasis, also known as Bilharzia, is a parasitic disease that is found in tropical and subtropical areas and is caused by a tapeworm called schistosome or bilharzias [1]. Schistosomes are parasites that have two phases of multiplication, one sexual in the definitive host, the human, and the other one asexual in the intermediate host, a freshwater snail (mollusc) [2]. Between the two hosts, the link is freshwater often shallow, quiet and grassy. This water is for snails the main development environment; for parasites the intermediate environment outside the hosts; and for humans a necessary resource. The meeting between these entities is the cause of infections. Indeed, the water is contaminated when it contacts parasite eggs that reside into the urine or feces of an infected human. After coming into contact with water, the eggs hatch and release a ciliated larval form called miracidia. This one whose life is short – only a few hours – swim to meet the snail to penetrate it. Inside the snail, larval development, which lasts about a month, leads to the release of thousands of larvae called cercariae. The liberation of cercariae is largely influenced by the nature of the climate. It is more important when the temperature in water is higher. Cercariae, with a lifetime of 4 to 7 weeks, are repented in water. When a human is in contact with fresh water infested, the cercaria can infect him. It penetrates the skin, releases its tail and becomes schistosomula [16].

Schistosomula spend several days in the skin before settling in the liver where they access through the blood vessels. They remain in the liver until sexual maturity and form pairs. The coupling step happens in the portal system of the liver and leads to lay egg either in the small and large intestines or in the rectal veins according to the different types of schistosomiasis. Much of the eggs are released from the body through the urine or feces and the parasite's life cycle resumes. The other part of the eggs is trapped in the body and makes the person sick [1] [16] [11].

With the description of the schistosome dynamics one can see that it is evolving within two hosts: a definitive host, which can be a human or livestock and an intermediate host which is a freshwater snail. It passes from one to the other through the water. This allows dividing the dynamics of the parasite in two parts: an intra-host part and an extra-host part called transmission dynamics, where the parasite interacts with other entities. For this purpose, the modeling of schistosomiasis occurs at two levels [6]:

- Intra-host models: they generally describe the dynamics of the larval stages of the parasite and its interactions with the organs of the body and the immune system [2] [3].
- Transmission dynamics models: they describe the evolution of infection in definitive population hosts [1] [4] [10] [11] [12] [14] [15] [16].

In this paper, we consider transmission dynamics models of the disease.

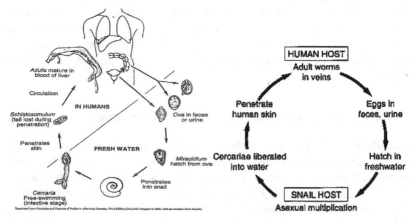

Fig. 1. Intra- and extra-host dynamics of schistosomiasis parasite

3   The Mathematical Modeling of Schistosomiasis

For many infectious diseases, including schistosomiasis, mathematical modeling has proved to be a suitable tool for the prediction of epidemic trends and the design of strategies [2]. Mathematical models of infectious diseases such as schistosomiasis are composed of two parts: the compartments and rules [10] [11].

Compartments are used to divide the population into subpopulations of homogeneous entities according to the different states of disease or infection, hence the name "compartmental model" used in mathematical modeling of epidemics in general. Considering the human population, for example, it can be partitioned into sub-populations of susceptible (S), infected individuals (I) and recovered individuals (R).

The rules give the proportions of migrant entities from one compartment to another. Example, the rule $r(S \rightarrow I)$ expresses the proportion of susceptible individuals becoming infected.

Each compartment is represented in the model by an equation and associated rules give the proportions of entities entering and exiting the compartment. Thus, the mathematical model is a set of equations and numerical simulation shows the quantitative evolution of each compartment according to the considered temporal or spatiotemporal scale. Most mathematical models of schistosomiasis are based on ordinary differential equations (ODE) and therefore take into account only the temporal dimension of infections. To illustrate this, we consider the following equation expressing the quantitative evolution of infected individuals over time [1]:

$$\frac{dI}{dt} = \frac{\beta}{1+\alpha P} PS - (\mu + \delta)I \tag{1}$$

The term $\frac{\beta}{1+\alpha P} PS$ involving the susceptible individuals, represented by S, and the cercariae, represented by P, gives the proportion of susceptible individuals becoming infected at a rate $\frac{\beta}{1+\alpha P}$. The term $(\mu + \delta)I$ gives the proportion of infected individuals leaving the compartment I by natural death (at rate μ) or mortality due to disease (at the rate δ).

The equations incorporate coefficients (β, μ, δ, etc.) called model parameters; expressing the rate of infection, incident, transmission, etc. those model parameters play a crucial role in the quality and the efficiency of a model. Their use makes the manipulation of mathematical models flexible. They also allow the development of a better understanding and an effective control of the schistosomiasis infections. Indeed, in the aim of assessing the impact of medical treatment on the dynamics of individual infections, a parameter τ, for example, can be introduced into the equation [1]:

$$\frac{dI}{dt} = \frac{\beta}{1+\alpha P} PS - (\mu + \delta + \tau)I \tag{2}$$

Thus, the proportion of individuals leaving the compartment I become more important. Similarly, the following equation gives the quantitative infected snails I_2 over time [1].

$$\frac{dI_2}{dt} = \frac{\beta_2}{M_0 + \varepsilon M^2} MS_2 - (\mu_2 + \delta_2 + \theta)I_2 \tag{3}$$

The term $\frac{\beta_2}{M_0 + \varepsilon M^2} MS_2$ involving the susceptible snails, represented by S_2, and the miracidiums, represented by M, gives the proportion of susceptible snails becoming infected. The term $(\mu_2 + \delta_2 + \theta)I_2$ gives the proportion of infected snails leaving the compartment I_2 by natural death (at rate μ_2) or mortality due to disease (at rate δ_2) or elimination (at rate θ).

Even if this kind of equations is widely used, it still has some conceptual limitations. Indeed, before running the simulation of an equation-based model, one has to give values to the parameters. And yet, for several reasons, these values usually contain a large amount of uncertainty. This can actually affect the accuracy of the simulation [1] [6]. Moreover, since infection depends on the involved species, the climate, the water, sanitation policies, etc. a model that does not take into account the spatial and environmental dimensions is improvable. These remarks, among others, lead us to use computational approaches – for modeling complex systems – to avoid some limits of mathematical models.

Thus, in this paper, we try to assess the spatial impact in the modeling of the transmission dynamics of schistosomiasis. We propose an agent-based model of the phenomenon from a mathematical model of reference established in [1]. The following section then describes the process of deriving the agent-based model from a mathematical model.

4   From the Mathematical Model to an ABM

The mathematical model is based on compartments and does not represent entities individually as perceptible objects. However, in agent-based modeling, it is necessary to explicitly represent the entities considered. Under these conditions, we represent as agent the entities of the same type, even if they are in different compartments in the mathematical model. For example, humans, who are susceptible or infected in the mathematical model, are represented by an agent "*human*". It is the same for snails. Only the status will allow us to distinguish both infected and susceptible human agents and snail agents. It is therefore reasonable to talk about "*human agent*" or "*snail agent*" because, they are reasonably counted. However, considering "*miracidium agent*" or "*cercaria agent*" is not without concerns regarding the uncountable number there may be in the water. Thus, we propose to represent, in the extended model, the water as an agent and link the existence of miracidia and cercariae with the water. We incorporate them in the "*agent water*" as properties and call them the "*content of miracidia in water*" and the "*content of cercariae in water*".

In the mathematical model, the proportion of susceptible humans becoming infected depends on the population of cercaria and the proportion of susceptible snails becoming infected depends on the population of miracidia. In the extended agent-based model, infection of humans will depend on the "*content of cercariae in water*" and infection of snails will depend on the "*content of miracidia in water*". Inversely, the "*content of miracidia in water*" will be determined by infected humans and the "*content of cercariae in water*" by infected snails.

We summarize it below with an UML state diagram of humans and snails, coupled with a class diagram of water, human and snail entities.

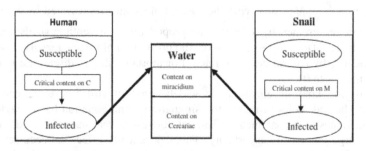

Fig. 2. State diagram of human and snail coupled with a class diagram of water, human and snail

5   A First Agent-Based Model Not Taking into Account the Spatial Dimension

In this model, snail and human agents are randomly distributed on space where they also move randomly. The space is composed of boxes (patches). A box can accommodate a human and / or snail agent and contains a concentration of miracidia and cercariae. A human agent (respectively, a snail agent) present in a box in which the concentration of cercariae (respectively, miracidia) reaches a threshold is contaminated. Similarly, an infected human agent (respectively, an infected snail agent) in a box produces a number of cercariae (respectively, miracidia).

A global agent, commonly called "world agent" in agent-based modeling [7], is created to oversee and synchronize the actions of agents. The interactions between agents are given in the sequence diagram in Figure 3. At each time step, the entire sequence defined in the sequence diagram is executed by the "world agent" using the parameters given in table 1.

Table 1. Values of parameters

Parameter	Value	Reference
Human's birth rate	8 daily	Estimated
Snail's birth rate	200 daily	1
Miracidia's Production rate	0.696 daily per infected person	1
Cercariae's Production rate	2.6 daily per infected snail	1
Miracidia's natural mortality rate	0.9 daily	1
Cercariae's natural mortality rate	0.004 daily	1
Infection rate of Human	$(\beta_1*C*S_1) / (1+\alpha*S_1)$	1
Infection rate of snail	$(\beta_2*M*S_2) / (M_0+\varepsilon*S^2_2)$	1
Human's natural mortality rate	0.0000384 daily	1
Snail's natural mortality rate	0.000569 daily	1
Human's mortality rate due to infection	0.0039 daily	1
Snail's mortality rate due to infection	0.0004012 daily	1
Snail's elimination rate for control	0.1 daily after 15 simulation days	1
Cercaria's elimination rate for control	0.05 daily after 15 simulation days	1
Human's treatment rate for control	0.03 daily after 15 simulation days	1
β_1	0.0000000306	Estimated
β_2	0.615	1
S_1: Susceptible humans visiting rivers	Calculated	
S_2: Susceptible snails	Calculated	
C: *content of cercariae in water*	Calculated	
M: *content of miracidia in water*	Calculated	
M_0: initial *content of miracidia in water*	10000	Estimated
ε	0.3	1

We have to recall that in this model, our goal is to show that we can, with agent-based modeling, reproduce similar results to those of the mathematical model with particularly the application of the same control policies used: controls consisting firstly to eliminate cercariaes, secondly to eliminate snails, thirdly to treat infected humans and fourth to combine the three control policies. Our simulation results are given in Figures 4, 5, 6, 7 and 8.

This first model does not take into account the spatial dimension of the phenomenon of the spread of schistosomiasis. Simulations of this model have produced results, especially in Figure 4 (left), which represent the temporal evolution of infections people. This figure shows that there is nearly 100% of infection after 100-days of simulations without any control policies. Our goal now is to show the

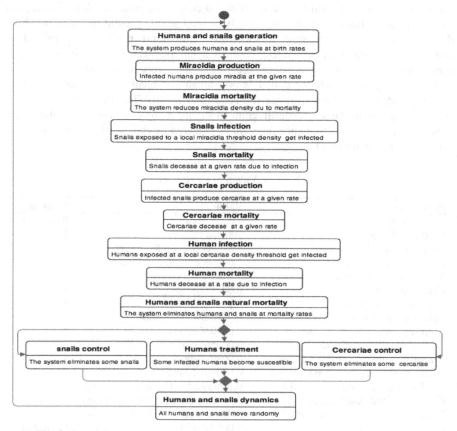

Fig. 3. Transition diagram in the model not taking into account spatial dimension

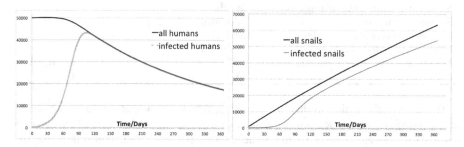

Fig. 4. Results of simulations without any control policies. (Left): The evolution of the human infection. (Right): The evolution of the snail infection.

impact of taking into account the spatial dimension in the spread of the disease. To do so, let us find out the time required to approach 100% infection, in a model taking into account spatial dimension, in order to compare the results with the first model.

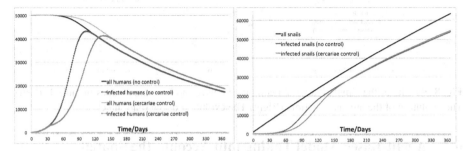

Fig. 5. Results of simulations with a cercariae control policy three time stronger than in table 1. (Left): The evolution of the human infection. (Right): The evolution of the snail infection.

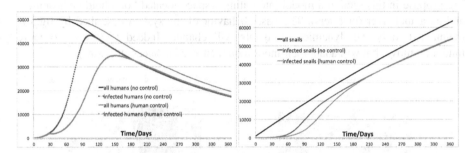

Fig. 6. Results of simulations with a human control policy three time stronger than in table 1. (Left) The evolution of the human infection. (Right) The evolution of the snail infection.

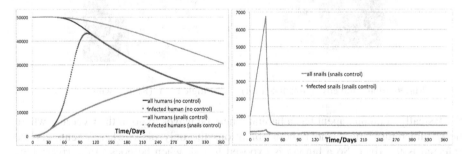

Fig. 7. Results of simulation with a snail control policy three time stronger than in table 1. (Left) The evolution of the human infection. (Right) The evolution of the snail infection.

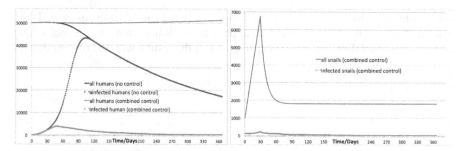

Fig. 8. Results of the simulation with a combined control policy, same values in table 1. (Left) The evolution of the human infection. (Right) The evolution of the snail infection.

6   Agent-Based Model Taking into Account the Spatial Dimension

We propose in this section a model integrating a space located household environment and an actual river (or rivers). The basic behavior of the system described by fig.3 is maintained, only the dynamics of agents will change. Indeed they will no more randomly move, but will adapt their motion on their need to access the water.

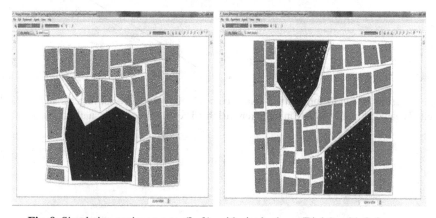

Fig. 9. Simulation environments: (Left): with single river. (Right): with 2 rivers.

Basically, a human agent has a habitat where he leaves to go to the river and where he should return after. Under these conditions the only place where agents can be infected is the river. So a human is exposed to risk when he goes to the river. This was not the case in the first model where the water is not located somewhere on space. That means in this extended model, everyone has not the same risk of exposure. This is the case in reality. We thus divide the population of human agents into three groups. One group, composed of 10% of the population, which has a regular attendance to the river (5 to 7 times per week), a second group, composed of 40% of the population, with an attendance to the river about 1 to 3 times a week and a third group, the rest of the population which has a rare attendance river (about 1 to 3 times every 15 days). To formalize this, each human agent, at the first beginning, has an attribute representing its attendance rate of the river. Those of the first group are more likely higher than those in the second group, also more likely higher than those in the third group. Each human agent also has a "decision mechanism" that assesses, according to the group membership of the agent, the conditions necessary for him to go to the river during the day (the time step).

The river and houses are here modeled as agents as well as human and snail agents. The river agent has attributes *miracidiums_content* and *cercaires_content* that allow infecting snails and humans visiting it.

6.1   Model with a Single River

To infect a person (respectively a snail), we use the same parameters (given in Table 1) of the mathematical model. In this case we can have the same inputs for both the models and be able to compare outputs. The only thing that we modified and adapted from the mathematical inputs is the rate of infection of individuals. Indeed, this rate is applied, at each time step, to all the susceptible persons in the mathematical model; this is not the case here. We apply it only to persons who may have visited the river. The simulation outputs are given in Figure 10 (left).

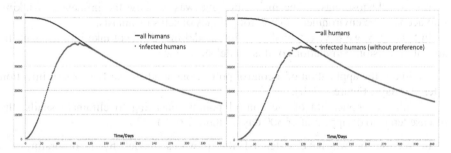

Fig. 10. (Left): Results of the simulation with a single river. (Right): Results of simulation with 2 rivers without specific properties.

6.2   Model with Two Rivers

We consider three scenarios:

Scenario 1: In this scenario, the two rivers do not have distinctive special properties. Our simulation outputs are shown in figure 8 (right). In this case, the disease spread is almost the same in the two environments; the first one with only one river and the second one with two rivers.

Scenario 2: Here, the two rivers are endowed with special properties making them attract a particular group of people. One river has properties attracting more humans with rare frequentation of rivers and the other attracts more the other groups. Under these conditions, a human agent, in its "decision mechanism", will choice before going into a river. In such a scenario, a river is frequented by almost the same people. Simulation results are given in Figure 11 (left).

Scenario 3: Compared to scenario 2, we add in this scenario control policy of eliminating snails. According to [1], this is the best control policy. Simulation results are given in Figure 11 (right).

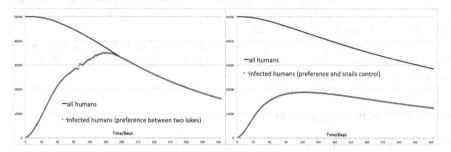

Fig. 11. (Left): Results of the simulation with 2 rivers where each river attract a specific group of persons. (Right): Applying the three control policies.

7   Discussion

In this paper, we assessed the environmental impact on the process of modeling the spread of schistosomiasis. The main objective was to show the influence of taking into account the environment on control policies of schistosomiasis.

In Section 5, we have established a first model that does not take into account the spatial dimension. Simulations of this model show that:

- Without the application of a control policy, one tends towards 100% of infection after 100 days: Figure 4 (left).
- With the application of the control policy consisting to eliminate snails, the disease tends to disappear after 300 days: Figure 7 (left).

In section 6, we have extended the previous model by incorporating the environment into the modeling process. We considered four scenarios.

In the first scenario, we have an environment with a single river. Here, our simulations produce approximately the same outputs as the model that does not take into account the spatial dimension: Figure 10 (left).

In the second scenario, we consider an environment with two rivers. Compared to the previous, our simulations show that without control, one tends towards 100% of infection after 120 days: Figure 10 (right).

In the third scenario, we still have an environment with two rivers. But this time, rivers have properties allowing them to attract a particular group of people. Our simulations show that without control, one approaches 100% of infection after 145 days: Figure 11 (left).

In the fourth scenario, we apply the control policy consisting to eliminate snails on the model of scenario 3. Compared to output in Figure 7 (left), simulations show here that the disease tends to disappear after 120 days: Figure 11 (right).

We can conclude from the simple implemented scenarios that taking into account the environment have a significant impact on the quality of the models of schistosomiasis spread. The "space-less" model and even the one with the only river give approximately the same result that equations-based model could give, especially on control policies. It is a good way to proceed. Nevertheless, when making some kind of realistic scenarios like river preference the outputs change a lot and so should the control policies. We can therefore state that control policies must be defined according to the environmental characteristics of each system that need to be modeled and the approach that we followed can help doing so. Agent-based modeling thus allows having general results and also going into details on some eventual cases study. Indeed, in our work, we use GAMA platform [13] which is an agent-oriented platform allowing GIS data loading to provide a graphical simulation environment faithfully representing the selected system (see figure 9).

References

1. Gao, S., Liu, Y., Luo, Y., Xie, D.: Control problems of a mathematical model for schistosomiasis transmission dynamics. Nonlinear Dyn. 63, 503–512 (2011)
2. Magombedze, G., Chiyaka, E.T., Mutimbo, L.: Modeling within host parasite dynamics of schistosomiasis. Computational and Mathematical Methods in Medicine 11(3), 255–280 (2010)
3. Feng, Z., Xu, D., Curtis, J., Minchella, D.J.: The Grid: On the role of schistosome Mating Structure in the maintenance of drug-resistant strains. Bulletin of Mathematical Biology 68, 209–229 (2006)
4. Allen, E.J., Victory Jr., H.D.: Modeling and Simulation of a schistosomiasis infection with biological control. ActaTropica 87, 251–267 (2003)
5. Remais, J., Liang, S., Spear, R.C.: Coupling Hydrologic and Infectious Disease Models to Explain Regional Differences in Schistosomiasis Transmission in Southwestern in China. Environ. Sci. Technol. 42, 2643–2649 (2008)
6. Rogier, C., Sallet, G.: Modélisation du Paludisme. Med. Trop. 64, 89–97 (2004)
7. Treuil, J.-P., Drogoul, A., Zucker, J.-D.: Modélisation et Simulation à base d'Agents. DUNOD, Paris (2008)
8. Le Moigne, J.-L.: La Modélisation des systèmes complexes, Paris, Dunod (1999)
9. Camara, G., Despres, S., Djedidi, R., Lo, M.: Modélisation Ontologique de processus dans le domaine de la veille épidémiologique. In: 18ème Congrès Francophone sur la Reconnaissance des Formes et l'Intelligence Artificielle, RFIA 2012 (2012)
10. Martcheva, M., Pilyugin, S.S.: An Epidemic Model Structured By Host Immunity. AMS Subject Classification: 92D30 (2005)
11. Chiyaka, E.T., Garira, W.: Mathematical Analysis of the Transmission Dynamics of Schistosomiasis in the Human-Snail Hosts. Journal of Biological Systems 17(3), 397–423 (2009)
12. Feng, Z., Eppert, A., Milner, F.A., Minchella, D.J.: Estimation of Parameters Governing the Transmission Dynamics of Schistosomes. Applied Mathematics Letters 17, 1105–1112 (2004)
13. GAMA platform, http://code.google.com/p/gama-platform/
14. Spear, R.C., Hubbard, A., Liang, S., Seto, E.: Disease Transmission Models for Public Health Decision Making: Toward and Approach for Designing Intervention Strategies for Schistosomiasis Japonica. Environ. Health Perspect. 110, 907–915 (2002)
15. Mangal, T.D., Paterson, S., Fenton, A.: Predicting the Impact of Long-Term Temperature Changes on the Epidemiology and Control of Schistosomiasis: A Mechanistic Model. PLoS ONE 3(1), e1438 (2008), doi:10.1371/journal.pone.0001438
16. Cohen, J.E.: Mathematical Model of Schistosomiasis. Annual Review of Ecology and Systematics 8, 209–233 (1977)

Behaviors of Actors in a Resource-Exchange Model of Geopolitics

Curtis S. Cooper[1], Walter E. Beyeler[1], Jacob A. Hobbs[1],
Michael D. Mitchell[1], Z. Rowan Copley[1,2], and Matthew Antognoli[1]

[1] Sandia National Laboratories,
Albuquerque, NM, 87185-1137 USA
[2] St. Mary's College of Maryland
{cscoope,jahobbs,webeyel,micmitc,mantogn}@sandia.gov,
zrcopley@smcm.edu

Abstract. We present initial findings of an ongoing effort to endow the key players in a nation-state model with intelligent behaviors. The model is based on resource exchange as the fundamental interaction between agents. In initial versions, model agents were severely limited in their ability to respond and adapt to changes in their environment. By modeling agents with a broader range of capabilities, we can potentially evaluate policies more robustly. To this end, we have developed a hierarchical behavioral module, based on an extension of the proven ATLANTIS architecture, in order to provide flexible decision-making algorithms to agents. A Three-Layer Architecture for Navigating Through Intricate Situations (ATLANTIS) was originally conceived for autonomous robot navigation at NASA's JPL. It describes a multi-level approach to artificial intelligence. We demonstrate the suitability of our reification for guiding vastly different types of decisions in our simulations over a broad range of time scales.

Keywords: Complex systems, software architecture, agent-based simulations, artificial intelligence, robot planning, emergent behaviors, policy analysis.

1 Introduction

There is growing interest in the social sciences in numerically modeling the key processes and interactions important for determining change in geopolitical systems. A major long-term goal is to develop models to critically evaluate U.S. foreign policy in a complex arena of multi-national corporations, military and economic rivalries, and long-term changes in the balance of power among nations. Eventually, the most useful models will capture the breadth of values and ideologies that guide nations of various sizes and political configurations.

The resource-exchange model of Beyeler et al. (2011) [1] has been adapted to be capable of representing both historical and contemporary economic interactions among key players in the geopolitical arena. The agents in the model can include national and sub-national entities (such as provinces) as well as corporations, whether

K. Glass et al. (Eds.): COMPLEX 2012, LNICST 126, pp. 70–82, 2013.
© Institute for Computer Sciences, Social Informatics and Telecommunications Engineering 2013

sponsored by a single nation to advance its interests (e.g., the East India Company) or a semi-autonomous organization with international affiliations (e.g., Microsoft). We often refer to the model as the "nation-state model" in this configuration, which is well-suited for geopolitical scenarios. Both terms, "resource-exchange model" and "nation-state model", will be used interchangeably throughout the text. Although it does not yet include prediction of military conflicts and their outcomes, the model nevertheless captures essential features of real economic systems. For example, it has been used successfully to represent economic interactions in the current Pax Americana, in which global-scale military conflicts are inhibited.

The nation-state model [1] is a hybrid model combining system dynamics and multi-agent based modeling. Differential equations representing the important rates of change in the system are integrated forward in time. The continuous timeline described by the rate equations is interspersed, however, with discrete resource-exchange events between agents (which the authors call entities). These exchanges occur through markets and constitute the fundamental interaction between entities. Money is represented in this framework as simply a resource which can be exchanged for any other, whereas all other resources are exchanged in specific ratios according to their relative value to different suppliers and consumers. Whether corporate, national, or sub-national, entities in the nation-state model [1] instantiate new processes and change parameters of processes already in operation in order to affect the flow of resources in and out of their boundaries. Processes represent what entities *can* do; what they *will* do is a separate question.

It is noteworthy that entities in the current implementation of the exchange model [1] do not have the ability to make choices or analyze their environments (e.g., to improve their performance). Their ability to adapt is limited to changing their valuations of different resources according to simple dynamical equations. Entity behavior is governed solely by the dynamics, which do not represent goal-seeking, internal world-model building, or any other intentional process. The behaviors exhibited are therefore quite limited. In assessing the strength of foreign-policy decisions, naïve models carry the risk of underestimating the capabilities of adversaries and therefore evaluating U.S. policies over-optimistically. Geopolitical players in credible models will be capable of innovative approaches, such as trade regulations and the initiation of conflicts, to promote their own goals and undermine those of their competitors.

It is the goal of this effort to implement agent behavior (with varying levels of skill) for the resource-exchange model using ideas from the artificial intelligence community. Our approach has been to design and develop reusable Java language [2] modules, which themselves depend minimally on the structure of the resource-exchange model, in order to facilitate future application of the behavioral layer to other agent-based simulations. Architecturally, the software is a reification of the ATLANTIS architecture developed by Erann Gat (see [3], [4]) as a solution to the complex problem of navigating NASA's ground rovers through rocky Martian terrain. We will discuss the advantages of their design for guiding entity behavior in the nation-state model.

2   Intelligent Agents

2.1   Degrees of Sophistication

Defining intelligent decision-making processes by agents in complex systems is critical to understanding the interactions between them, which can be both cooperative and competitive.[1] The algorithms that accomplish this are expected to be broadly applicable. By not tying the behavioral layer to a particular complex-system model, we confer the ability for it to be used to inform decision-making for a wide range of agent-based systems. For example, flocking and steering behaviors can be encapsulated as common algorithms of certain animals. Once a particular behavior for an animal has been implemented, other animals exhibiting similar behaviors can be modeled using existing algorithms as a foundation.

Beyond intelligent agents, it is desirable to be able to provide certain agents with decision-making capabilities that are not cognitive, knowledge-based, or even particularly intelligent. For example, to assess the performance of a particular entity and its decisions, it may make sense to script the specific actions of its competitors. Similarly, in systems such as ant colonies, inter-agent cooperation is possible without cognitive decision-making by any one member. Individual ants behave reflexively to specific chemical stimuli they receive from others in the colony. Such autonomous agents are well-adapted to their problem domain and achieve success without manifesting human-like thought processes.

The first-order, rudimentary approach to modeling behavior, which has been used with some success in video games, is scripting. Scripted actors have a set strategy they follow. For example, a nation-state in our model could be told to always spend 10% of its GDP on military products, unless that action engenders a response from others. This approach, which is straightforward to implement, can certainly instill nations and corporations with behavior. It is important to recognize, however, that scripted, predictable strategies, in which the behaviors of agents cannot adapt in time to new input and new situations, will almost certainly fail in competitive environments against adversaries that learn and adapt.

The next level of sophistication for agent behavior is to instill actors with reflexes. For example, even animals we would not consider to be particularly intelligent have the ability to blink an eye to avoid being hit by a pebble. This rule-based approach to behavior is sufficient for making good decisions in some situations. For example, contemplation is not required (or a good idea) to avoid a car crash while driving. Braking and/or steering actions must be taken in a timely manner to preserve the driver's health.

The most sophisticated artificial intelligence systems, designed to solve challenging problems, are deliberative in nature. That is, the decision-maker must evaluate multiple options and make choices based on imperfect or distorted information about its surroundings, making outcomes of certain choices very difficult to predict. Although challenging to implement (and potentially intractable),

---

[1] The ideas we present in this section come from internal discussions as well as a review of the artificial intelligence literature, especially the excellent textbooks that have been written on the subject; see [5] and [6].

deliberative decision-making may be needed in order to achieve the ultimate goals of this research: to begin evaluating computationally the performance of diverse policy choices in a multitude of hypothetical scenarios.

2.2   Components of Intelligent Agents

We assume here that the agents we will be dealing with in practice are neither omniscient nor omnipotent. They are limited in their knowledge (and their ability to acquire knowledge), and those agents that can perform actions have a discrete, limited set of capabilities to affect their surroundings. Some but not all agents will also have memories and the ability to learn from past experiences. For agents that learn and adapt to their surroundings, we do not expect repeatable output for a given sequence of inputs. Intelligent agents exhibit dynamic behaviors; they are capable of improving their performance through repetition.

We consider as a reasonable starting point a functional model for decision-making, which has the flexibility to encapsulate at least some common elements of the range of problems that require agents to make decisions and perform actions. Note we conceive of decisions and actions (or plans of action) as separate concepts. The latter are clearly the output of any decision model, but the implementation of a plan is time-dependent and therefore susceptible to dynamical obstacles in the environment or disruption by the actions of other agents. In other words, things don't always go according to plan. Dynamic, intelligent agents should be capable of perceiving obstacles and rethinking plans as unforeseen circumstances arise.

Before discussing our proposed functional decomposition of decision-making problems, other temporal questions should be mentioned. It seems likely that complex agents will perform multiple tasks simultaneously. How often do decisions about these tasks need to be reconsidered? What level of analysis or effort is appropriate for each decision type? How much time is available to make each decision before action is required for the well-being (even survival) of the agent? How is the timing of decisions related to the rate of sensory input and the agent's ongoing accumulation of knowledge about its environment? Answers to these questions are likely to be highly application-dependent. We note here only that how decisions are made is distinct from when decisions are executed. An application attempting to model the behaviors of intelligent agents must consider both problems in turn.

If the output of a decision-making model (or algorithm) is a plan of action, what is the input? To determine this, we must first separate elements of the problem intrinsic to the agent from extrinsic elements. The latter are the agent's environment, from which it receives input through its sensors and receptors in the form of *percepts*. Environmental awareness is limited by the agent's sensors (which poll for new percepts) and receptors (which receive signals from the environment). The time scales for updating the agent about the environment depend on how these sensory mechanisms operate (e.g., ears and eyes, through which animals gather information at different frequencies).

The agent must have a self-model of its sensors, actuators, and world view, which specify the inputs to various decision-making problems. Note also that complex agents may have different approaches to selecting actions in different situations. For example, human eyes blink reflexively to prevent damage to them. Such low-level tasks can be addressed effectively by quick responses to reflexive impulses. Humans are also capable, however, of solving difficult problems through knowledge-based, cognitive thought processes.

Hence, for agents that acquire and accumulate knowledge, decisions are not always made in reaction to percepts directly. Rather, sensory information is acquired, filtered, processed, and stored in the form of memories. Cognitive decisions are made based on knowledge–i.e., some sort of world model–comprising an agent's fragmented memories about its environment and its experiences with the outcomes of past decisions. Certain percepts to be sure (such as an imminent car crash) will trigger immediate decisions at specific times. Intelligent agents, however, will also make decisions at various (in principle unpredictable) times, depending on the importance of the topic, the agent's self-confidence about its current plans, and the availability of new, relevant information.

2.3   Decision Problems and Thinking Agents

As mentioned in Section 2.2, we hypothesize in this discussion that many important decision-making problems can be embodied in a *decision function*. The decision function can be represented abstractly with a decision model, which evaluates different actions against whatever (presumably domain-specific) criteria are appropriate for the problem at hand. From an object-oriented programming (OOP) standpoint, the decision model is likely to be an aspect of the agent attached using some sort of strategy pattern [7]. This design makes it possible for new ways of making decisions (for different types of problems) to be dynamically plugged into the agent. We elaborate here on the rationale for the simplified form of the decision function that we have adopted in these interfaces.

The decision function clearly must take in the sequence of percepts from the agent's sensors that have been acquired since the last time the agent processed sensory information (e.g., by acting on it or incorporating the new information into its world model). In addition to the environmental information (what the agent knows or can detect), we also will need a description of the agent's actuators; i.e., those things that the agent can do. A plan of action, which is the output of a decision function, will naturally consist of a sequence of actions within the set of actions possible (e.g., according to the agent's physics). The set of allowed actions limits the scope of each decision problem to a finite set of possible choices, although the correct plan for the agent might in some cases consist of an infinitely repeating sequence of actions.

With suitable time-dependent decision functions, engineered agents in models can mimic intelligent behavior (at least in specific problem domains). Agents must combine their knowledge to solve problems in their task environments, with the potential to improve their performance with experience. In practice, decision-making algorithms can consider multiple choices for actions (or plans of actions) against various goodness criteria. The challenge is then optimizing the agent's path through the decision tree, considering tradeoffs and likely events in the environment and responses from allies, neutrals, and adversaries.

Significant progress can be made towards self-aware, thinking agents using a simplistic model of human psychology, the so-called *rational-actor* hypothesis. Rational-actor models assume intelligent agents always pursue their own interests, or what they believe to be their own interests based on past experiences, limited only by their imperfect ability to predict the future outcomes of their actions. That is, they are goal-oriented and seek effective solutions to the problems they encounter to accomplish their goals. The filtering of information by sensory mechanisms in the agent leads to skewed perceptions or incomplete knowledge about the system.

The rational-actor hypothesis may be flawed (in that a rational actor will not always mimic human behaviors, which are not strictly rational), but it nevertheless affords great breadth of freedom to entities in a framework such as the nation-state model [1]. The notion that judgments are subjective can be captured by differences in values between entities; i.e., those concepts the entities view as priorities. The value sets can differ greatly between entities and are in principle time-dependent, varying with a given entity's perception of its surroundings. In rational decision-making, all options for an entity to pursue its interests are on the table.

3   The ATLANTIS Agent

3.1   Description of the Architecture

Many approaches have been tried since the inception of artificial intelligence to address the complexities involved in programming agents to solve problems (for example, see the discussion in [8]). In perusing various proposed solutions in the literature, we came to the conclusion that a multi-layered approach would provide the flexibility sought for governing the behaviors of entities in the nation-state model, for reasons we discuss in detail in Section 3.2.

We eventually settled on the three-layered architecture described by ATLANTIS and shown in Figure 1. We also show in Figure 2 the key interfaces in our Java reification of ATLANTIS derived from an object-oriented design process, and how the agents interact with the other objects in the system. In particular, state information about the world is encapsulated as "projection" objects. The intent of Figure 2 is not to befuddle the reader with unnecessary implementation details but to show how the ATLANTIS architecture translates in practice to a statically typed language like Java (note Gat's original implementations of ATLANTIS were written in LISP; see [3], [4]).

The primary attraction of the three-layered approach for guiding agent behavior is its modularity. By dividing the decision-making process of an actor (our term for the subclass of agents that make decisions) into three separate layers—control, sequencing, and deliberative (see Figure 1)—the architecture supports implementations of increasing (and ultimately arbitrary) levels of sophistication. We will first describe the control layer and deliberative layer, which are the most intuitive conceptually.

The control layer defines all interactions with the actor's task environment. Actions must be performed by specific actuators attached to the agent, and actuators can only perform one task at a time. This is a key constraint that greatly simplifies the system and allows for local processing of information by the actor. As the only tools available to actors to affect their environment in constrained ways, actuators greatly

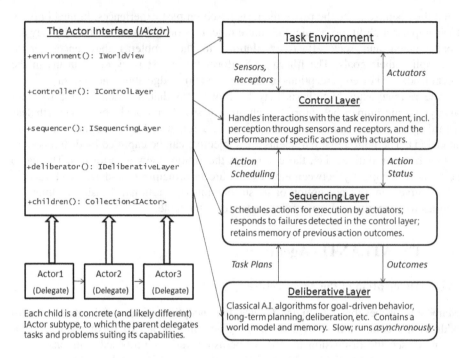

Fig. 1. As shown in this high-level overview of the ATLANTIS architecture [3], actors aggregate three interacting layers to make decisions and accomplish tasks. The control and deliberative layers send messages to and from the mediating sequencing layer to coordinate tasks. The control layer is close to the environment and operates on short time scales, whereas the deliberative layer, which runs asynchronously, solves complex problems and suggests action sequences for queuing by the sequencing layer.

limit the scope of the agent, the decisions it must make, and the information required to make those decisions. Reflexes by actuators on short time scales are possible in the control layer (e.g., blinking eyes), thus permitting basic agent behaviors. At this level of the system, however, the actor does not manifest "intelligence" (or problem-solving). Rather, it performs only the low-level, primitive actions defined by its actuators.

Note in Figure 1 that an actor's 'environment()' method returns a "world view" object, not the world itself. The task environment (for the most general types of acting agents) is itself a function of the agent's perceptions. Actors interact with an imperfect model of their local environment; they do not know the full state of system. Any information actors have about their surroundings must go through their sensors or receptors. They have no access to other information about the world. This "fog" is another key constraint simplifying the construction of a true, problem-solving agent: they are not omniscient and require specific tools to gather and process information.

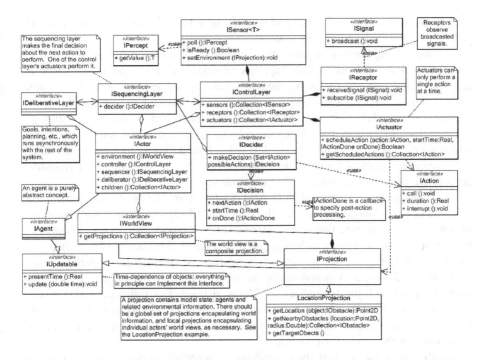

Fig. 2. In this class diagram (see [7] for more information about UML) of the key objects involved in our Java-based reification of the ATLANTIS architecture (for intelligent decision-making), the task environment is represented with composite projections (i.e., IProjection implementations). We represent the notions of perceptions and beliefs with these projection objects. Hence, the world view of an actor can represent a distorted or incomplete description of the state of the system. All direct interaction of the actor with its environment occurs through the control layer.

In ATLANTIS, the most sophisticated problem-solving for the actor occurs in the deliberative layer, which notably runs asynchronously with the rest of the system, so it does not reduce the responsiveness of the control layer. The bulk of the memory and processing power of the agent is expected to be consumed by deliberation, which includes development of a world model, time management, action selection and planning, and achieving goals. The learning also occurs in the deliberative layer by retaining memories of past experiences. In a robotics system, the control layer is a robust, real-time simulator, whereas the deliberative layer is not real-time but rather must partition its computational resources to achieve long-term goals prioritized by their importance to the agent.

The sequencing layer is the core innovation of ATLANTIS and three-layer A.I. architectures in general (see Gat, 1998 [8]). By allowing agent control and deliberation to operate asynchronously, with the controller always having real-time priority, the agent can potentially perform both simple and complicated tasks well. In early intelligent-agent architectures, the deliberator was in complete control. Such an architecture is much more limited because action selection is often only occurring on long timescales compared to the polling time of a typical agent's sensors (e.g., of order 10 Hz for the human vision system).

The sequencing layer is in principle complicated. It must sort out the priorities of different actions, provide scheduling, interruption, and rescheduling of tasks as necessary, and respond to failure messages sent by the control layer. For simple agents, this is likely to be fairly straightforward. Difficulties emerge for more complex agents, however, in which goals and plans can potentially conflict or compete for use of actuators. Much complexity is hidden in the sequencing layer, which we consider a good design. By delegating final authority for action scheduling to a mediating layer, a good implementation can allocate computational resources appropriately and balance the agent's short-term and long-term needs.

Note another key feature of ATLANTIS shown in Figure 1: the hierarchical structure of the actors themselves. In object-oriented design terms, actors are naturally organized in a composite pattern [7]. Hence, specific tasks either too complicated or too low-level for the agent can be delegated to a set of child actors (the delegates in the diagram) under the supervision of the parent. Very different implementations might be used for decision-making by an actor's children, allowing the intelligent agent to solve multiple specialized problems at once. For instance, a nation-state might delegate discrete information-gathering to an espionage agency. To make life interesting, child actors may not always act in the best interests of the parent (depending on the level of oversight).

Finally, we point out that the control and sequencing layers together, without a deliberative layer, are sufficient to capture certain types of agent behaviors, such as reflexes occurring over short time spans, in which deliberation and planning are not necessary to make reasonable decisions. Initial implementations should therefore focus on constructing a robust control layer supervised by a simple sequencing layer, which does not need to do much work because actions are simple and the time scales for their completion are short. Such a configuration can exhibit behaviors based on simple rules, such as an expert system. Behaviors in such a system can be quite interesting in their own right (e.g., emergent behaviors observed in cellular automata), although complex problem-solving and learning still require the deliberative layer.

3.2   Strengths and Weaknesses

The solution offered by the ATLANTIS architecture offers numerous advantages over many others we investigated (e.g., a blackboard architecture (briefly described in [5], p. 369-70). A full Java implementation of the components (even ignoring domain-specific objects and/or algorithms) is likely to involve a lot of code. But the architecture affords a sufficiently fine-grained division of labor amongst the components (without sacrificing flexibility) to allow the system to be built from the bottom up. Bottom-up construction is highly desirable because the individual pieces of the software are reasonably simple and testable. Furthermore, simple agent behaviors can be examined in detail before attempting to address challenging domain problems.

The ATLANTIS architecture stays close to the objects in the problem domain itself and is therefore more intuitive than many other architectural descriptions for intelligent agents. It lends itself to development of a Java version without having detailed knowledge of the original version (or access to its source code, which was written in LISP by Erann Gat while at Virginia Tech and NASA's JPL; see [3, 4]).

The architecture also strongly decouples the parts of the problem that are challenging (deliberation in order to solve domain-specific problems) from aspects that are more straightforward (agent control through sensors and actuators). In software-engineering terms, ATLANTIS describes objects with weak coupling but strong cohesion, which is a key principle of building maintainable software (as discussed at length in [7]). This decoupling of deliberation and control extends not just to code complexity but computational resources, since the deliberative layer runs asynchronously from the rest of the system and is allowed to consume more memory and processor time than the other layers.

We furthermore consider it a major advantage of ATLANTIS that dealing with error conditions and reporting unknown situations is well-specified at the architectural level. It is the control layer's responsibility to detect the end state of the actions it attempts (through actuators) and report this information to the sequencing layer. Agents in ATLANTIS are *failure cognizant* [4]. This implies that in a fully functional system, there is natural cohesion between the information-gathering objects (sensors and receptors) and the effector objects (the actuators).

It is worth noting that the intermediate sequencing layer is potentially complicated, depending on the application. Indeed, part of the architecture's design is to hide complexity in the sequencing layer. However, for a first pass-implementation, its operation can be approximated with a thread-safe priority queue (thread-safe because the deliberative layer runs asynchronously), in which the control layer has the option to interrupt in-progress actions at any time. We admit, however, that the sequencing layer is the main potential disadvantage to three-layered architectures. For complicated agents, it is conceivable that a robust implementation of the sequencing layer will prove elusive.

Nevertheless, a solution based on ATLANTIS has been deployed to solve a sophisticated robotics problem: how to autonomously steer a robot on the surface of another planet. Hence, the basic ideas are thought to be sound and compare favorably to alternative approaches to intelligent agent architectures (see [7]). Due to the generality of ATLANTIS and its dissection of the key interacting objects into manageable pieces, we expect the architecture to have broad applicability to complex systems problems, even though many of the anticipated applications are quite different from the robotics problem domain in which it was conceived.

4   Application of ATLANTIS

We have used the ATLANTIS architecture in designing the behavioral layer for entities in the nation-state model. As shown in Figure 3, it is within the behavioral layer that entities process information and perform actions, thereby manifesting complex behaviors that potentially can defy prediction. Without the behavioral layer, behaviors of entities are limited to simple local functions of their internal resource levels [1]. The inclusion of the behavior layer provides entities, such as nations and corporations, with capabilities to alter model structure in more sophisticated ways in order to promote their policies and weaken competitors. Implementations will enumerate these possible actions (examples of which are shown in Figure 3), from which decisions will be made according to different strategies and algorithms encapsulated in the behavior layer (see Figure 4).

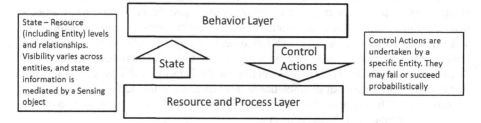

Kinds of Control Actions	
Concrete Description	**Abstract Description**
Instantiate a new process	
Connect to/Disconnect from a Market	Add/Delete a relationship between an Entity and a Market Entity
Instantiate Entities	Trigger a specific kind of production process
Adjust parameters on processes	
Join/Leave a compound Entity	Add/Delete a relationship
Set defense policy	Change boundary rules
Collect information	Receive messages (of a specific kind)

Fig. 3. The connection between the behavioral layer of a resource-exchange model [1] entity (configured for geopolitical scenarios) and the model structure itself, comprising the external environment with which entities interact. Some examples are given in the table of the types of actions entities must choose from in order to succeed and increase their health.

Sensations and capabilities of entities, shown in Figure 4, as well as discrete action execution by actuators, comprise the control layer of the ATLANTIS architecture. The control layer is the hard boundary between entities and the world. This structural locality is built-in intentionally to confine the scope of entities' knowledge, awareness, and ability to affect their surroundings. Entities are part of the world (and may understand it structure) but only act locally. Internal thought processes of the most intelligent entities (e.g., national leaders) might include a world model with memory, values and objectives, contemplation of future possibilities, and action planning in order to accomplish goals. The black box in Figure 4 captures these ideas of scoring choices, which are made in the deliberative layer. Final arbitration of the actions to be performed by an agent when its goals or needs are conflicting is handled in the sequencing layer.

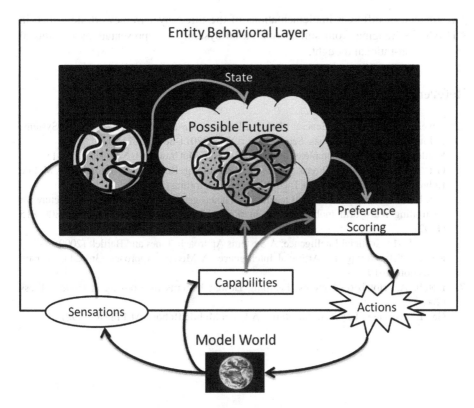

Fig. 4. The behavioral layer will provide intelligent decision-making capabilities to entities in the nation-state model [1]. We discuss in the text how these components fit into our reification of the versatile three-layer architecture (ATLANTIS) in Java, as shown in Figure 2.

The ATLANTIS architecture was originally conceived to provide autonomous navigation for vehicles, but its applicability is not specific to that problem domain because it captures fundamental aspects of intelligent agents. The discrete character of agents' interactions with their surroundings, embodied in the control layer, is a general feature of A.I. problems. Also, the general problems of action ordering and prioritization, along with failure response, are described well by the architecture's sequencing layer.

Furthermore, the deliberative layer's internals are not specified by the architecture. The deliberative layer can be as simple or complex as needed to achieve adequate performance in the problem domain; it is the agent's brain, and it can be tailored to solve a broad range of problems. It is in the deliberative layer that the agent "comes to life". Although there is coupling to the sequencing layer, the deliberative layer is fully separated in ATLANTIS from the control layer. The control layer code can therefore be reused for multiple implementations of an agent's deliberative layer. Once the capabilities of the entities in the nation-state model are defined, their level of sophistication can be developed incrementally. In future work, we will compare the performance of different implementations of the three-layer framework described in ATLANTIS, ranging from simple rule-based systems to representations of rational and even non-rational thought.

References

1. Beyeler, W.E., et al.: A General Model of Resource Production and Exchange in Systems of Interdependent Specialists. Sandia Report SAND2011-8887 (2011)
2. Schildt, H.: Java: The Complete Reference, 8th edn. McGraw-Hill, New York (2011)
3. Gat, E.: Reliable Goal-Directed Reactive Control for Real-World Autonomous Mobile Robots, Ph. D. thesis, Virginia Tech, Blacksburg, Virginia (1991)
4. Gat, E.: Integrating planning and reacting in a heterogeneous asynchronous architecture for controlling real-world mobile robots. In: Proc. of the 10th AAAI Conf., pp. 809–815 (1992)
5. Jones, T.M.: Artificial Intelligence: A Systems Approach. Jones and Bartlett (2009)
6. Russell, S.J., Norvig, P.: Artificial Intelligence: A Modern Approach, 3rd edn. Pearson Education (2010)
7. Holub, A.: Holub on Patterns: Learning Design Patterns by Looking at Code. APress (2004)
8. Gat, E.: On Three-Layer Architectures, A.I. and Mobile Robots. AIII Press (1998)

Portfolio Optimization and Corporate Networks: Extending the Black Litterman Model

Germán Creamer

Stevens Institute of Technology, Hoboken, NJ 07030
gcreamer@stevens.edu
http://www.creamer-co.com

Abstract. The Black Litterman (BL) model for portfolio optimization combines investors' expectations with the Markowitz framework. The BL model is designed for investors with private information or with knowledge of market behavior. In this paper I propose a method where investors' expectations are based on accounting variables, recommendations of financial analysts, and social network indicators of financial analysts and corporate directors. The results show promise when compared to those of an investor that only uses market price information. I also provide recommendations about trading strategies using the results of my model.

Keywords: Link mining, social network, machine learning, computational finance, portfolio optimization,boosting, Black Litterman model.

1 Introduction

Contemporary investment literature is significantly influenced by [28, 29]'s portfolio optimization approach that suggests an optimal allocation of assets that maximizes expected return and minimizes volatility. The problem is that this mean-variance portfolio optimization process may lead to the selection of few top assets, it is very sensitive to small changes in inputs, it is based on past price history, and investors can not formally input their own knowledge of the market. As a reaction to these limitations, [4] proposed a mean-variance portfolio optimization model that included investors' expectations. This methodology creates views that represent investors' market expectations with different confidence levels, and uses these views as inputs for the selection of the optimal portfolio.

A different aspect that has not been deeply explored in the literature is the application of link mining to solve finance problems. Link mining is a set of techniques that uses different types of networks and their indicators to forecast or to model a linked domain. Link mining has had several applications [31] to different areas such as money laundering [26], telephone fraud detection [16], crime detection [32], and surveillance of the NASDAQ and other markets [26, 21]. One of the most important business applications of link mining is in the area of viral marketing or network-based marketing [15, 30, 27, 22], and more

K. Glass et al. (Eds.): COMPLEX 2012, LNICST 126, pp. 83–94, 2013.

recently, in finance. [1] have applied network analysis to quantify the flow of information through financial markets. [12] have applied a link mining algorithm called CorpInterlock to integrate the metrics of an extended corporate interlock (social network of directors and financial analysts) with corporate fundamental variables and analysts' predictions (consensus). CorpInterlock used these metrics to forecast the trend of the cumulative abnormal return and earnings surprise of US companies.

In this paper, I propose PortInterlock which is a variation of the CorpInterlock algorithm that uses the return forecast to complement or substitute the investors' view of the BL approach. This methodology may help investors to incorporate qualitative and quantitative factors into their investment decisions without the intervention of financial management experts.

2   Methods

2.1   The Black Litterman Model

The BL model is one of the most extended tactical allocation models used in the investment industry. The BL model calculates the posterior excess return using a mean variance optimization model and the investors' view. Since the introduction of the BL model [4], many authors have proposed several modifications. [25] extends the BL model including a factor uncorrelated with the market; [3] and [2] substitute the investors' views by analysts' dividend forecast and by GARCH derived views, respectively.

I follow [5, 6] in describing the BL model. Additional useful references about the BL model are [23, 33].

The excess returns of n assets over the risk free rate R_f are normally distributed and are represented by the n-vector μ. Using a Bayesian framework, the prior distribution of excess returns is $\mu \sim N(\Pi, \Sigma)$ where Π is a n-vector of implied equilibrium excess return and Σ is the nxn variance covariance matrix of excess return.

Σ can be obtained by the historical excess return and the equilibrium excess return $\Pi = \lambda \Sigma w$ is the solution to the following unconstrained return maximization problem:

$$max_w w'\Pi - \frac{\lambda w' \Sigma w}{2} \qquad (1)$$

where λ is the risk aversion parameter. The vector of optimal portfolio weights can be derived from the Π formula:

$$w = (\lambda \Sigma)^{-1} \Pi \qquad (2)$$

In equilibrium, the market portfolio w_{mkt} derived from the capital asset pricing model (CAPM) should be the same as the mean variance optimal portfolio w. So, the prior expected excess return should be the equilibrium expected excess return:

$$\Pi = \lambda \Sigma w_{mkt} \qquad (3)$$

where $w_{mkt} = \frac{M_i}{\sum_i M_i}$ and M_i is the market capitalization value of asset i.

The main innovation of the BL model is that the investor may specify k absolute or relative scenarios or "views" about linear combinations of the expected excess return of assets. The views are independent of each other and are also independent of the CAPM. They are represented as:

$$P\mu = Q - \epsilon \tag{4}$$

P is a $k x n$ matrix where each row represents a view. Absolute views have weights for the assets that will outperform their expected excess return and their total sum is one; relative views assign positive and negative weights to assets that over- or underperform respectively and their total sum is zero. Q is a k-vector that represents the expected excess return of each view, τ is a k-vector that represents the confidence indicator of each view, and $\epsilon \sim N(0, \Omega)$ is an error term normally distributed that represents the uncertainty of the views where Ω is a $k x k$ diagonal covariance matrix of error terms of the views.

The posterior distribution of excess returns $\hat{\mu}$ combines the prior excess return Π and the investors' views P:

$$\hat{\mu} = N([(\tau\Sigma)^{-1} + P'\Omega^{-1}P]^{-1}[(\tau\Sigma)^{-1}\Pi + P'\Omega^{-1}P], [(\tau\Sigma)^{-1}\Pi + P'\Omega^{-1}P]^{-1}) \tag{5}$$

An alternative expression of the expected excess return is:

$$\hat{\mu} = \Pi + \tau\Sigma P'[P\tau\Sigma P' + \Omega]^{-1}[Q - P\Pi] \tag{6}$$

and the optimal portfolio weights on the unconstrained efficient frontier using the posterior distribution is:

$$\hat{w} = (\lambda\Sigma)^{-1}\hat{\mu} \tag{7}$$

2.2   Boosting

Adaboost is a machine learning algorithm invented by [19] that classifies its outputs by applying a simple learning algorithm (weak learner) to several iterations of the training set where the misclassified observations receive more weight. [18] proposed a decision tree learning algorithm called an *alternating decision tree* (ADT). In this algorithm, boosting is used to obtain the decision rules and to combine them using a weighted majority vote.

[20], followed by [8] suggested a modification of AdaBoost, called LogitBoost. LogitBoost can be interpreted as an algorithm for step-wise logistic regression. This modified version of AdaBoost–known as LogitBoost–assumes that the labels $y_i's$ were stochastically generated as a function of the $x_i's$. Then it includes $F_{t-1}(x_i)$ in the logistic function to calculate the probability of y_i, and the exponent of the logistic function becomes the weight of the training examples. Figure 1 describes Logitboost.

2.3   PortInterlock: A Link Mining Algorithm

CorpInterlock is a link mining algorithm proposed by [12] to build a bipartite social network with two partitions: one partition includes members of board of

$$F_0(x) \equiv 0$$
$$\text{for } t = 1 \ldots T$$
$$w_i^t = \frac{1}{1 + e^{y_i F_{t-1}(x_i)}}$$
$$\text{Get } h_t \text{ from } weak\ learner$$
$$\alpha_t = \frac{1}{2} \ln \left(\frac{\sum_{i:h_t(x_i)=1, y_i=1} w_i^t}{\sum_{i:h_t(x_i)=1, y_i=-1} w_i^t} \right)$$
$$F_{t+1} = F_t + \alpha_t h_t$$

Fig. 1. The Logitboost algorithm [20]. y_i is the binary label to be predicted, x_i corresponds to the features of an instance i, w_i^t is the weight of instance i at time t, h_t and $F_t(x)$ are the prediction rule and the prediction score at time t respectively.

directors and another partition consists of financial analysts representing companies that they cover. This social network is converted into a one-mode network where the vertices are the companies and the edges are the number of directors and analysts that every pair of companies have in common. This is the extended corporate interlock. The basic corporate interlock is calculated in the same way using only directors. The algorithm selects the largest strongly connected component of a social network and ranks its vertices using a group of investment variables presented in the appendix 1 and a group of social network statistics obtained from the basic or extended corporate interlock. Finally, the algorithm predicts the trend of a financial time series using a machine learning algorithm such as boosting. I propose the PortInterlock algorithm, an extension of CorpInterlock, to be used for portfolio optimization. This algorithm uses the trend of the financial time series predictions as the view of the investors and the prior asset excess returns to define the optimal portfolio weights (Figure 2).

Forecasting Earnings Surprise. I used the definition of earnings surprise or forecast error proposed by [14]:

$$FE \doteq \frac{\text{CONSENSUS}_q - \text{EPS}_q}{|\text{CONSENSUS}_q| + |\text{EPS}_q|}$$

where CONSENSUS_q is the mean of earnings estimate by financial analysts for quarter q, and EPS_q is the actual earnings per share for quarter q. FE is a normalized variable with values between -1 and 1. Additionally, when CONSENSUS_q is close to zero and EPS_q is not, then the denominator will not be close to zero.

The increasing importance of organizational and corporate governance issues in the stock market suggests that the integration of indicators from the corporate interlock with more traditional economic indicators may improve the forecast of FE and CAR.

The following indicators obtained by the PortInterlock algorithm captures the power relationship among directors and financial analysts:

Input: Two disjoint nonempty sets V_{11} and V_{12}, a matrix ER of historical excess returns of each asset i, a financial time series Y to be predicted, the covariance matrix Ω of error terms of the views of investors, the vector τ that represents the confidence indicator of each view, the risk factor λ, a vector M with market capitalization values of each asset i, and additional exogenous variables.

1. Build a bipartite graph $G_1(V_1, E_1)$ where its vertex set V_1 is partitioned into two disjoint sets V_{11} and V_{12} such that every edge in E_1 links a vertex in V_{11} and a vertex in V_{12}.

2. Build a one-mode graph $G_2(V_2, E_2)$ in which there exist an edge between v_i and v_j: $v_i, v_j \in V_2$ if and only if v_i and v_j share at least a vertex $u_i \in V_{12}$. The value of the edge is equal to the total number of objects in V_{12} that they have in common.

3. Calculate the largest strongly connected component of G_2 and call it $G_3(V_3, E_3)$.

4. Calculate the adjacency matrix A and geodesic distance matrix D for G_3. a_{ij} and d_{ij} are the elements of A and D respectively.

5. For each vertex $v_i \in V_3$ calculate the following social network indicators:
   - Degree centrality: $deg(v_i) = \sum_j a_{ij}$
   - Closeness centrality (normalized): $C_c(v_i) \doteq \frac{n-1}{\sum_j d_{ij}}$
   - Betweenness centrality: $B_c(v_i) = \sum_i \sum_j \frac{g_{kij}}{g_{kj}}$, where g_{kij} is the number of geodesic paths between vertices k and j that include vertex i, and g_{kj} is the number of geodesic paths between k and j.
   - Clustering coefficient: $CC_i = \frac{2|\{e_{ij}\}|}{deg(v_i)(deg(v_i)-1)} : v_j \in N_i, e_{ij} \in E$
   - Normalized clustering coefficient: $CC_i' = \frac{deg(v_i)}{MaxDeg} CC_i$, where MaxDeg is the maximum degree of vertex in a network

6. Merge social network indicators with any other relevant set of variables for the population under study such as analysts' forecasts and economic variables and generate test and training samples.

7. Run a machine learning algorithm with above test and training samples to predict trends of Y.

8. Define a matrix P where each row k is the multiplication of the confidence of the prediction and the prediction of the trends of Y for each asset. P represents the absolute view of the investors.

9. Obtain Q as a k-vector that represents the expected excess return of each asset or each view k, Σ as the variance covariance matrix of ER, and $\Pi = \lambda \Sigma w_{mkt}$ as the equilibrium expected excess return where $w_{mkt} = \frac{M_i}{\sum_i M_i}$

10. Optimize the portfolio using the Black Litterman model where the expected excess return is:
$\hat{\mu} = \Pi + \tau \Sigma P'[P\tau\Sigma P' + \Omega]^{-1}[Q - P\Pi]$
and the vector of optimal portfolio weights is $\hat{w} = (\lambda \Sigma)^{-1}\hat{\mu}$.

Output: Optimal portfolio weights (w).

Fig. 2. The PortInterlock algorithm

1. Degree centrality: directors and analysts of a company characterized by a high degree or degree centrality coefficient are connected through several companies.
2. Closeness centrality: directors and analysts of a company characterized by a high closeness centrality coefficient are connected through several companies that are linked through short paths.
3. Betweenness centrality: directors and analysts of a reference company characterized by a high betweenness centrality coefficient are connected through several companies. Additionally, the reference company mentioned above has a central role because it lies between several other companies, and no other company lies between this reference company and the rest of the companies.
4. Clustering coefficient: directors and analysts of a company characterized by a high clustering coefficient are probably as connected amongst themselves as is possible through several companies.

Each of the measures above show a different perspective of the relationship between directors and analysts. Hence, I could include them as features in a decision system to forecast FE and CAR. Because the importance of these features combined with a group of financial variables to predict FE may change significantly in different periods of time, I decided to use boosting, specifically Logitboost, as the learning algorithm. Boosting is well-known for its feature selection capability, its error bound proofs [19], its interpretability, and its capacity to combine continuous and discrete variables. [9, 10, 11] have already applied boosting to forecast equity prices and corporate performance showing that Logitboost performs significantly better than logistic regression, the baseline algorithm. [14] have also compared tree-induction algorithms, neural networks, naive Bayesian learning, and genetic algorithms to classify the earnings surprise before announcement.

3 Experiments

The asset price and return series are restricted to the US stock market. They are from the Center for Research in Security Prices (CRSP), the accounting variables from COMPUSTAT[1], the list of financial analysts and earnings forecast or consensus from IBES, and the annual list of directors for the period 1996 - 2005 is from the Investor Responsibility Research Center. The number of companies under study changes every year. The minimum and maximum number of companies included in my study are 3,043 for 2005 and 4,215 for 1998.

I implemented the PortInterlock algorithm (Figure 2) with the software Pajek [13] to obtain the basic (social network of directors) and extended corporate interlock. I computed the investment signals as described in appendix 1 and the social network statistics introduced in the previous section of the basic and extended corporate interlock. I merged the accounting information, analysts'

---

[1] COMPUSTAT is an accounting database managed by Standard & Poor's.

predictions (consensus) and social networks statistics using quarterly data and selected the last quarter available for every year.[2] I forecasted the trend of FE and CAR. CAR is calculated using the cumulative abnormal return of the month following the earnings announcement. Every instance has the label 1 if the trend was positive and -1 otherwise. CAR is calculated as the return of a specific asset minus the value weighted average return of all assets in its risk-level portfolio according to CRSP. FE is based on the predictions of the analysts available 20 days before the earnings announcement as fund managers may suggest [14]. Fund managers take a position, short or long [3], a certain number of days before the earnings announcement and, according to their strategy, they will liquidate the position a given number of days after the earnings announcement. Investors profit when the market moves in the direction expected and above a certain threshold, even though the market movement might not be in the exact amount forecasted.

I restricted my analysis to trading strategies using FE because the prediction of FE (test error of 19.09%) outperformed the prediction of CAR (test error of 47.56%). According to [12], the long-only portfolio is the most profitable strategy when it is compared with a long-short, a long-short for the most precise decile, and a long only strategy when analysts predict that earnings will be larger than consensus. Based on these results, the weights of the long-only portfolio multiplied by the confidence of the prediction are used as the investors' views of the BL model. This portfolio is compared against a market portfolio where the weight of each asset is based on its market capitalization.

4   Results

Table 1 compares the result of several views based on a portfolio completely generated by the PortInterlock algorithm with an equally weighted portfolio and the market portfolio. The PortInterlock portfolio and the investors' view based on the PortInterlock show the largest Sharpe ratio (risk-adjusted return). When the confidence in this view decreases (lower Ω and τ), the Sharpe ratio deteriorates. The Sharpe ratio decreases even more when the risk parameter (λ) increases. The difference of Sharpe ratios between these scenarios and the market portfolio is significant according to the heteroskedasticity and autocorrelation robust (HAC) estimation test.

In the simulations, a portfolio based on social networks and fundamental indicators with high confidence in the investors' perspective has an annual Sharpe ratio of 6.56, while the market portfolio with 20% confidence has an annual Sharpe ratio of 1.415.

---

[2] Most of the fundamental and accounting variables used are well-known in the finance literature and [24] demonstrated that these variables are good predictors of cross-sectional returns.

[3] Long or short positions refer to buy a specific asset or to sell a borrowed asset based on the expectation that price of the asset will increase or decrease respectively.

Table 1. Annual Sharpe ratio, risk and return by portfolio.
Ω is the covariance and τ is the confidence indicator in a particular view (according to equation 6). λ is a risk factor. BL is the Black Litterman model that includes investors' views. Sharpe ratio is the ratio of mean and standard deviation of excess return over the risk free rate. *,**: 95% & 99% confidence level of the Sharpe ratio difference between each scenario and the market portfolio.

Portfolios/Views	Sharpe	Risk	Return
BL, PI, $\tau=1,\Omega=0.000001$, $\lambda=0.00001$	6.563 **	0.49	31.36% *
BL, CI, $\tau=1,\Omega=0.0001,\lambda=0.0001$	6.563 **	0.49	31.36% *
BL, CI, $\tau=1,\Omega=0.001,\lambda=0.001$	6.561 **	0.49	31.36% *
BL, CI, $\tau=1,\Omega=0.001,\lambda=0.0025$	6.558 **	0.49	31.35% *
BL, CI, $\tau=1,\Omega=0.001,\lambda=0.005$	6.552 **	0.49	31.34% *
BL, CI, $\tau=1,\Omega=0.001,\lambda=0.01$	6.542 **	0.49	31.32% *
BL, CI, $\tau=1,\Omega=0.001,\lambda=0.5$	5.438 **	0.51	29.26% *
BL, CI, $\tau=1,\Omega=0.001,\lambda=1$	4.402 **	0.53	27.53%
BL, CI, $\tau=0.5,\Omega=0.001,\lambda=1$	4.063 **	0.55	27.34%
BL, CI, $\tau=0.01,\Omega=0.001,\lambda=1$	1.604 *	0.61	14.06%
BL, CI, $\tau=0.005,\Omega=0.001,\lambda=1$	1.517 *	0.62	13.59%
BL, CI, $\tau=0.0025,\Omega=0.001,\lambda=1$	1.468 *	0.62	13.31%
BL, CI, $\tau=0.001,\Omega=0.001,\lambda=1$	1.437 *	0.63	13.13%
PortInterlock (PI): soc.network	6.563 **	0.49	31.36%
Equally weighted	2.840 *	0.47	14.01%
Market portfolio	1.415	0.63	13.00%

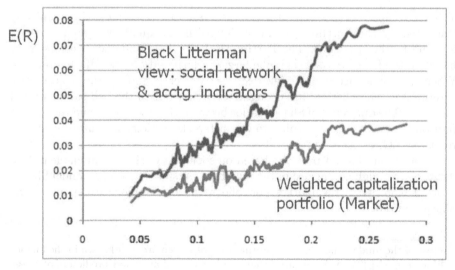

Fig. 3. Abnormal return and risk by portfolio type

Graph 3 indicates that the inclusion of social network and accounting indicators (red line) generates a portfolio with a higher level of accumulated expected return than a portfolio that uses the current market capitalization (blue line) as input. The inclusion of corporate social network indicators might capture interactions among directors and financial analysts that improve the prediction of earnings surprise. This effect, combined with the predictive capacity of selected accounting indicators, explains why a portfolio with a social network perspective outperforms the market portfolio.

5 Conclusions

This paper shows that a modified BL model that includes a forecast based on social networks and fundamental indicators as investors' view outperforms the market portfolio. Even though the BL model includes the investors' subjective views, these views can be substituted or enriched by forecasts based on the optimal combination of social networks and accounting indicators.

Acknowledgment. The author would like to thank Ionut Florescu, Maria Christina Mariani, Frederi G. Viens and participants of the Financial Mathematics session of the American Mathematical Society meeting 2010, the Workshop on Information in Networks (WIN)-NYU 2010, the IEEE CEC2011 Workshop on Agent-Based Economics and Finance, the Sixth Rutgers-Stevens Workshop Optimization of Stochastic Systems 2011, and the Eastern Economics Association meeting 2012 for their valuable comments. The opinions presented are the exclusive responsibility of the author.

References

[1] Adamic, L., Brunetti, C., Harris, J., Kirilenko, A.: Information flow in trading networks. Paper presented at 1st Workshop on Information in Networks, NYU, NY (2009)

[2] Beach, S., Orlov, A.: An application of the Black-Litterman model with EGARCH-M derived views for international portfolio management. Financial Markets Portfolio Management 21, 147–166 (2007)

[3] Becker, F., Gurtler, M.: Quantitative forecast model for the application of the Black-Litterman approach. In: 2009 Finance International Meeting AFFI - EUROFIDAI (December 2010)

[4] Black, F., Litterman, R.: Asset allocation: Combining investor views with market equilibrium. Fixed Income Research. Goldman Sachs & Co. (September 1990)

[5] Black, F., Litterman, R.: Global portfolio optimization. Financial Analysts Journal 48(5), 28–43 (1992)

[6] Black, F., Litterman, R.: The intuition behind Black-Litterman model portfolios. Investment Management Research. Goldman Sachs & Co. (December 1999)

[7] Borgatti, S.P., Everett, M.: A graph-theoretic perspective on centrality. Social Networks 28(4), 466–484 (2006)

[8] Collins, M., Schapire, R.E., Singer, Y.: Logistic regression, AdaBoost and Bregman distances. Machine Learning 48(1-3), 253–285 (2004)

[9] Creamer, G., Freund, Y.: Predicting performance and quantifying corporate governance risk for Latin American ADRs and banks. In: Proceedings of the Second IASTED International Conference Financial Engineering and Applications, pp. 91–101. Acta Press, Cambridge (2004)

[10] Creamer, G., Freund, Y.: Using AdaBoost for equity investment scorecards. In: Machine Learning in Finance Workshop in NIPS 2005, Whistler, B.C. (2005)

[11] Creamer, G., Freund, Y.: A boosting approach for automated trading. The Journal of Trading 2(3), 84–95 (2007)

[12] Creamer, G., Stolfo, S.: A link mining algorithm for earnings forecast and trading. Data Mining and Knowledge Discovery 18(3), 419–445 (2009)

[13] de Nooy, W., Mrvar, A., Batagelj, V.: Exploratory social network analysis with Pajek. Cambridge University Press, New York (2005)

[14] Dhar, V., Chou, D.: A comparison of nonlinear methods for predicting earnings surprises and returns. IEEE Transactions on Neural Networks 12(4), 907–921 (2001)

[15] Domingos, P., Richardson, M.: Mining the network value of customers. In: KDD 2001: Proceedings of the Seventh ACM SIGKDD International Conference on Knowledge Discovery and Data Mining, pp. 57–66. ACM, New York (2001)

[16] Fawcett, T., Provost, F.: Activity monitoring: Noticing interesting changes in behavior. In: Proceedings of the Fifth ACM SIGKDD International Conference on Knowledge Discovery and Data Mining (KDD 1999), San Diego, CA, USA, pp. 53–62. ACM, New York (1999)

[17] Freeman, L.: Centrality in social networks conceptual clarification. Social Networks 1(3), 215–239 (1979)

[18] Freund, Y., Mason, L.: The alternating decision tree learning algorithm. In: Machine Learning: Proceedings of the Sixteenth International Conference, pp. 124–133. Morgan Kaufmann Publishers Inc., San Francisco (1999)

[19] Freund, Y., Schapire, R.E.: A decision-theoretic generalization of on-line learning and an application to boosting. Journal of Computer and System Sciences 55(1), 119–139 (1997)

[20] Friedman, J., Hastie, T., Tibshirani, R.: Additive logistic regression: A statistical view of boosting. The Annals of Statistics 38(2), 337–374 (2000)

[21] Goldberg, H.G., Kirkland, J.D., Lee, D., Shyr, P., Thakker, D.: The NASD securities observation, news analysis and regulation system (SONAR). In: IAAI 2003, Acapulco, Mexico (2003)

[22] Hill, S., Provost, F., Volinsky, C.: Network-based marketing: Identifying likely adopters via consumer networks. Statistical Science 21(2), 256–276 (2006)

[23] Idzorek, T.: A step-by-step guide to the Black-Litterman model. Working paper (January 2009)

[24] Jegadeesh, N., Kim, J., Krische, S.D., Lee, C.M.C.: Analyzing the analysts: When do recommendations add value? Journal of Finance 59(3), 1083–1124 (2004)

[25] Khrishnan, H., Mains, N.: The two-factor Black-Litterman model. Risk, 69–73 (2005)

[26] Kirkland, J.D., Senator, T.E., Hayden, J.J., Dybala, T., Goldberg, H.G., Shyr, P.: The NASD Regulation Advanced Detection System (ads). AI Magazine 20(1), 55–67 (1999)

[27] Leskovec, J., Adamic, L.A., Huberman, B.A.: The dynamics of viral marketing. In: EC 2006: Proceedings of the 7th ACM Conference on Electronic Commerce, pp. 228–237. ACM, New York (2006)

[28] Markowitz, H.: Portfolio selection. The Journal of Finance 7(1), 77–91 (1952)

[29] Markowitz, H.: Portfolio Selection: Efficient Diversification of Investments. John Wiley & Sons, New York (1959)

[30] Richardson, M., Domingos, P.: Markov logic networks. Machine Learning 62(1-2), 107–136 (2006)

[31] Senator, T.E.: Link mining applications: Progress and challenges. SIGKDD Explorations 7(2), 76–83 (2005)

[32] Sparrow, M.K.: The application of network analysis to criminal intelligence: An assessment of the prospects. Social Networks 13(3), 251–274 (1991)

[33] Walters, J.: The Black-Litterman model in detail. Working paper (February 2009)

[34] Watts, D., Strogatz, S.: Collective dynamics of small world networks. Nature 393, 440–442 (1998)

Appendix 1. Investment Signals Used for Prediction

I do not include firm-specific subscripts in order to clarify the presentation. Subscript q refers to the most recent quarter for which an earnings announcement was made. The fundamental variables are calculated using the information of the previous quarter (SUE,SG,TA,and CAPEX). My notation is similar to the notation used by [24].

Variable	Description	Calculation detail
SECTOR	Two-digit sector classification according to the Global Industrial Classification Standards (GICS) code.	Energy 10, Materials 15, Industrials 20, Consumer Discretionary 25, Consumer Staples 30, Health Care 35, Financials 40, Information Technology 45 Telecommunication Services 50, Utilities 55
Price momentum:		
CAR1	Cumulative abnormal return for the preceding six months since the earnings announcement day	$[\Pi_{t=m-6}^{m-1}(1+R_t)-1]-[\Pi_{t=m-6}^{m-1}(1+R_{tw})-1]$, where R_t is return in month t, R_{tw} is value weighted market return in month t, and m is last month of quarter
CAR2	Cumulative abnormal return for the second preceding six months since the earnings announcement day	$[\Pi_{t=m-12}^{m-7}(1+R_t)-1]-[\Pi_{t=m-6}^{m-1}(1+R_{tw})-1]$
Analysts variables:		
ANFOR (ANFOR-LAG)	Number of analysts predicting that earnings surprise increase (lagged value)	
CONSENSUS	Mean of earnings estimate by financial analysts	
FELAG	Lagged forecast error	$\frac{\text{CONSENSUS}_q-\text{EPS}_q}{|\text{CONSENSUS}_q|+|\text{EPS}_q|}$ [14] where EPS is earnings per share
Earnings momentum:		
FREV	Analysts earnings forecast revisions to price	$\sum_{i=0}^{5}\frac{\text{CONSENSUS}_{m-i}-\text{CONSENSUS}_{m-i-1}}{P_{m-i-1}}$ where P_{m-1} is price at end of month $m-1$, and i refers to the previous earnings revisions

| SUE | Standardized unexpected earnings | $\frac{(EPS_q - EPS_{q-4})}{\sigma_t}$ where EPS is earnings per share, and σ_t is standard deviation of EPS for previous seven quarters |

Growth indicators:

| LTG | Mean of analysts' long-term growth forecast | |
| SG | Sales growth | $\frac{\sum_{t=0}^{3} Sales_{q-t}}{\sum_{t=0}^{3} Sales_{q-4-t}}$ |

Firm size:

| SIZE | Market cap (natural log) | $ln(P_q\, shares_q)$ where $shares_q$ are outstanding shares at end of quarter q |

Fundamentals:

| TA | Total accruals to total assets | $\frac{\Delta C.As._q - \Delta Cash_q - (\Delta C.Lb._q - \Delta C.Lb.D_q) - \Delta T_q - D\&A_q}{\frac{(T.As._q - T.As._{q-4})}{2}}$ where $\Delta\, X_q = X_q - X_{q-1}$ and C.As., C.Lb.,C.Lb.D.,T,D&A,and T.As. stands for current assets, current liabilities, debt in current liabilities, deferred taxes, depreciation and amortization, and total assets respectively. |
| CAPEX | Rolling sum of capital expenditures to total assets | $\frac{\sum_{t=0}^{3} capital\ expenditures_{q-t}}{(T.As._q - T.As._{q-4})/2}$ |

Valuation multiples:

| BP | Book to price ratio | $\frac{book\ value\ of\ common\ equity_q}{market\ cap_q}$, where $market\ cap_q = P_q\, shares_q$ |
| EP | Earnings to price ratio (rolling sum of EPS of the previous four quarters deflated by prices) | $\frac{\sum_{t=0}^{3} EPS_{q-t}}{P_q}$ |

Social networks:

$deg(v_i)$	Degree centrality or degree: number of edges incidents in vertex v_i	$\sum_j a_{ij}$, where a_{ij} is an element of the adjacent matrix A
$C_c(v_i)$	Closeness centrality (normalized): inverse of the average geodesic distance from vertex v_i to all other vertices	$\frac{n-1}{\sum_j d_{ij}}$, where d_{ij} is an element of the geodesic distance matrix D [17, 7]
$B_c(v_i)$	Betweenness centrality: proportion of all geodesic distances of all other vertices that include vertex v_i	$\sum_i \sum_j \frac{g_{kij}}{g_{kj}}$, where g_{kij} is the number of geodesic paths between vertices k and j that include vertex i, and g_{kj} is the number of geodesic paths between k and j [17]
CC_i	Clustering coefficient: cliquishness of a particular neighborhood or the proportion of edges between vertices in the neighborhood of v_i divided by the number of edges that could exist between them [34]	$\frac{2|\{e_{ij}\}|}{deg(v_i)(deg(v_i)-1)}$: $v_j \in N_i$, $e_{ij} \in E$, where each vertex v_i has a neighborhood N defined by its immediately connected neighbors: $N_i = \{v_j\} : e_{ij} \in E$.
CC_i'	Normalized clustering coefficient	$\frac{deg(v_i)}{MaxDeg} CC_i$, where MaxDeg is the maximum degree of vertex in a network [13]
C (not used for forecasting)	Mean of all the clustering coefficients	$\frac{1}{n} \sum_{i=1}^{n} CC_i$
SW (not used for forecasting)	"Small world" ratio [34].	$\frac{C}{L} \frac{L_{random}}{C_{random}}$, where $L_{random} \approx \frac{ln(n)}{ln(k)}$ and $C_{random} \approx \frac{k}{n}$

Labels:

| LABELFE | Label of forecast error (FE) | 1 if $CONSENSUS \geq EPS$ (current quarter), -1 otherwise |
| LABELCAR | Label of cumulative abnormal return (CAR) | 1 if $CAR_{m+1} \geq 0$, -1 otherwise, where CAR_{m+1} refers to the CAR of the month that follows the earnings announcement |

Modeling Decision-Making Outcomes in Political Elite Networks

Michael Gabbay

Applied Physics Laboratory, University of Washington, Seattle, WA 98105, USA
gabbay@uw.edu

Abstract. A methodology for modeling group decision making by political elites is described and its application to real-world contexts is illustrated for the case of Afghanistan. The methodology relies on the judgments of multiple experts as input and can improve analysis of political decision making by elucidating the factional structure of the group of elites and simulating their interaction in a policy debate. This simulation is performed using a model of small group decision making which integrates actor policy preferences and their inter-relationship network within a nonlinear dynamical systems theory framework. In addition to the basic nonlinear model, various components required to implement the methodology are described such as the analyst survey, structural analysis, and simulation. Implementation and analysis results are discussed for both the government and insurgent sides of the current conflict in Afghanistan.

Keywords: political networks, social networks, computational social science, nonlinear dynamics, Afghanistan.

1 Introduction

This paper describes a methodology for quantitatively modeling group decision making by political elites. The methodology involves the use of expert judgment as input, structural analysis, and computational simulation using a nonlinear model of small group decision making which can address questions involving the outcome and level of dissent in a given policy debate. The methodology can aid analysis of group decision making by providing both a quantitative and qualitative framework. Quantitative implementation affords a systematic framework for assessing the interaction of member policy preferences and inter-relationships. This is difficult to do on a purely qualitative level as the structure of the group's social network and distribution of policy preferences may be complex — a difficulty that is compounded by the nonlinear nature of the interaction between group members. As a qualitative framework, the model of group decision-making dynamics can provide guidance as to when one should be on guard for the possibility of "nonlinear behaviors" that can lead to sudden and dramatic changes in policy or group discord or to unanticipated, perhaps counterintuitive dynamics.

This paper proceeds as follows: Section 2 presents the nonlinear model of group decision-making dynamics. In Sec. 3, the implementation methodology is

K. Glass et al. (Eds.): COMPLEX 2012, LNICST 126, pp. 95–110, 2013.
© Institute for Computer Sciences, Social Informatics and Telecommunications Engineering 2013

described. Section 4 illustrates the application of the methodology for the current conflict in Afghanistan for both Afghan government and insurgent leadership groups.

2   Nonlinear Model of Group Decision Making

This section describes the nonlinear model of small group decision making which is used to simulate the evolution of group member policy or ideological positions [8,7]. The theoretical basis of the model draws from social psychology theories of attitude change and small group dynamics and theories of foreign policy decision making [1,17,15]. The model is concerned with the evolution of group member positions for a given policy issue or broader ideological axis. The group member policy positions are arrayed along a one-dimensional continuum known as the *position spectrum*. A group member's position along the position spectrum is subject to change under the influence of three separate forces: (i) the self-bias force; (ii) the group influence force; and (iii) the information flow force. Only the first two forces will be discussed in this paper but information flow force has been used to model interactions between two rival decision-making groups and as a stochastic forcing representing random flow of incoming information.

2.1   Self-Bias Force

For a given policy decision episode, each member comes to the debate with his own preferred position called the *natural bias*. It is a reflection of the member's underlying beliefs, attitudes, and worldview of relevance to the matter at hand. If a member's position is shifted from his natural bias due to group pressures, he will experience a psychological force that resists this change. This self-bias force can be viewed as a form of cognitive dissonance [1]. Denoting the ith member's current policy position by x_i and his natural bias as μ_i, then i's self-bias force $S_i(x_i)$ is proportional to the difference between his current position and natural bias,

$$S_i(x_i) = -\gamma_i(x_i - \mu_i).\tag{1}$$

The proportionality constant γ_i is called the *commitment*.

2.2   Group Influence Force

The group influence force is the total force acting to change a member's position due to the other members of the group. The influence of member j upon member i is assumed to be a function of the difference in their current positions, denoted by $H_{ij}(x_j - x_i)$ and called the *coupling force*. In general, the reciprocal coupling forces between two members will not be of equivalent strength, $|H_{ij}| \neq |H_{ji}|$. If there are N members in the group, the total group influence force on member i, denoted by $G_i(x_i)$, is given by the sum

$$G_i(x_i) = \sum_{j=1}^{N} H_{ij}(x_j - x_i).\tag{2}$$

The coupling force, depicted in Fig. 1, is taken to have the form,

$$H_{ij}(x_j - x_i) = \kappa_{ij}(x_j - x_i) \exp\left(-\frac{(x_j - x_i)^2}{2\lambda_i^2}\right), \tag{3}$$

where κ_{ij} is the *coupling strength* and λ_i is i's *latitude of acceptance*. κ_{ij} gives the strength of the influence of j upon i given their personal relationship and is equivalent to a tie strength in a weighted adjacency matrix ($\kappa_{ij} \geq 0$, $\kappa_{ii} = 0$). It is useful to define a *coupling scale* α which is equal to the average coupling strength, $\alpha = \sum_{i,j} \kappa_{ij}/N$. The coupling scale can be used to represent the overall group cohesion stemming from factors such as the frequency of communications between members, their camaraderie and dedication to the group, and the overall threat to the group.

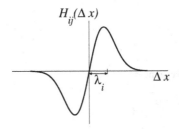

Fig. 1. Plot showing the nonlinear dependence of the coupling force on the inter-member opinion difference. $\Delta x = x_j - x_i$.

2.3   Equation of Motion

The sum of the self-bias and group influence forces determines the rate of change of the ith member's opinion so that $dx_i/dt = S_i(x_i) + G_i(x_i)$. Using the expressions (1)–(3) then yields the following equation of motion for each of the group members:

$$\frac{dx_i}{dt} = -\gamma_i(x_i - \mu_i) + \sum_{j=1}^{N} \kappa_{ij}(x_j - x_i) \exp\left(-\frac{(x_j - x_i)^2}{2\lambda_i^2}\right). \tag{4}$$

With regard to formal models of group decision making, this model is most similar to "social influence network theory," a linear model in which the force producing opinion change in a dyad is always proportional to the level of disagreement [5]. The nonlinear model of Eq. (4), however, has both a "linear" regime at low disagreement levels in which the behavior is intuitive and a "nonlinear" regime at high disagreement levels in which behaviors can run counter to initial intuition. The linear regime is characterized by: gradual changes in policy outcomes and the level of equilibrium group discord as parameters such as the coupling scale are varied; only one equilibrium for a given set of parameter values; lower group discord for higher network tie densities; and symmetric conditions of opinions and couplings always lead to symmetric final states. The nonlinear regime can exhibit the opposite behaviors: discontinuous transitions between deadlock and consensus as parameters are varied; multiple equilibria for a given set of parameter values; greater discord reduction in less dense networks; and asymmetric outcomes of majority rule even for symmetric conditions [8,7,13].

3   Implementation

This section describes the methodology for implementing the model on real-world, ongoing political contexts based on input obtained from analysts with expertise on the situation of concern. An overview of this methodology is depicted in Figure 2.

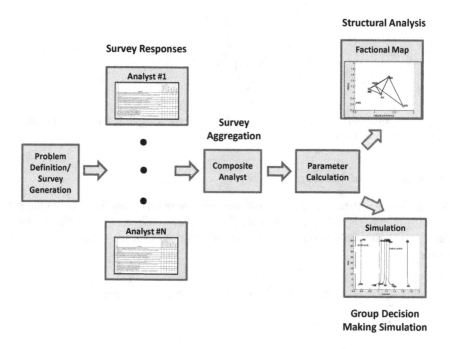

Fig. 2. Overview of Methodology

3.1   Problem Definition and Actor Selection

Problem definition concerns identifying the policy issue(s) of concern and the actors who will comprise the members of the decision-making group that will be modeled. The model assumes that relationships are stable during the course of the decision-making episode. It also assumes that group members are on the same "team" in that they have important common goals which can be furthered by joint, coordinated action and that their fates are tied together by the success or failure of that action. Accordingly, the policy issue to be modeled should be one in which the achievement of common goals is at stake. This speaks to choosing issues which are core to the success of the group. This can also be achieved by using a broader ideological axis which represents a combination of multiple issues for the position spectrum.

Typically, the members of the decision-making group are individual elites whose policy stances and relationships are critical to the decision-making process. The use of individuals is consistent with the basis of the model in social psychology, although there is no reason based purely on the model formalism which precludes the use of groups or organizations as actors.

Selecting the political elites to include in the model is often difficult given the need to limit the number of actors. This limit does not stem from computational demands of the model but rather practical demands on analyst time for survey completion. A limit of twenty actors seems reasonable based on having a survey that can be completed within a few hours. Another practical factor limiting the size of the group is that it appears to be rare for analysts to have knowledge of a large number of actors at the resolution required by the survey. In addition, a large number of actors can also excessively complicate model interpretation and visualization without significantly improving the analytical value.

Actor selection is most straightforward in situations where there is a formally-constituted small group for making decisions such as the Politburo Standing Committee in China or the General Secretariat of the FARC rebel group in Colombia. In cases where there is no such group, it may be helpful to include actors on some common basis such as having an independent power base external to the group, e.g., bureaucracies, political parties, militias, religious institutions, and tribes (see Sec.4.1).

3.2 Analyst Survey

This section describes the components of the survey which elicits expert judgment on the political group under study. Not all of the components below need to be included in every survey but the Ideologies and Strategic Attitudes and Influence Matrix components are essential.

Ideologies and Strategic Attitudes: This component of the survey is designed to assess the attitudes of the group members relevant to the policy issues of concern. It is used to calculate member natural biases and latitudes of acceptance and to set the intervals along the position spectrum corresponding to different policies. For each member, analysts are asked to estimate the member's level of agreement/disagreement with a series of statements on a scale ranging from 1 (Strongly Disagrees) to 5 (Strongly Agrees). The instructions direct analysts to evaluate agreement with the statements on the basis of the private beliefs of the members if thought to be at odds with their public rhetoric. The statements cover a range of issues, goals, identities, and specific policies. Examples are shown in Table 3.

Influence Matrix: For the influence matrix, analysts are asked to estimate the strength of each person's direct influence, i.e., that resulting from direct verbal or written communications (perhaps via trusted intermediaries), upon each of the other members in the group. The influence strength depends on factors such as the frequency of communications, status within the group, common or rival factional membership, and personal relationships of friendship or animosity. The influence strength is scaled on a range from 0 (None) to 4 (Very Strong). Each pair of members is represented by two cells in the matrix: one corresponding to influence of i upon j and one for j upon i. The influence matrix values are used to calculate the coupling strengths and commitments.

Status: Analysts are asked to rate the "status" of each group member on a scale from 1 to 10. Status is an estimate of the power of the elite in terms of his ability to influence others within the group. It depends on factors such as his formal rank within the group, the strength and nature of his power base, the amount of resources he controls, and the respect accorded to him. It is used in calculating the policy that emerges from the weighted majority and consensus decision rules and in the factional maps.

Group Affinity: A member's group affinity refers to the extent to which his allegiance resides within the leadership group as opposed to something outside the group such as the organization that he commands or to his ideology. It gives a measure of the degree to which the member will put aside his own personal policy preferences for the sake of preserving group unity. The group affinity is akin to the concept of "primary group identity" used in the decision units framework for foreign policy analysis [16]. The group affinity is scaled from 0 to 1 where 0 signifies total disregard for the opinions of the other group members and 1 signifies that the member is completely concerned with the positions of the others and ignores his own natural bias. Group affinity can be used to calculate the coupling scale.

Decision Rule: The decision rule is the way in which the final positions of the group are combined into a policy decision. Three possible choices are used:

- *Leader Choice:* The chosen policy is the final position of the group leader.
- *Weighted Majority:* The policy supported by the highest status subset of group members wins.
- *Consensus:* All group members must support the final policy. If no consensus policy exists, the status quo policy is the default.

Confidence Level: This component asks the analysts to assess their level of confidence in their knowledge of each of the actors with respect to the information solicited by the survey. A scale of 1 to 4 is used where 1 is "minimal confidence" and 4 is "high confidence." These scores are used in aggregating the analyst surveys to form the composite analyst.

3.3   Survey Aggregation

A composite analyst can be formed by averaging the survey responses of the individual analysts. If desired this can be done in a weighted fashion so that an analyst's answers are weighted by her confidence level for each actor. The aggregation of individual surveys allows for analyst judgments to be synthesized independently of each other, thereby minimizing the chances of social pressures altering individual judgment as can happen if the modeler elicits inputs in an oral discussion with a group of analysts. Note also that results can be generated on the basis of individual surveys as well. This allows for the comparison of the results from individual analysts with the composite analyst and with each other, thereby providing a way of stimulating debate about differences between analyst viewpoints.

3.4   Parameter Calculation

Some parameters can be essentially taken straight from the survey whereas others involve more elaborate calculation. Only the natural bias and latitude of acceptance calculation are noted here.

The natural bias for a given issue is the overall attitude score of a member for that issue which is obtained by averaging the member's responses to the relevant statements for that issue (after flipping those statements phrased to indicate a negative attitude). The attitude scores are put on a scale from -2 (strongly unfavorable) to +2 (strongly favorable). If a linear combination of a number of different issues is used as the position spectrum (e.g., via PCA, see Sec. 3.5), then the natural bias is the linear combination of the attitude scores for the different issues. It is important to remark that this method of placing group members on a position spectrum does not demand of the analyst the task of directly abstracting the range of policy options into a mathematical axis as do some spatial models of group decision making [3,18] — a task for which they may be ill-suited to perform. Rather, it asks for analyst assessments of the level of member agreement/disagreement on the more elemental and concrete aspects of the situation presented in the individual attitude statements.

The latitude of acceptance is calculated as the standard deviation of the natural biases obtained from the individual analysts. This makes the assumption that analyst differences with respect to the member's natural bias reflects genuine ambiguity or uncertainty in his position which in turn affects how open he is to different opinions. Other techniques are possible as well.

3.5   Structural Analysis

Independently of the ultimate simulation of the group interaction dynamics, the survey data can be analyzed to glean insight into the structure of the group of actors with respect to issues, the network of relationships, and actor power.

Structure of Issue Space: Actor positions on the attitude statements and issues can be investigated to understand relationships between different issues and factional divisions among actors as defined by their positions on the issues. Matrix decomposition techniques such as Principal Component Analysis (PCA) can be used to investigate correlations between issues and actors and the effective dimensionality of the system [2]. PCA decomposes the matrix of actor attitudes on issues into orthogonal principal components. These principal components are ranked in descending order according to the variance of the data along each component. If the first principal component carries the bulk of the variance, then the system is effectively one dimensional. This would be the case, for instance, if one faction of actors consistently takes similar positions on distinct issues whereas another faction takes opposing positions on those issues. In such a situation, the differences between actors on a number of issues can be approximately reduced to a one-dimensional axis in accordance with the assumption of the nonlinear model. The position spectrum can be constructed in such a manner although interpretation is complicated by the fact that it is now a linear combination of a number of issues, rather than a single issue.

Network Structure: The network structure and actor roles as defined by the influence matrix can be analyzed using standard social network analysis methods. This is a distinct picture from that provided by the issue space structure although one would expect there to be similarities in the factional structures exhibited by both under the assumption that birds of a feather flock together, i.e., homophily. As with the issue space, PCA can be used to analyze and visualize the network [4]. If there is a strong factional breakdown in the network, this should be evident in the PCA visualization; those actors with a similar set of relationships should be found near each other in the visualization. For assessing individual roles and influence, metrics such as degree and betweenness centralities can be calculated. Weighted out and in-degree centralities reflect, respectively, the influence going out from and coming into the actor. These can be compared with the direct assessment of actor status from the survey; typically, the correlation between them is high. While the correlation between high status and high out-degree centrality would be expected for a leader, the correlation between high power and high in-degree centrality might be less expected. This stems from the larger number of actors that leaders are connected to and to whom they must be responsive if they seek to maintain the cohesion of the group; one would particularly expect leaders who are interested in consensus-building to have high in-degree.

Factional Maps: Actor issue positions, relationships, and power can be jointly visualized using a "factional map." The actor natural biases for the issue of concern are plotted on the horizontal axis, actor status on the vertical axis, and the relationships are plotted as links between the actors. Examples are shown in Fig. 3. The factional map provides an integrated representation of issue and network-based factional structure. Potential alliances can be identified as well as actors who could play key roles such as brokers or swing players. As an example, factional maps of Iraqi insurgent groups constructed directly from their rhetoric (rather than analyst judgments) reflected alliances that eventually formed and showed the role of the Islamic Army in Iraq as a bridge between different ideological wings of the insurgency [10].

Another way of integrating ideologies and relationships is via the use of an ideology-weighted centrality metric. Here the tie strengths from the influence matrix are further weighted by a function that decreases with ideological distance, a gaussian for instance. This metric was used to analyze potential successors to Putin in 2007 [6].

3.6   Model Simulation

Model simulation is used to investigate potential results of the group decision-making process with respect to the policy outcome, the level of discord associated with that policy, and which group members sign on to the policy and which dissent. Group members are typically initialized at their natural biases and the model is run until equilibrium. (Currently, the time units are arbitrary given the difficulty of estimating the actual rates implicit in the commitment and coupling parameters.) The decision rule is used to aggregate the final member positions along the policy axis and the members of the winning coalition and dissenters are calculated. Sensitivity and scenario analyses can then be conducted to more fully assess the implications of the model.

The decision rule used to aggregate the group member final positions can be taken as the one chosen by the majority of analysts or it can be varied as well. For leader choice, the leader's final position is the policy. For weighted majority, the policy that has the most status-weighted support is the outcome; the support that each member provides to a prospective policy position decreases as a gaussian function of the distance between the prospective policy and his final position. This method allows for the policy outcome in a case of majority rule to reside within the range of positions of the majority. Otherwise, if a simple status-weighted linear combination of member positions were used then the chosen policy could lie somewhere between the majority and minority positions and, hence, would not correspond to majority rule at all. All those within their latitude of acceptance of the final policy are said to be in the winning coalition and those further away are deemed dissenters. The policy for a consensus decision rule is calculated in the same way as for a weighted majority but there can be no dissenters in order for the policy to be chosen.

Both sensitivity and scenario analysis involve varying parameters but their goals and the manner in which they are accomplished can differ. Sensitivity analysis involves running the simulation while sweeping over a parameter(s). This is used to judge the range of potential outcomes that can result due to uncertainty in model parameters. The selected parameter might be: (1) an intrinsically important one such as the coupling scale which is hard to pin down precisely and could significantly affect the results; or (2) one for which there is a large variance in analyst estimates indicating that there is substantial uncertainty in its value. The ability of an individual to sway the simulation outcome by changing his natural bias can be assessed using an "outcome centrality" metric which can serve as a sensitivity analysis measure for addressing the importance of uncertainties in the preferences of individual group members [6].

Scenario analysis entails changing parameters to correspond with a hypothesized change in the situation, e.g., a particular member(s) dramatically shifts his position, a member's status increases, a member leaves the group or dies, or a tie between two members is severed. The scenario analysis can be run using natural bias initial conditions or from the equilibrium positions that resulted prior to the changes effected in the scenario.

4   Afghanistan Application

This section illustrates application of the methodology for the case of the ongoing conflict in Afghanistan. Both Afghan government elites and insurgent leaders were included as separate decision-making groups in the analyst survey. Analytical questions focused on the prospects of a negotiated solution between the two sides, continued U.S. presence and influence, the degree of centralization of the Afghan state, and ethnic tensions. Survey responses from analysts with expertise on Afghan politics and the insurgency were obtained in the spring of 2011. Analysis and simulation were conducted in Fall 2011. Some of the implications of this modeling exercise were incorporated into the analysis of Taliban strategy and Afghan government vulnerability presented in Ref. [12].

4.1   Elite Actors

The set of Afghan Government elite actors is listed in Table 1 and the Insurgent elite actors in Table 2. For actor selection purposes, an elite actor was considered to be an independently powerful individual who has communication with other members of his group and should have a power-base independent of his title or position. An actor's power base can be tribal, ethnic, regional, military, religious, or organizational in nature and the constituent members of the power base should hold more allegiance to the individual actor than to the elite actor group (Government or Insurgent) to which he belongs. For inclusion in the Afghan Government group, an elite had to (1) generally support the concept of an Afghanistan arranged along the lines of the current constitution; and (2) not use his influence or constituents to incite large-scale violence against Afghan government or Coalition forces. Insurgent elites had to be marked by the opposites of (1) and (2).

After the surveys were completed but prior to the analysis of results, two major events affected the composition of the actors in these groups: (1) Osama Bin Laden was killed by U.S. commandos in May 2011; (2) Burhanuddin Rabbani was killed in September 2011 by a suicide bomber posing as a Taliban peace emissary meeting with him in his capacity as chairman of the High Peace Council. The use of the Al Qaida core leadership as an actor rather than Bin Laden himself meant that the actor was not lost but his death clearly would be expected to have an impact on Al Qaida's status and relations with the other insurgent actors not

Table 1. Afghan government actors included in survey

Actor	Symbol	Ethnicity	Position
Hamid Karzai	KRZ	Pashtun	President
Mohammed Qasim Fahim	FHM	Tajik	Vice President
Karim Khalili	KAL	Hazara	Vice President
Burhanuddin Rabbani	RAB	Tajik	Chairman, High Peace Council; Head, Jamiat-e-Islami party
Abdul Rashid Dostum	DOS	Uzbek	Founder, Junbesh party; Armed Forces Chief of Staff (ceremonial)
Atta Mohammed Nur	NUR	Tajik	Governor, Balkh
Gul Agha Sherzai	SHZ	Pashtun	Governor, Nangarhar
Mohammed Mohaqiq	MOQ	Hazara	Head, Wahdat-e-Mardum party; Member of Parliament
Ismail Khan	IK	Tajik	Energy Minister
Abdul Rasul Sayyaf	SAY	Pashtun	Member of Parliament

Table 2. Afghan insurgent actors included in survey

Actor	Symbol	Organization	Role/Notes
Mullah Omar	MO	Afghan Taliban	Supreme Leader
Mullah Baradar	MB	Afghan Taliban	Former First Deputy (detained)
Mawlawi Abdul Kabir	AK	Afghan Taliban	Military Commander, Eastern Region
Haqqanis	HQN	Haqqani Network; Afghan Taliban	Amalgam of leaders Jaluluddin & Sirajuddin; also in Taliban Leadership Council
Gulbuddin Hekmatyar	HIG	Hezb-e-Islami	Leader
Al Qaida Leadership	AQ	Al Qaida	Amalgam of core leaders, e.g., Bin Laden, Zawahiri, Abu Yahya al-Libi

reflected in the original survey. However, simulations in which the corresponding parameters were reduced had little impact on the results. Rabbani's death meant the total loss of an actor. Simulations were conducted mostly with him removed from the data but the analysis helps reveal the potential motive behind his assassination as discussed in Sec. 4.4.

4.2   Survey Attitude Statements

The Ideology and Strategic Attitudes component of the survey contained 40 statements for the Insurgent side and 37 for the Government side. The statements explored a number of actor policy issues, ideological attitudes, and social identities such as insurgent political power, state centralization, U.S. influence, Pakistani influence, and Afghan vs. ethnic identities. Table 3 shows a selection of statements for Afghan Government actors bearing on the issues of state centralization and accommodation of insurgent political power.

4.3   Structural Analysis

Factional maps for both sides are displayed in Fig. 3. The policy issue concerns insurgent political power, which entails, on the Insurgent side, how much political

Table 3. Selected attitude statements for Afghan Government actors

1. Partition of Afghanistan should be considered to end the conflict, if necessary.
2. Afghanistan should have a federal system of government where regions have effective autonomy to govern themselves.
3. Karzai's efforts to concentrate power in the presidency show that the Afghan Constitution should be changed to institute a parliamentary-centered system of government.
4. A strong central government is needed in order to hold Afghanistan together.
5. The insurgents are criminals, terrorists and rebels who must be put down militarily, not negotiated with.
6. If the insurgents were to halt their armed struggle and disarm, they could legitimately represent their constituents as a political party.
7. It would be acceptable for the insurgents to openly join the political process without disarming if a permanent ceasefire is agreed to.
8. A coalition government with members including insurgent leaders would be the best way to represent the Afghan population and end the conflict.
9. The best way to achieve peace is to cede effective control of some parts of Afghanistan to the insurgents.

power they are striving for and on the Government side the degree of political power they should be accommodated. It is plotted on the same scale for both sides so that the Government actors mostly have negative scores indicating less accommodation of insurgent power and the Insurgents have positive scores (specific policy labels are noted in Sec. 4.4).

The Afghan Government map shows the non-Pashtun ethnic groups on the hawkish side of the spectrum and Pashtuns on the dovish side. Karzai is the most powerful actor and his network ties show him as a bridge between Pashtuns and non-Pashtuns. Importantly, Rabbani is seen to occupy a pivotal position as the least hawkish of the non-Pashtuns and having strong ties with Karzai and most of the other non-Pashtun actors. This indicates why Karzai may have selected him as chairman of the High Peace Council — to help bring non-Pashtuns onboard with the process of reconciliation with insurgents. For the Insurgents, Mullah Omar is the most powerful and is on the hawkish side of the spectrum. The other Taliban-affiliated actors are less hawkish. Al Qaida is seen to be on the extreme hawkish end of the spectrum but having the least status. Hekmatyar is on the dovish extreme of the spectrum but has relatively little power and has poor relationships with the other insurgent actors.

4.4 Simulations

Simulations of the nonlinear decision-making model of Eq. (4) are shown in Fig. 4 for the insurgent political power issue. The intervals along the position spectrum corresponding to different qualitative policies are indicated: "no power" — no insurgent political power is to be accommodated; "unarmed party" — insurgents can participate in politics after disarming; "coalition" — insurgent leaders should be brought into a national coalition government; "armed party" — insurgents can retain their arms, control some territory, and participate as a political party if they end their violence against the government; "central control" — insurgents seek to conquer the central state and control Afghanistan. The dashed lines which bracket the policies serve as rough guides rather than hard boundaries.

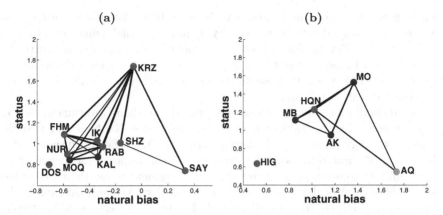

Fig. 3. Factional maps for insurgent political power issue. (a) Afghan Government actors, ethnicities — Pashtun (red), Tajik (green), Hazara (blue), Uzbek (purple). (b) Insurgent actors — those formally part of the Taliban organization in blue or green. Tie strength is proportional to link thickness; weak ties have been thresholded.

For the Afghan Government, the policy under a leader choice decision rule is seen to be "unarmed party," a policy which would support negotiations with insurgents and attempts to bring them into the political process. Dissenters include Nur, Mohaqiq, and Dostum on the hawkish side and Sayyaf on the dovish side. Note that although Rabbani is left out of the simulation shown in Fig. 4(a), none of the other model parameters were changed to account for the effects of his death, but scenario analyses aimed at doing so were conducted. For instance, to model hardened stances of anti-Taliban hawks in response to his killing, the above non-Pashtun dissenters had their commitments set to one, i.e., their positions are fixed, which has the effect of bringing Karzai to the "no power" policy interval. In addition, a sensitivity analysis shows that if the coupling scale were increased, due perhaps to an increased sense of threat to the government, then Karzai also would swing toward a more hardline policy closer to that favored by non-Pashtuns. In the immediate aftermath of Rabbani's assassination, Karzai did indeed become more hardline although since then he appears to have drifted back to a more dovish position, at least on a rhetorical level. In general, scenario analyses show that it is extremely difficult to forge a consensus policy on this issue and that if Karzai moves significantly to the left or right he will lose either Pashtun or non-Pashtun support respectively. This indicates his vulnerability to being isolated from one of these two key constituencies.

The Taliban are seen to coalesce around a "central control" policy which is Mullah Omar's choice. The only Insurgent dissenter is Hekmatyar who does not move significantly from his "armed party" natural bias given his weak links with the other actors. This solid support for a policy of seizing the central state indicates that Taliban negotiations overtures toward the United States in late 2011/early 2012 did not reflect a sincere desire to seek a peace deal with the Afghan government, as argued in Ref. [12]. In perhaps a confirmation of this conclusion, a recent article states that the U.S. government, previously hopeful, has largely given up on negotiations with the Taliban [19]. Both sensitivity and scenario analyses indicate great difficulty in moving Mullah Omar from the "central control" policy to the "armed party" policy. For example, no matter how much Mullah Baradar were to move toward a dovish position (which might be a condition of his release), it would still not be sufficient to shift Mullah Omar into the "armed party" zone.

These simulations along with insight from the Afghan Government factional map suggest why the Taliban may have assassinated Rabbani and also their broader strategy toward the Afghan government [12]. The conclusion that the Taliban are dedicated to the goal of "central control" implies that they must pursue a military solution vis a vis the Afghan government rather than a negotiated one. Rabbani's pivotal position within the network of Afghan Government elites noted above suggests that his killing would serve to exacerbate ethnic tensions between Pashtun and non-Pashtun government elites and heighten the divide over how to deal with the Taliban; both through the loss of his direct influence as well as the shock of the act itself. This in turn would make it more difficult for Karzai to effectively act as a bridge between Pashtuns and non-Pashtuns as seen in Fig. 3(a) and increases his potential to be isolated from one of those groups. An isolated Karzai decreases the sense of national unity among Afghan government elites and the population at large. This weakened national unity and drop in cohesion within the Afghan government would in turn decrease support for the Afghan National Security Forces — the primary obstacle to a Taliban military victory given the planned U.S. force drawdown.

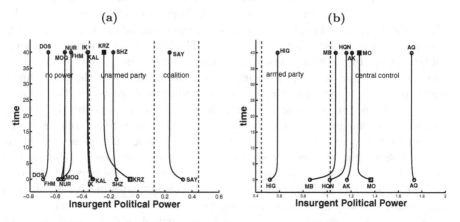

Fig. 4. Simulations of insurgent political power issue. (a) Afghan Government (w/o Rabbani), Karzai choice decision rule. (b) Insurgents, Mullah Omar choice decision rule. Open circles are actor initial positions, solid circles are final positions. Lines are actor position trajectories. Solid square indicates the final policy position; open square would be decision in absence of debate.

5   Conclusion

The nonlinear model employed to simulate decision-making outcomes synthesizes attitude change theory, social network structure, and nonlinear dynamical systems mathematics and so represents an innovative approach to the formal modeling of political decision making. The combination of the policy preference distribution in the group and its social network can form a complex structure whose complexity is further compounded by the nonlinear nature of the interactions between members in which member opinions need not move in simple proportion to their differences. The model provides a framework wherein these elements are integrated in a self-consistent manner that is not readily done by qualitative analysis alone, and allows for the controlled testing of the effects of changes or uncertainties in group variables. The nonlinear aspect of the model gives rise to the fact that the group dynamics can change *qualitatively* — and not merely as a matter of degree — as a function of the level of disagreement.

The associated analyst survey provides a systematic way of obtaining analyst judgment on the substantive aspects of the decision making group that enter into the model. The survey's use of attitude scale methodology to assess and calculate the ideological and policy positions of group members is natural given the nonlinear model's foundations in attitude change theory. This combination of attitude scaling and a formal model of elite decision making is another innovative aspect of the methodology outlined in this paper. It elicits analyst expertise on actor policy preferences without demanding that they perform the abstraction needed to create a policy axis or space itself, — a task which instead is left to the modeler.

As an alternative to implementation with analyst input, the use of rhetoric-based methods of obtaining actor ideologies and networks has been explored at the individual and organizational levels and used to inform policy analysis of ongoing situations [10,9,11,14]. A comparison of rhetoric-based Afghan Government and Insurgent actor ideologies with analyst assessments from the survey yielded good correlations for major issue dimensions. Other potential items for further research include: modeling multi-dimensional issue space dynamics; incorporating stochastic modeling and forecasting; a co-evolution model in which policy positions and actor relationships can evolve simultaneously; and integration with game-theoretic approaches.

References

1. Eagly, A., Chaiken, S.: The Psychology of Attitudes. Harcourt, Fort Worth (1993)
2. Elden, L.: Matrix Methods in Data Mining and Pattern Recognition. SIAM, Philadelphia (2007)
3. Feder, S.: Factions and Policon: New ways to analyze politics. Studies in Intelligence 31, 41–57 (1987); Reprinted in Westerfield, H.B. (ed.) Inside CIA's Private World. Yale University Press, New Haven (1995)
4. Freeman, L.C.: Graphic techniques for exploring social network data. In: Carrington, P.J., Scott, J., Wasserman, S. (eds.) Models and Methods in Social Network Analysis. Cambridge University Press, Cambridge (2005)

5. Friedkin, N.E., Johnsen, E.C.: Social Influence Network Theory: A Sociological Examination of Small Group Dynamics. Cambridge University Press, Cambridge (2011)
6. Gabbay, M.: Application of a social network model of elite decision making. Paper Presented at the Annual Meeting of the International Studies Association, Chicago (2007)
7. Gabbay, M.: A dynamical systems model of small group decision making. In: Avenhaus, R., Zartman, I.W. (eds.) Diplomacy Games. Springer, Heidelberg (2007)
8. Gabbay, M.: The effects of nonlinear interactions and network structure in small group opinion dynamics. Physica A: Statistical Mechanics and its Applications 378(1), 118–126 (2007)
9. Gabbay, M.: The 2008 U.S. elections and Sunni insurgent dynamics in Iraq. CTC Sentinel 1(10), 13–16 (2008)
10. Gabbay, M.: Mapping the factional structure of the Sunni insurgency in Iraq. CTC Sentinel 1(4), 10–12 (2008)
11. Gabbay, M.: Quantitative analysis of Afghan government actor rhetoric. Tech. rep., Applied Physics Laboratory, Univ. of Washington (2011)
12. Gabbay, M.: The Rabbani assassination: Taliban strategy to weaken national unity? CTC Sentinel 5(3), 10–14 (2012)
13. Gabbay, M., Das, A.K.: Majority rule in nonlinear opinion dynamics. In: Proceedings of the International Conference on Applications in Nonlinear Dynamics, Seattle, WA (2012) (to appear)
14. Gabbay, M., Thirkill-Mackelprang, A.: A quantitative analysis of insurgent, frames, claims, and networks in Iraq. Paper Presented at the Annual Meeting of the American Political Science Association, Seattle (2011)
15. 't Hart, P., Stern, E., Sundelius, B. (eds.): Beyond Groupthink. University of Michigan Press, Ann Arbor (1997)
16. Hermann, C., Stein, J., Sundelius, B., Walker, S.: Resolve, accept, or avoid: Effects of group conflict on foreign policy decisions. In: Hagan, J., Hermann, M. (eds.) Leaders, Groups, and Coalitions. Blackwell, Malden (2001)
17. Hermann, M.: How decision units shape foreign policy: A theoretical framework. In: Hagan, J., Hermann, M. (eds.) Leaders, Groups, and Coalitions. Blackwell, Malden (2001)
18. Bueno de Mesquita, B.: A decision making model: Its structure and form. International Interactions 23(3-4), 235–266 (1997)
19. Rosenberg, M., Nordland, R.: U.S. abandoning hopes for Taliban peace deal. New York Times (October 1, 2012)

Cooperation through the Endogenous Evolution of Social Structure

David Hales[1] and Shade T. Shutters[2]

[1] The Open University, Milton Keynes, UK
dave@davidhales.com
http://davidhales.com
[2] Center for Social Dynamics and Complexity and School of Sustainability,
Arizona State University, Tempe, AZ, USA
sshutte@asu.edu
http://www.public.asu.edu/~sshutte

Abstract. A number of recent models demonstrate sustained and high levels of cooperation within evolutionary systems supported by the endogenous evolution of social structure. These dynamic social structures co-evolve, under certain conditions, to support a form of group selection in which highly cooperative groups replace less cooperative groups. A necessary condition is that agents are free to move between groups and can create new groups more quickly than existing groups become invaded by defecting agents who do not cooperate.

Keywords: evolution of cooperative, agents, group selection, prisoner's dilemma, cultural evolution.

1 Introduction

Human society is pervaded by groups. Some are formal, such as corporations, educational institutions and social clubs. Others are informal such as youth tribes, collections of old men in a town square who discuss politics and play chess, and more recently various online forums. Some last for a long time, replenishing their membership over many generations and others are ephemeral and fleeting. Some have distinct and clear boundaries others are more diffuse – formed of overlapping networks of relationships. It would appear hard to make sense of human social behaviour without some reference to groups. Indeed, if individuals are asked to describe themselves then it is highly likely that they will refer to the groups that they hold membership of.

1.1 The Danger of Intuition

Although it seems intuitively clear that humans (and even other species) benefit from organising, coordinating and cooperating within groups, understanding how these might evolve from individual behaviour poses major puzzles for both political economy and evolutionary theory. If we start from individual self-interest or selfish replicators then why would individuals behave for the benefit of the group

K. Glass et al. (Eds.): COMPLEX 2012, LNICST 126, pp. 111–126, 2013.

if they can "get away" with free riding on the group - extracting the benefits of group membership without making a contribution?

One of the dangers of intuition about group processes is that it can lead one to ascribe agency to a group where none exists. Indeed the idea that if a group has interests - in the sense of something that would benefit all members - then rational or evolutionary individuals will behave in a way that promotes those interests has been debunked by careful analysis. Olson's famous work clearly describes the folly of this intuition when considering rational agents [23] and biologists have also challenged this idea from an evolutionary perspective [39,14].

Given these results there has been a desire to understand the highly groupish phenomena observed in human societies which appears to be altruistic from the point of view of the the individual. One way to tackle this is to attempt to capture the kinds of cultural evolutionary processes that might support learned behaviour that does not conform to self-interest [4].

In this paper we follow this line by discussing recent evolutionary models that rely on a dynamically evolving population structure - that constrains interaction possibilities - that produce remarkable groupish phenomena. With only minimal assumptions these models support the endogenous formation of groups and high levels of cooperation within the groups even when there are significant incentives for individuals to free ride.

1.2   Recent Models

Recent evolutionary models demonstrate novel forms of group selection based on simple learning rules. They function via the spontaneous formation and dissolution of groupings of selfish agents such that altruistic behaviour evolves within their in-groups.

These social dynamics, offering an alternative to rational action theories, demonstrate several notable features of human systems such as seemingly irrational altruism, highly tribal or "groupish" behaviour, and complex dynamics of social structures over time. We overview several classes of such models some based on evolving network structures and others based on different forms of population structures indicating their key features and potential applications.

Recent agent-based computational simulation models have demonstrated how cooperative interactions can be sustained by simple imitation rules that dynamically create simple social structures [28,27,11,13,12,7,20,37,30] . These classes of models implement agents as adaptive imitators that copy the traits of others and, occasionally, adapt (or mutate) them. Although these models bear close comparison with biologically inspired models – they implement simple forms of evolution – the interpretation can be of a minimal cultural, or social, learning process in which traits spread through the population via imitation and new traits emerge via randomised, or other kinds of, adaption.

Often agent-based models represent social structures such as groups, firms or networks of friends, as external and a priori to the agents – see so-called "network reciprocity" results [19]. In the models we discuss in this paper the social structures are endogenous such that agents construct, maintain and adapt them through on-going behaviour. A subset of traits supports the formation and maintenance of simple social structures [16].

As will be seen, it is the non-equilibrium processes of dynamic formation and dissolution of these structures over time that drives, or incentivises, the agents to behave cooperatively. Yet, as we will show, it is not necessary for the individual agents to prefer socially beneficial structures or outcomes, rather they emerge through a self-organising process based on local information and adaption.

1.3   When in Rome

In the models we present here agents are assumed to have incomplete information and bounded processing abilities (bounded rationality). Given these relaxed assumptions agents use social learning heuristics (imitation) rather than purely individual learning or calculation. It has been argued by Herbert Simon [32,33] that complex social worlds will often lead to social imitation (or "docility" in Simon's terminology) because agents do not have the information or cognitive ability to select appropriate behaviours in unique situations. The basic idea is "imitate others who appear to be performing well" which might also be captured by the famous quote of Saint Ambrose: "when in Rome, do as the Romans do".

The models we present demonstrate that simple imitation heuristics can emerge social behaviours and structures that display highly altruistic in-group behaviour even though this is not part of the individual goals of the agents and, moreover, may appear irrational from the point of view of the individual agents. Agents simply wish to improve their own individual condition (or utility) relative to others and have no explicit conception of in- or out-group. Yet a side effect of their social learning is to sustain group structures that constrain the spread of highly non-social (selfish) or cheating behaviour such as freeloading on the group. They can be compared to models of group selection [40,26].

We could replace the term "side effect" with the term "invisible hand" or "emergent property". We can draw a lose analogy with Adam Smith's thoughts on the market [34]. The difference is that there is no recognisable market here but rather a dynamic evolution of social structure that can transform egotistical imitative behaviour into socially beneficial behaviour.

1.4   Tribal Systems

One might term these kinds of models "tribal systems" to indicate the grouping effects and tendency for intra-group homogeneity because individuals joining a group often join this group via the imitation of others who are already a member of the group. We do not use the term "tribal" to signify any relationship between these kinds of models and particular kinds of human societies but rather to indicate the tribal nature of all human organisations i.e. that individuals almost always form cliques, gangs or other groupings that may appear arbitrary and may be highly changeable and ephemeral yet have important effects on inter-agent dynamics and behaviour.

In these kinds of tribal systems individual behaviour cannot be understood from a standpoint of individual rationality or equilibrium analysis but rather only with reference to the interaction history and group dynamics of the system as a whole. The way an individual behaves depends on their history and relationship to the groups or tribes that they form collectively.

2   Situating the Models

Diverse models of cultural group selection have been proposed from a wide range of disciplines [40]. More recently attempts to formalise them through mathematical and computer based modelling have been made.

We wish to situate the models we will discuss in this chapter with reference to the more traditional game theory [3] approach that assumes agents are rational, in the *homo economicus* sense, and have perfect information, common knowledge and no social structures to constrain interactions.

Our aim in this section is to give the non-modelling expert a sense of the relation between rational action approaches (game theory) and the more bio- and socially- inspired approaches of cultural group selection by presenting a number of broad dimensions over which they differ. It is of course the case that the boundaries between approaches is never as clean or distinct as simple categories suggest, however, to the extent that a caricature can concisely communicate what we consider to be key points that distinguish approaches it can be of value.

Figure 1(a) shows two dimensions along which game theory and cultural group selection approaches may be contrasted. Traditionally game theory models have focused on agents with unbounded rationality (i.e. no limit on computational ability) and complete information (i.e. utility outcomes can be calculated for all given actions). The cultural group selection models presented here focus on highly bounded rationality (agents just copy those with higher utility) and highly limited information (agents can not calculate a priori utility outcomes). The benefit that game theory gains by focusing on the bottom left-hand region are analytic tractability by proving equilibrium points such as Nash Equilibrium for given games [21]. Given incomplete information and bounded rationality it generally becomes more difficult to find tractable solutions and hence (agent-based) computer simulation is often used.

Figure 1(b) shows another two dimensions, learning and utility, along which a broad distinction can be made. Game theory models tend to focus on individual utility maximisation and action or strategy selection (a kind of learning) at the individual level via deduction (bottom-left). Cultural group selection focuses on social learning based on imitation in combination with rare innovation events (comparable to mutation in biological models). The emergent result is increase in social utility even though the agents themselves use a heuristic based on trying to improve their own individual utility. Hence cultural group selection could also be placed in the bottom-right quadrant.

Figure 1(c) shows another two dimensions, interaction and social structure, that distinguish the cultural group selection models and game theory. The cultural group selection models presented here represent interactions within dynamic social structures whereas game theory has tended towards static mean

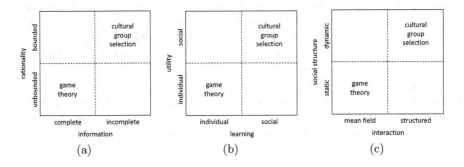

Fig. 1. Six qualitative dimensions distinguishing traditional game theory models and many cultural group selection models

field structures, by which we mean that game interactions are often assumed to occur stochastically, with equal probability, between agents over time. In the cultural group selection models (as will be seen later) a key aspect that drives the evolution of cooperation and increases in social utility is the dynamic formation of in-groups of agents that interact together exclusively, excluding interactions with the out-group.

3   Three Kinds of Models

Historically group selection has been seen as controversial within both biological and social sciences due to the difficulty in advancing a plausible theory and the inability of identifying such processes empirically in the field. Also certain kinds of non-formalised group selection approaches were exposed as naive by biologists. However these objections have been challenged due to recent advances in the area due to extensive use of computational (often agent-based) modelling and a theoretical shift that accepts that selection operating at the individual level can, under broad conditions, emerge group level selection at a higher level. The historical debate from a group selectionist perspective is well covered elsewhere [40].

We will not retread old ground here but will concentrate on presenting a specific class of group selection models that have recently emerged in the literature. These models may be interpreted as cultural evolutionary models in which imitation allows traits to move horizontally. We do not concern ourselves here with the biological interpretation of such models but rather the cultural interpretation.

Group selection relies on the dynamic formation and dissolution of groups. Over time individual entities may change groups by moving to those that offer better individual performance. Interaction between entities that determine performance is mainly restricted to those sharing the same group. Essentially then, in a nutshell, groups that support high performance for the individuals that comprise them grow and prosper whereas exploitative or dysfunctional groups dissolve as individuals move away. Hence functional groups, in terms of satisfying individual goals, are selected over time. It should be noted, as will be seen in the network rewire models discussed below, that clearly defined group boundaries are not required so long as interactions between agents are sufficiently localised through an emergent structure [41].

Key aspects that define different forms of group selection are: How group boundaries are formed (or interaction is localised); the nature of the interactions between entities within each group; the way that each entity calculates individual performance (or utility) and how entities migrate between groups.

3.1 Pro-social or Selfish Behaviour

In almost all proposed social and biological models of group selection, in order to test if group selection is stronger than individual selection, populations are composed of individuals that can take one of two kinds of social behaviour (or strategy). They can either act pro-socially, for the good of their group, or they can act selfishly for their own individual benefit at the expense of the group. This captures a form of commons tragedy [15].

Often this is formalised as a Prisoners Dilemma (PD) or a donation game in which individuals receive fitness payoffs based on the composition of their group. In either case there is a fitness cost c that a pro-social individual incurs and an associated fitness benefit b that individuals within a group gain. A group containing only pro-social individuals will lead each to gain a fitness of $b-c$ which will be positive assuming $b - c > 0$. However, a group containing only selfish individuals will lead each to obtain no additional fitness benefit or cost. But a selfish individual within a group of pro-socials will gain the highest fitness benefit. In this case the selfish individual will gain b but the rest will gain less than $b-c$. Hence it is always in an individual's interests (to maximise fitness) to behave selfishly. In an evolutionary scenario in which the entire population interacts within a single group then selfish behaviour will tend to be selected because this increases individual fitness. This ultimately leads to an entire population of selfish individuals and a suboptimal situation in which no individual gains any fitness. This is the Nash Equilibrium [21] and an Evolutionary Stable Strategy for such a system [35].

There have been various models of cooperation and pro-social behaviour based on reciprocity using iterated strategies within the PD [1]. However, we are interested in models which do not require reciprocity since these are more generally applicable. In many situations, such as large-scale human systems or distributed computer systems, repeated interactions may be rare or hard to implement due to large population sizes (of the order of millions) or cheating behaviour that allow individuals (or computer nodes) to fake new identities.

3.2 Tag Models

In [11] a tag model of cooperation was proposed which selected for pro-social groups. It models populations of evolving agents that form groups with other

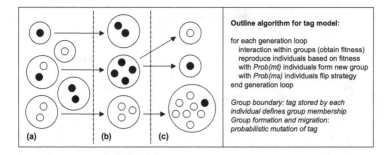

Fig. 2. Schematic of the evolution of groups in the tag model. Three generations (a-c) are shown. White individuals are pro-social, black are selfish. Individuals sharing the same tag are shown clustered and bounded by large circles. Arrows indicate group linage. Migration between groups is not shown.

agents who share an initially arbitrary tag or social marker. The tag approach was discussed by Holland [16] and developed by Riolo [28,27]. The tag is often interpreted as an observable social label (e.g. style of dress, accent etc.) and can be seen as a group membership marker. It can take any mutable form in a model (e.g. integer or bitstring). The strategies of the agents evolve, as do the tags themselves. Interestingly this very simple scheme structures the population into a dynamic set of tag-groups and selects for pro-social behaviour over a wide range of conditions. Figure 2 shows a schematic diagram of tag-group evolution and an outline algorithm that generates it.

In general it was found that pro-social behaviour was selected when $b > c$ and $mt \gg ms$, where mt is the mutation rate applied to the tag and ms is the mutation rate applied to the strategy. In this model groups emerge from the evolution of the tags. Group splitting is a side effect of mutation applied to a tag during reproduction. Figure 3 shows typical output from a tag simulation model visualising a portion of the tag space. Each line on the vertical axis represents a unique tag value (i.e. a possible group). Groups composed of all cooperative agents are shown in light grey, mixed groups of cooperators and defectors (mixed) are dark grey and groups composed of all defectors are black. The size of each group is not shown. Notice that over time new groups form, persist for some period and then become invaded by some defectors and quickly die. This cyclical process continues persistently. Since mixed groups die quickly the number of defectors in the entire population at any time instant are small. Hence at any given time the highest scoring agents are those who are defecting but this is not a sustainable strategy because they destroy the groups they are situated in.

A subsequent tag model [27] produced similar results with a different tag and cooperation structure. Tags are defined as single real values [0..1] and cooperation is achieved through a donation game in which a related tolerance value [0..1] specifies a range around the tag value to which agents will donate if they find other agents with tags within this range. The tolerance and tag are traits that evolve based on fitness (payoff). This provides the potential for overlapping group

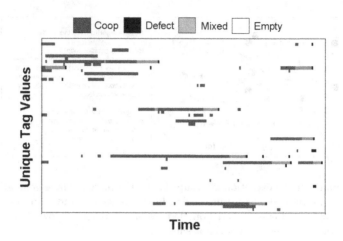

Fig. 3. Evolution of tag groups over time. Here 200 cycles from a single typical simulation run are shown for a portion of the tag space. Cooperative groups come into to existence, persist for some time, and then die.

boundaries. Also the model works through random sampling of the population for game partners rather than only sampling within a group. However It has been argued that this approach cannot be applied to pro-sociality in general because it does not allow for fully selfish behaviour between identically tagged individuals [29,6]. Put simply, those agents which share the same tag must cooperate and can not evolve a way to defect since there is no separate trait defining the game strategy.

Recent work by Traulsen and Nowak examined the tag process both analytically and in simulation deriving the necessary conditions under which high cooperation can be sustained [38]. They derived an elegant mathematical description of the stochastic evolutionary dynamics of tag-based cooperation in populations of finite size. They found that in a cultural model with equal mutation rates between all possible phenotypes (tags and strategies), the crucial condition for high cooperation is $b/c > (K+1)/(K-1)$, where K is the number of tags. A larger number of tags requires a smaller benefit-to-cost ratio. In the limit of many different tags, the condition for cooperators to have a higher average abundance than defectors becomes $b > c$. Hence this result indicates that it is not necessary to have higher mutation rates applied to tags per se but rather to have enough tag space relative to the benefit-to-cost ratio.

Finally, more recent work rigorously replicates and compares several existing tag models from the literature and introduces the notion of weak and strong cheating [31]. It was found that the way that agents implement their cheating mechanism can have dramatic effects of the results obtained from some previous tag models. This has implications for both cultural and biological interpretations of the models.

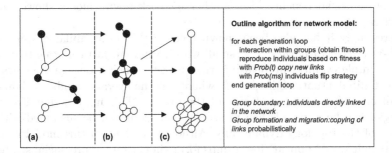

Outline algorithm for network model:

for each generation loop
    interaction within groups (obtain fitness)
    reproduce individuals based on fitness
    with Prob(t) copy new links
    with Prob(ms) individuals flip strategy
end generation loop

Group boundary: individuals directly linked in the network
Group formation and migration:copying of links probabilistically

Fig. 4. Schematic of the evolution of groups (cliques) in the network-rewiring model. Three generations (a-c) are shown. White individuals are pro-social, black are selfish. Arrows indicate group linage. Notice the similarity to the tag model in figure 2.

3.3 Network Rewiring Models

Network rewiring models for group selection have been proposed with direct application to peer-to-peer (P2P) protocol design [7,8]. In these models, which were adapted from the tag model described above, individuals are represented as nodes on a graph. Group membership is defined by the topology of the graph. Nodes directly connected are considered to be within the same group. Each node stores the links that define its neighbours. Nodes evolve by copying both the strategies and links (with probability t) of other nodes in the population with higher utility than themselves. Using this simple learning rule the topology and strategies evolve promoting pro-social behaviour and structuring the population into dynamic arrangements of disconnected clusters (where $t = 1$) or small-world topologies (where $0.5 < t < 1$). Group splitting involves nodes disconnecting from all their current neighbours and reconnecting to a single randomly chosen neighbour with low probability mt. As with the tag model pro-social behaviour is selected when $b > c$ and $mt >> ms$, where ms is the probability of nodes spontaneously changing strategies. Figure 4 shows a schematic of network evolution (groups emerge as cliques within the network) and an outline algorithm that implements it. Figure 5 shows output from a typical simulation run where $t = 1$. Four snapshots of the network structure are shown at different key time cycles. As can be seen the system self-organises into a dynamic ecology of cooperative components. The evolution of the network is therefore the outcome of a coevolutionary process between nodes which effects both the strategy and the network structure based on local node information. This can be compared to global forms of network optimisation, that interestingly, also implement localised mutation and rewiring [18].

In this model we have translated dynamics and properties similar to the tag model into a graph. In [8,7] the rewiring approach was suggested as a possible protocol that could be applied in an online peer-to-peer system. In [12] the same fundamental rewiring protocol was applied to a scenario requiring nodes to adopt specialised roles or skills within their groups, not just pro-social behaviour alone, to maximise social benefit. Also another network rewire model shows similar properties [30].

Interestingly it has also been shown recently [22] in a similar graph model tested over *fixed topologies* (e.g. small-world, random, lattice, scale-free) that under a simple evolutionary learning rule pro-social behaviour can be sustained in some limited situations if $b/c > k$, where k is the average number neighbours over all nodes (the average degree of the graph). This implies that if certain topologies can be imposed then pro-social behaviour can be sustained without rewiring of the topology dynamically. Although analysis of this model is at an early stage it would appear that groups form via clusters of pro-social strategies forming and migrating over the graph via nodes learning from neighbours. This reinforces the insight that localisation of interaction rather than strict group boundaries is sufficient to produce cooperation [41].

3.4   Group-Splitting Model

In [37] a group selection model is given that sustains pro-social behaviour if the population is partitioned into m groups of maximum size n so long as $b/c > 1 + n/m$. In this model group structure, splitting and extinction is assumed *a priori* and mediated by exogenous parameters. Splitting is accomplished by explicitly limiting group size to n, when a group grows through reproduction beyond n it is split with (high) probability q into two groups by probabilistically reallocating each individual to one of the new groups. By exogenously controlling n and m a detailed analysis of the model was derived such that the cost / benefit condition is shown to be *necessary* rather than just sufficient. The model also allows for some migration of individuals between groups outside of the splitting process. Significantly, the group splitting model can potentially be applied recursively to give multilevel selection – groups of groups etc. However, this requires explicit splitting and reallocation mechanisms at each higher level. Figure 6 shows a schematic of group-splitting evolution and an outline algorithm that implements it. We include this example here for comparison but it should be noted that it is not a fully endogenous group formation model.

4   Discussion

What do the kinds of models presented in this paper actually tell us? They do not aim to capture any particular phenomena but rather to demonstrate certain kinds of possible process under given assumptions. Hence they should be seen as an aid to intuition. Models such as these deduce logically, through a computer program, outcomes of given assumptions. As we have discussed previously unchecked intuition can be unreliable when applied to complex systems such as these. They can be viewed, then, as thought experiments with the aid of a computer.

Given an understanding of these possible processes it may then be possible to identify real world phenomena that may be evidencing them. This is a much

(a) Grouping before cooperation

(b) Cooperation spreading

(c) Giant cooperative component breaks apart

(d) Cooperative groups are formed

Fig. 5. Evolution of network structure in a network rewire model. From an initially random topology (not shown) composed of all nodes playing the defect strategy (dark shaded nodes), components quickly evolve, still containing all defect nodes (a). Then a large cooperative component emerges in which all nodes cooperate (b). Subsequently the large component begins to break apart as defect nodes invade the large cooperative component and make it less desirable for cooperative nodes (c). Finally an ecology of cooperative components dynamically persists as new components form and old components die (d). Note: the cooperative status of a node is indicated by a light shade.

harder task requiring extensive empirical work. However, the models can suggest the kinds of data that would be required and the kinds of hypotheses that can be tested against that data.

4.1   Institutional Economics

The idea of an evolutionary and cyclical group competition process goes at least as far back as Ibn Khaldun writing in the 14th Century [17]. Khaldun, who is considered one of the founding fathers of sociology and economics, developed a

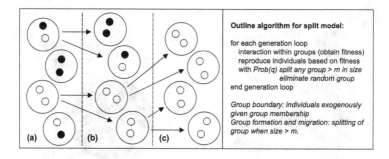

Fig. 6. Schematic of the evolution of groups in the group-splitting model. Three generations (a-c) are shown. White individuals are pro-social, black are selfish. Individuals sharing the same group are shown clustered and bounded by large circles. Arrows indicate group linage. Migration between groups is not shown.

concept of "asabiyyah" (roughly meaning social cohesion) to explain how groups arise, become powerful and ultimately are conquered by other groups. We do not claim that the models presented here capture the sophistication of Kaldun's model - indeed he discusses groups with highly sophisticated internal power structures with roles for a ruling elite, artisans and the like - however, we believe, that some of the essential intuitive dynamics are captured in them in a minimal and formal way.

Essentially we have a model of perpetual social conflict – an arms race – in which groups die from within after going through several stages in which social cohesion is successively weakened.

More generally we argue that kinds of models presented here can be viewed as an initial (and minimal) way to establish a link between a line of work called "institutional economics" and the more individualistic evolutionary approaches which have been heavy influenced by biology. The aim is to find the kinds of simple mechanisms that can support institutional-like dynamics. We might view the groups in our models as highly abstract and simple "photo-institutions". We have shown elsewhere that similar models can support limited forms of specialisation within groups – where different agents perform different functions for the good of the group [12].

Ostrom's famous work describes in detail how groups with known boundaries – and other "design principles" – can self-organise their own solutions to Common Pool Resource dilemmas without market mechanisms or central control from government [24]. We believe that the kinds of mechanism evidenced within the models presented could be sufficient to support the evolution of more sophisticated cooperation mechanisms rather than simply cooperate or defect. i.e. policing and punishment strategies.

4.2   Applications

The models presented here could potentially have applications in both understanding real existing social systems and engineering new tools that support new kinds of social systems – particularly in online communities. Increasingly online Web2.0 and other communities allow for the tracking and measurement of the dynamics of groups overtime [25]. Massive clean datasets can now be collected and it might be possible for (versions of) the models presented here to be calibrated and validated (or invalidated). However, to our knowledge, this has so far not been achieved for full-blown cultural group selection, however recent highly detailed work shows great promise in this area [36].

In addition, as has already been discussed above, peer-to-peer (P2P) systems composed of millions of online nodes (running client software on users machines) could benefit from the application of group selection techniques by applying them directly to the algorithms (or protocols) used by nodes to self-organise productive services for users. There has already been some deployed on-going experiments with a P2P file-sharing client called "Tribler"[1] which applies a so-called "group selection design pattern" to self-organise cooperative communities for media content sharing [10,9].

These two kinds of application of the models are not independent because by increasing our understanding of productive human social processes we can automate aspects of those processes into computer algorithms to increase their speed and reach - one might consider the success of online social networking as an example of this.

5   Conclusion

The models presented here show how simple agent heuristics based on imitation directed towards individual improvement of utility can lead to behaviour in which agents behave "as if" there is a motivating force which is higher than self-interest: the interests of the group or tribe. This higher force does not need to be built-in to agents but rather emerges through time and interactions - a historical process. The formation of social structures, overtime, creates conditions that favour pro-social behaviour. Agents receive utility by interacting in tribes (simple social structures that localise game interactions). Tribes that cannot offer the agent a good "utility deal" relative to other tribes will disband as agents "vote with their feet" by joining other better tribes based on their individual utility assessment. Of course movement between tribes, here, is not interpreted as a physical relocation but rather a social and mental one. By copying the traits of others who have higher utility the appropriate social structures emerge. Increasingly in electronic and virtual communities the cost of such movement is converging towards zero or very low individual cost. It could be conjectured it is this low cost, and consequent freedom from geographical and organisational constraints,

---

[1] See: http://tribler.org

which is a major factor in the recent success of online communities, virtual social networks and other peer-production communities such as Wikipedia [2].

However, this process would not preclude agents with explicit group level utility preferences i.e. incorporating "social rationality" functions or such like. Such agents could potentially improve general welfare through a modicum of explicit planning and encouragement of pro-social group formation. The models presented here rely on random trial and error to find cooperative pro-social "seeds" which then are selected and grow via the evolutionary process as other agents join the seed. We speculate that an agent with a correctly aligned internal model of what would make a successful seed could pro-actively recruit others from the population. However, this introduces issues such as explicit recruitment processes, explicit internal social models and, potentially, transferable utility. Here we begin to see the formation of something that resembles a market or an institutional strategy. In this context the models we have presented could be seen as "pre-market" mechanisms or "proto-institutions" in which value is not separated from the social structures that produce it because it cannot be stored, accumulated, transferred or controlled.

The models we have presented mainly focus on social dilemma scenarios - situations in which individuals can improve their own utility at the expense of the group or tribe they interact with. Often the application of the market in these situations does not resolve the dilemma in a socially equitable way (i.e. does not lead to cooperation) but rather can incentivise non-cooperation. This is such a serious issue that game theory explicitly addresses it within the emerging area of Mechanism Design [5]. However, often these models rely on standard Rational Action assumptions and a high degree of central control that enforce the rules of the game.

As previously discussed, the kinds of models presented here might be viewed as the first steps on a bridge between simple evolutionary – often biologically inspired – cultural models and some of the puzzles and finding of institutional economics. Rather than starting from the assumption of existing and complex institutions we attempt to grow them from individuals following simple copying heuristics.

References

1. Axelrod, R.: The evolution of cooperation. Basic Books, NY (1984)
2. Benkler, Y.: The Wealth of Networks: How Social Production Transforms Markets and Freedom. Yale University Press (2006)
3. Binmore, K.: Game Theory and the Social Contract. Playing Fair, vol. 1. The MIT Press, Cambridge (1994)
4. Boyd, R., Richarson, P.: Culture and cooperation. In: Beyond Self-Interest, Chicago University Press, Chicago (1990)
5. Dash, R., Jennings, N., Parkes, D.: Computational-Mechanism Design: A Call to Arms. IEEE Intelligent Systems, 40–47 (November 2003)
6. Edmonds, B., Hales, D.: Replication, Replication and Replication - Some Hard Lessons from Model Alignment. Special Issue on Model-2-Model Comparison, Journal of Artificial Societies and Social Simulation 6(4) (2003)

7. Hales, D., Arteconi, S.: SLACER: A Self-Organizing Protocol for Coordination in P2P Networks. IEEE Intelligent Systems 21(2), 29–35 (2006)
8. Hales, D., Edmonds, B.: Applying a socially-inspired technique (tags) to improve cooperation in P2P Networks. IEEE Transactions in Systems, Man and Cybernetics - Part A: Systems and Humans 35(3), 385–395 (2005)
9. Hales, D., et al.: QLECTIVES Report D2.1.1: Candidate theory models for cooperation algorithms (2010), http://davidhales.com/qlectives/D2.1.1.pdf
10. Hales, D., Arteconi, S., Marcozzi, A., Chao, I.: Towards a group selection design pattern. In: Meyer, F. (ed.) Proceedings of the Final Workshop on The European Integrated Project "Dynamically Evolving, Large Scale Information Systems (DELIS)", Barcelona, February 27-28, Heinz Nixdorf Institute, University of Paderborn, Band 222 (2008)
11. Hales, D.: Cooperation without Memory or Space: Tags, Groups and the Prisoner's Dilemma. In: Moss, S., Davidsson, P. (eds.) MABS 2000. LNCS (LNAI), vol. 1979, pp. 157–166. Springer, Heidelberg (2001)
12. Hales, D.: Emergent Group-Level Selection in a Peer-to-Peer Network. Complexus 3(1), 108–118 (2006)
13. Hales, D.: From Selfish Nodes to Cooperative Networks Emergent Link-based Incentives in Peer-to-Peer Networks. In: Proceedings of the Fourth IEEE International Conference on Peer-to-Peer Computing (p2p 2004). IEEE Computer Society Press (2004)
14. Hamilton, D.W.: Geometry for the selfish herd. Journal of Theoretical Biology 31, 295–311 (1971)
15. Hardin, G.: The tragedy of the commons. Science 162, 1243–1248 (1968)
16. Holland, J.: The Effect of Labels (Tags) on Social Interactions. Santa Fe Institute Working Paper 93-10-064. Santa Fe, NM (1993)
17. Khaldun, I.: The Muqaddimah – An Introduction to History, 9th edn. Princeton University Press (1989)
18. Kaluza, P., Vingron, M., Mikhailov, A.S.: Self-correcting networks: Function, robustness, and motif distributions in biological signal processing. Chaos 18, 1–17 (2008)
19. Konno, T.: A condition for cooperation in a game on complex networks. Journal of Theoretical Biology 269(1), 224–233 (2011)
20. Marcozzi, A., Hales, D.: Emergent Social Rationality in a Peer-to-Peer System. Advances in Complex Systems (ACS) 11(4), 581–595 (2008)
21. Nash, J.F.: Equilibrium Points in N-Person Games. Proc. Natl. Acad. Sci. USA 36, 48–49 (1950)
22. Ohtsuki, H., et al.: A simple rule for the evolution of cooperation on graphs and social networks. Nature 441(25), 502–505 (2006)
23. Olson, M.: The Logic of Collective Action. Harvard University Press, Cambridge (1970)
24. Ostrom, E.: Governing the Commons – The evolution of institutions for collective action. Cambridge University Press, UK (1990)
25. Palla, G., Barabasi, A.-L., Vicsek, T.: Quantifying social group evolution. Nature 446, 664–667 (2007)
26. Richerson, P., Boyd, R.: The Biology of Commitment to Groups: A Tribal Instincts Hypothesis in Evolution and the Capacity for Commitment. Russell Sage Press, New York (2001)
27. Riolo, R., Cohen, M.D., Axelrod, R.: Cooperation without Reciprocity. Nature 414, 441–443 (2001)

28. Riolo, R.: The Effects of Tag-Mediated Selection of Partners in Evolving Populations Playing the Iterated Prisoners Dilemma. Santa Fe Institute Working Paper 97-02-016. Santa Fe, NM (1997)
29. Roberts, G., Sherratt, N.T.: Similarity does not breed cooperation. Nature 418, 449–500 (2002)
30. Santos, F.C., Pacheco, J.M., Lenaerts, T.: Cooperation prevails when individuals adjust their social ties. PLoS Comput Biol. 2(10), e140 (2006)
31. Shutters, S.T., Hales, D.: Tag-mediated altruism is dependent on how cheaters are defined. Journal of Artificial Societies and Social Simulation (expected pub. date November 2012) (in press)
32. Simon, H.A.: A Mechanism for Social Selection and Successful Altruism. Science 250 (1990)
33. Simon, H.A.: Models of Bounded Rationality, Empirically Grounded Economic Reason, vol. III. MIT Press, Cambridge (1997)
34. Smith, A.: An inquiry into the nature and causes of the wealth of nations, 1st edn. William Clowes and Sons, England (1776, 1836)
35. Smith, J.M.: Evolution and the Theory of Games (1982)
36. Szell, M., Thurner: Measuring social dynamics in a massive multiplayer online game. Soc. Networks 32, 313–329 (2010)
37. Traulsen, A., Nowak, M.A.: Evolution of cooperation by multilevel selection. Proceedings of the National Academy of Sciences 130(29), 10952–10955 (2006)
38. Traulsen, A., Nowak, M.A.: Chromodynamics of Cooperation in Finite Populations. Plos One 2(3), e270 (2007)
39. Williams, G.C.: Adaptation and Natural Selection: A Critique of Some Current Evolutionary Thought. Princeton University Press, Princeton (1966)
40. Wilson, D.S., Sober, E.: Re-introducing group selection to the human behavioural sciences. Behavioural and Brain Sciences 17(4), 585–654 (1994)
41. Wilson, D.S., Wilson, E.O.: Rethinking the theoretical foundation of sociobiology. Q. Rev. Biol. 82, 327–348 (2007)

The Role of Community Structure
in Opinion Cluster Formation

Ryan J. Hammer, Thomas W. Moore, Patrick D. Finley, and Robert J. Glass

Complex Adaptive Systems of Systems (CASoS) Engineering
Sandia National Laboratories
Albuquerque, New Mexico USA
{rhammer,tmoore,pdfinle,rjglass}@sandia.gov
http://www.sandia.gov/CasosEngineering/

Abstract. Opinion clustering arises from the collective behavior of a social network. We apply an Opinion Dynamics model to investigate opinion cluster formation in the presence of community structure. Opinion clustering is influenced by the properties of individuals (nodes) and network topology. We determine the sensitivity of opinion cluster formation to changes in node tolerance levels through parameter sweeps. We investigate the effect of network community structure through rewiring the network to lower the community structure. Tolerance variation modifies the effects of community structure on opinion clustering: higher values of tolerance lead to less distinct opinion clustering. Community structure is found to inhibit network wide clusters from forming. We claim that advancing understanding of the role of community structure in social networks can help lead to more informed and effective public health policy.

Keywords: Social networks, community structure, complex networks, opinion dynamics, agent-based models, complex systems.

1   Introduction

Network science plays an important role in public health policy. Social, contact and organizational networks have been shown to affect various aspects of health. Public health researchers increasingly apply social network research to population health problems [1] in areas of infectious disease propagation [2], obesity [3], smoking [4], and even happiness [5].

Understanding the effects of social network topology is essential to crafting optimally effective public health policy. Social network research is increasingly focusing on the presence and effects of community structure within various networks. Community structure refers to heterogeneous degree distributions that result in groups of nodes more densely connected to each other than to the rest of the network [6]. This understanding of community structure can be traced back to Granovetter's research which found that two nodes having friendship relationships with a third node are more likely to be connected to each other [7]. Community structure is also a factor

K. Glass et al. (Eds.): COMPLEX 2012, LNICST 126, pp. 127–139, 2013.

in homophily, where nodes with similar characteristics tend to be more highly connected to one another than nodes with dissimilar characteristics [8]. Community structure has been shown to play an important role in the dynamics of networks through social diffusion [9].

Opinion Dynamics (OD) is a powerful social-network modeling technique integrating research findings from sociology and statistical physics. Conceptually OD traces its origins to Cartwright and Harary's structural balance theory which posits that an individual's opinion regarding another person or idea is influenced by those with whom he/she shares positive social ties [10]. The mathematics and algorithms of OD are derived from the Ising spin model which captures spin alignment of adjacent particles within a lattice. OD models extend the particle interaction concepts of statistical physics to include structural balance theory to produce a generalizable method of modeling the flow of ideas, opinions and concepts through social networks [11].

Several different implementations of OD models have been proposed; for a comprehensive review, see Castellano, Fortunato and Loreto [11]. The basic assumptions of opinion dynamics can be extended to capture various facets of social dynamics. Some of these modifications include continuous-valued states, bounded confidence models using tolerance, and network moderated interactions to introduce social structure into the model. Opinion dynamics initially provided insight into the fundamental dynamics of information spread in networks and well-mixed populations. More recently, OD has been used to investigate the diffusion of agricultural practices in farming communities [12], the formation of extremist groups in larger communities [13], and the effects of influence-based interventions on differential social structures, including gendered networks [14].

Opinion dynamics is useful in studying the formation of opinion clusters. Opinion clusters are an emergent result of the collective interactions of people within a network influenced by their individual characteristics and the network structure. Various groups in nature display clustering behavior including flocks of birds [15], bacterial colonies [16], fish schools [17], and human behavior such as walking patterns [18]. Recent work that has highlighted the effect of opinion clusters on a system includes studies that show how clusters of unvaccinated individuals can lead to a dramatic increase in disease outbreak probability [19]. In a study of individual characteristics affecting opinion cluster formation, Schelling presented an exhaustive study using a spatial proximity model [20]. Studies on opinion cluster formation within adaptive networks [21] have elucidated network influences on opinion clusters.

In this paper, we address the role of individual node properties and the role of network topology on opinion cluster formation within and among communities. Section 2 documents the model formulation used in this investigation. Section 3 describes the experiments we performed and the results from these experiments. Section 4 presents a discussion of potential applications of our findings to public health.

2   Model Formulation

An agent-based model is used to investigate the influence of individual and network characteristics on the formation of opinion clusters. We use a modified version of the opinion dynamics model of Deffuant and Weisbuch to model the flow of opinions in a network of heterogeneous agents. The Deffuant-Weisbuch model simulates the spread of opinions in a well-mixed population [12]. We modify the original model by mapping it to a directed network of agents. This directed network represents relationship ties such as friendship nominations where the directionality indicates non-reciprocal nominations. Incorporating edge directionality is supported by studies showing that friendship networks are often directed [22]. In addition, empirical studies on network-based properties of tobacco use have identified correlations incorporating directionality [23].

Agents have an individual opinion, modeled as a continuous variable in the range [0.0, 1.0], and a tolerance value indicating how open an agent is to the opinions of others. The tolerance value constrains opinion-changing interactions to agents whose opinions are within a tolerance bound. If the difference in opinion between two agents is less than the tolerance value, the agents can influence one another, incrementally changing opinions to become more similar to each other. Henceforth, the term node shall refer to an agent. At each time step the opinion of each node is updated using the following equation:

$$x_i(t+1) = x_i(t) + \frac{1}{|S_i|} \sum_{j \in S_i} \mu_{ij} [x_j(t) - x_i(t)] . \tag{1}$$

In equation (1), $x_i(t+1)$ represents the opinion of node i at the next time step. The opinion is updated by adding to node i's current opinion the average difference between node i's opinion and that of every one of its neighbors, $x_j(t)$, at the current time step t. If the difference in opinion between node i and a neighbor is greater than the tolerance bound, the two do not influence one another. An edge weight, μ_{ij}, allows for giving certain friendships more influence as might be the case for a family member or a close friend.

Tolerance is an important feature of the model to study opinion cluster formation. As the model execution proceeds, opinions of nodes in various portions of the network tend to converge to common local mean values with the number and average size of the clusters determined by tolerance [12]. Higher tolerance values generate fewer, larger clusters. Opinion clusters emerge from the network as a result of the constraint that tolerance places on the number of interactions that can take place. These clusters consist of groups of neighboring nodes with similar opinions. To be considered as part of a given opinion cluster, a node must both be reachable by every node in the cluster and hold an opinion within tolerance bounds of the most proximal nodes in the cluster to which it is connected.

Opinion cluster formation is also influenced by network structure: node interactions only occur if an edge exists. Random networks, such as Erdős-Rényi graphs, exhibit less distinct clustering properties than do scale free networks (Moore, et al, 2012 in review).

The algorithm employed for cluster detection is a variant of the DBSCAN algorithm [24]. We have modified the algorithm to map it onto a network. The algorithm operates by mapping nodes to clusters if they are within the tolerance bounds of the initial cluster node and reachable via an edge of the node. This algorithm also removes the need to know a priori the number of clusters involved, which is most suitable for our application.

We investigate the influence of tolerance-constrained interactions in a network containing community structure by generating 250-node networks comprised of five communities. We vary tolerance over the range [0, 0.5], and examine the formation of opinion clusters within and among communities. To investigate the contribution of community structure in the network, we decrease the community structure by increasing the number of edges between communities, adding random edges between nodes in different communities.

3   Experiments

We conducted two experiments to study the effects of individual-level constraints and network-level constraints on the formation of opinion clusters: one to study the effects of tolerance level and one to study the effects of network topology.

We use a similar network topology in both experiments. Our primary criterion for the network structure is that distinct community structure exists. We create our network by generating 5 communities using the Erdős–Rényi model [25]. Each community consists of 50 nodes connected using an edge probability p = 0.1633 resulting in 400 expected edges within each community.

Once constructed, we randomly connect each individual community to every other community with a specified number of edges depending on the experiment (Figure 1). We don't claim that this graph formation process generates networks representative of those in the real world, only that the generated networks contain distinct community structure, the condition under which we are interested in studying opinion clusters. Future studies will elicit the effects of using other network formation models more demonstrative of real-world topology.

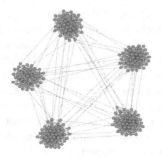

Fig. 1. An example network displaying community structure in which five densely connected communities are connected by more sparse inter-group edges

3.1   Tolerance Experiment

The OD model is run using networks with community structure. The tolerance of each node is varied to determine the effect of tolerance on opinion cluster formation. For this experiment networks are generated with five distinct communities (Figure 1). Each community is connected to every other community by adding 25 edges at random between each community for a total of 250 additional edges. Using the modularity metric proposed by Newman and Girvan [26] to provide a measurement of community structure, this network produces a modularity of ~0.72 indicating a high level of community structure. Initial opinion values of the nodes are uniformly distributed at the outset of the model run. Tolerance is increased over the range [0, 0.5] in a series of simulations. For each simulation, a different stochastic realization of the example network is generated. For each tolerance value, the model is run with 100 stochastically-generated networks.

We investigate two things. First, as tolerance increases, at what tolerance value does each community form a single majority-opinion cluster? Second, at what tolerance value does the entire network form a single majority-opinion cluster? In this context, a single-majority cluster contains a large percentage of the nodes in the relevant community or network. It is rare that clusters are entirely defined by communities as network topology can cause certain nodes to be drawn into a different community.

3.2 Tolerance Results

As illustrated in Figure 2, at a tolerance value of 0.0 every node maintains its baseline opinion; no opinion adjustment can take place. At this value each node forms its own opinion cluster. As tolerance increases, the number of clusters decreases while the average cluster size increases.

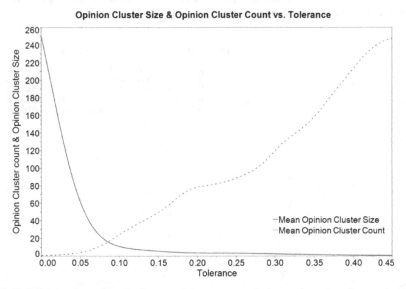

Fig. 2. Tolerance values affect both mean cluster count and cluster size. As tolerance increases the number of opinion clusters falls and the size of the clusters rises.

Once tolerance has increased to a certain level, each of the five communities emerges as a single opinion cluster, as seen in Figure 3. Individual node opinions for a single community are plotted relative to tolerance values. As tolerance increases, node opinions begin to draw closer together. Finally, at a tolerance value of ~0.27 the nodes in the community converge to a single majority-opinion cluster.

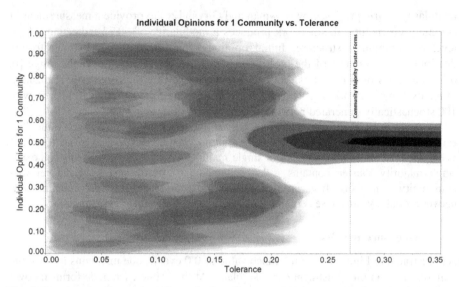

Fig. 3. Density plot shows the effect of tolerance on equilibrium distribution of opinion in a single community. Opinion distributions develop clustering patterns at a tolerance value of 0.05, become bimodally distributed at approximately 0.15 and coalesce to a single mean-value cluster at a tolerance of approximately 0.27.

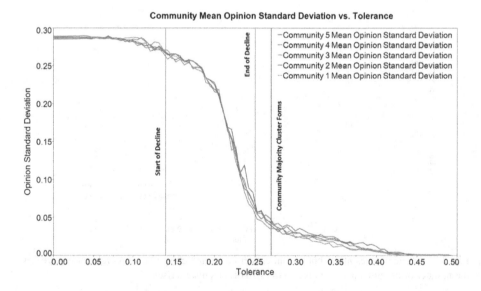

Fig. 4. Standard deviation of opinion distribution for each community. Note the transition in the rate of opinion deviation decline at 0.25. The tolerance value at which a community majority-opinion cluster forms (0.27) is highlighted by the blue line.

Figure 4 shows a decrease in the standard deviation of node opinions for each community as tolerance increases. A sharp drop can be seen between tolerance values of ~0.14 to ~0.25, indicating minimal node interaction and opinion adjustment at low tolerance levels.

At progressively higher tolerance values, the clusters begin to merge together into larger clusters until eventually the entire network coalesces into a single-opinion cluster. Each community reaches a single majority-opinion cluster at a tolerance value of ~0.27, while coalescence into a single network-wide opinion cluster occurs at a tolerance value of ~0.45 (Figure 5). The top portion of the two-part figure plots each individual opinion in the entire network at tolerance values from 0.0 to 0.5. The bottom portion of the figure shows a plot of the standard deviation in opinion for the entire network across the same range of tolerance values. Two rough transitions can be seen in the top portion. A transition occurs at the tolerance value of ~0.27, the point at which communities form single-majority clusters as demonstrated in previous figures. The final transition to a single majority-opinion cluster occurs at a tolerance of ~0.45, a finding verified by the plot illustrated in the bottom half of Figure 5 in which the standard deviation reaches a minimum at a tolerance value of ~0.45.

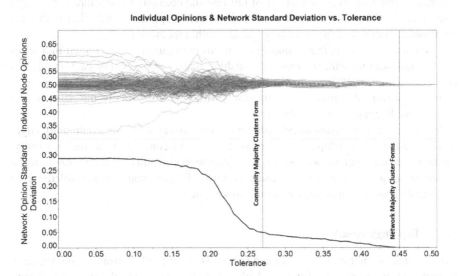

Fig. 5. Individual node opinions for the entire network plotted relative to tolerance values (above) are shown in the context of the standard deviation of opinion for the entire network (below). The individual node opinions come into consensus at tolerance ~0.45 indicating the formation of the single majority-opinion network cluster.

3.3   Topology Experiment

We examine the role of topology in opinion cluster formation, specifically in regard to the structure of communities. The same process of generating a network with five distinct communities is utilized. Based on our findings from the previous experiment, every node is assigned a constant tolerance value of 0.27, the value at which distinct community-wide opinion clusters form. Unlike the uniform distribution used in the previous experiment, here nodes in each community draw an initial opinion from one of five distinct opinion intervals ranging from low (0.00, 0.12) to high (0.88, 1.00). These intervals are defined in Table 1.

Table 1. Opinion ranges assigned to each community for topology experiment

Community 1	[0.00, 0.12]
Community 2	[0.22, 0.34]
Community 3	[0.44, 0.56]
Community 4	[0.66, 0.78]
Community 5	[0.88, 1.00]

The heterogeneous assignment of community opinion among the different communities and a homogenous assignment within each community serves to illustrate model operation in community networks which more realistically reflect the homophily to be expected in social networks. Where communities hold distinctly separate opinion intervals, the effect of blurring the community structure can be seen more clearly. One side effect of community-specific opinion distribution is that the communities come to consensus very quickly within themselves.

The number of edges that connect the various communities is increased in a series of 100 runs starting with 0 edges and finishing with 250 edges. For each number of edge connections, 100 stochastic network realizations are modeled. Each edge increment lowers the amount of community structure until finally the network is a single densely connected community.

We examine edge ratios of between-community edges to within-community edges to identify the relationship at which communities will converge to a single majority-opinion cluster. That is, we ask to what degree community structure needs to be degraded to allow a majority cluster to form. Rather than varying tolerance to investigate cluster formation, we use network topology.

3.4   Topology Results

The modularity metric of Newman and Girvan is again used to measure the degree of degradation of community structure. A plot of modularity versus the ratio of between-community edges to within-community edges, presented in Figure 6, demonstrates that as the ratio increases the community structure decreases. In light of the degradation of community structure, both the modularity and the edge ratio can be analyzed at the point at which the single majority network cluster forms. Two sharp transitions in modularity can be seen in this figure: the first steep drop occurs in the edge ratio range of ~0.11 to ~0.20, the second in the edge ratio range of ~0.27 to ~0.31. This provides an understanding of what is happening to community structure as edges are added and at what edge ratio the community structure is obscured, giving us a clear picture of the modularity of the network when the communities converge.

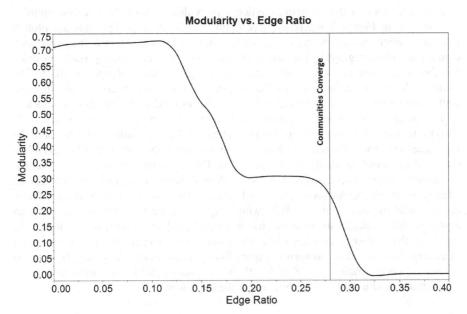

Fig. 6. Modularity plotted versus the ratio of between-community edges to within-community edges. As edges are added between the communities, the community structure fades. At the ratio of ~0.31, community structure has been eliminated.

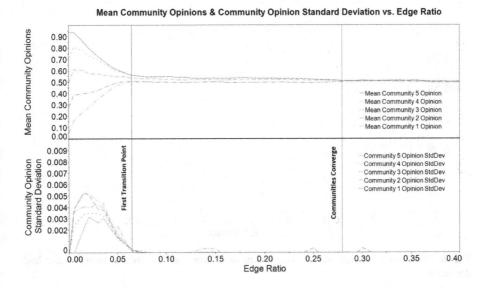

Fig. 7. Effect of between-community to within-community edge ratio on mean community opinion and standard deviation. The mean community opinion (upper) plot shows consensus among the five communities commencing at ~0.065 with near complete consensus at ~0.28. The standard deviation (lower) plot shows the communities at consensus amongst themselves at the onset.

A broader look at the changing relationships within each of the five communities can be seen in Figure 7 where the five communities' average opinions are plotted relative to edge ratios (top graph) as are the standard deviation of opinion for each community (lower graph). The mean opinion for each community starts to draw together at an edge ratio of ~0.065 and moves together more sharply at ~0.28. The standard deviation graph indicates that the communities are in consensus from the start based on the assigned tolerance value and the initial distribution range for each.

The standard deviation in opinion for the network can be observed in Figure 8. Two transition points can be seen: the first occurs at the edge ratio of ~0.08 where the first steep drop ends; the second transition can be seen at the edge ratio of ~0.30 when the standard deviation finally levels out near 0. This provides further evidence that the communities have converged and the network has formed a majority cluster near an edge ratio of ~0.28. As seen earlier in Figure 6, the last steep drop of modularity concludes at the edge ratio of ~0.31 which highlights the fact that the network can converge once community structure has been obscured to a high enough degree. In this case, the convergence takes place when structure has been obscured to the point of an edge ratio of ~0.28 (as seen in Figure 7) and a modularity value near 0. In terms of this network, a ratio of 0.28 indicates that a total of 2000 edges exist within the communities and a total 560 edges connect the communities.

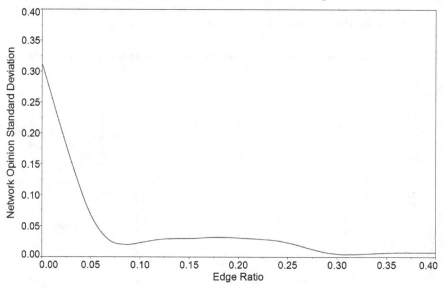

Fig. 8. Effect of in-community to inter-community edge ratio on overall network standard deviation

The opinion cluster size should also level out at ~0.28. We observe this in Figure 9. Additionally it is of note in Figure 9 how quickly cluster size grows with just a small increase in the edge ratio. However, in order to achieve a majority cluster very close to the total number of nodes, an edge ratio ~0.28 must exist.

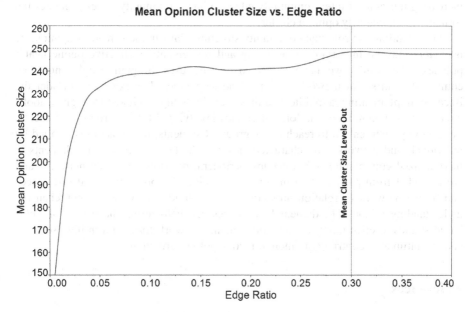

Fig. 9. Effect of between-community to within-community edge ratio on mean opinion cluster size. Mean cluster size can be seen to level out in the vicinity of an edge ratio of ~0.30, very near the point at which community opinions converged in the previous results (0.28).

These results indicate that community structure plays a role in the attainment of network-wide consensus. Even though communities are able to come to internal consensus, the community structure must be diminished to a certain degree in order to overcome the network-level constraint placed on opinion cluster formation. The results shown indicate that once an edge ratio of ~0.28 has been reached, the communities converge to form a single majority opinion cluster.

4   Discussion

We have presented an opinion dynamics model highlighting the role that community structure plays in the formation of opinion clusters. Using the individual constraint of tolerance, communities form single majority-opinion clusters at the relatively low tolerance value of ~0.27. A majority network-wide cluster forms at a tolerance value of ~0.45. We also found that community structure can inhibit the formation of a majority network opinion cluster when a constant tolerance value is imposed on a level which constrains the communities to single majority-opinion clusters. To overcome this network-level constraint, the community structure must be obscured to a certain degree by the establishment of inter-community relationships. In the case of the given topology, an edge ratio of between-community edges to within-community edges needed to be ~0.28 in order to overcome network constraints. This indicates that tolerance can be driven by means other than individual characteristics. Simply increasing the ratio of inter-community edges to intra-community edges promotes the formation of a majority opinion cluster.

Understanding the role that community structure plays in the spread of opinions or behaviors may be important to designing and implementing effective public health policies. The results we have presented can be used to gain insight into how communities in social networks will respond to policies. For example, consider the issue of implementing a public health policy in a high school. In high schools, community structure takes the form of cliques involving different groups of students. For a policy intervention to reach every group of students, the constraints implied by individual- and network-level characteristics must be taken into account. Students in marginalized communities often occupy peripheral positions on social networks and are excluded from participation in core communities by both individual differences and by the network of relationships connecting individuals. A more comprehensive understanding of these fundamental influences can help refine the design of field-based studies to elicit empirical data on dynamic network formation in high schools, and can ultimately contribute to more effective policy formation.

References

1. Smith, K.P., Christakis, N.A.: Social networks and health. Annual Review of Sociology 34, 405–429 (2008)
2. Glass, L.M., Glass, R.J.: Social contact networks for the spread of pandemic influenza in children and teenagers. BMC Public Health 8(1), 61 (2008)
3. Christakis, N.A., Fowler, J.H.: The spread of obesity in a large social network over 32 years. New England Journal of Medicine 357(4), 370 (2007)
4. Moore, T.W., Finley, P.D., Linebarger, J.M., Outkin, A.V., Verzi, S.J., Brodsky, N.S., Cannon, D.C., Zagonel, A.A., Glass, R.J.: Extending Opinion Dynamics to Model Public Health Problems and Analyze Public Policy Interventions. In: Eighth International Conference on Complex Systems, Quincy, MA (2011)
5. Fowler, J.H., Christakis, N.A.: Dynamic spread of happiness in a large social network: longitudinal analysis over 20 years in the Framingham Heart Study. British Medical Journal 337(dec04 2), a2338 (2008)
6. Girvan, M., Newman, M.E.J.: Community structure in social and biological networks. Proceedings of the National Academy of Sciences 99(12), 7821–7826 (2002)
7. Granovetter, M.: The strength of weak ties: A network theory revisited. Sociological Theory 1, 201–233 (1983)
8. McPherson, M., Smith-Lovin, L., Cook, J.M.: Birds of a feather: Homophily in social networks. Annual Review of Sociology 27(1), 415–444 (2001)
9. Colbaugh, R., Glass, K.: Predictive analysis for social diffusion: The role of network communities. Arxiv preprint arXiv:0912.5242 (2009)
10. Cartwright, D., Harary, F.: Structural Balance: A Generalization of Heider's Theory. Psychological Review 63(5) (1956)
11. Castellano, C., Fortunato, S., Loreto, V.: Statistical physics of social dynamics. Reviews of Modern Physics 81(2), 591–646 (2009)
12. Weisbuch, G., Deffuant, G., Amblard, F., Nadal, J.P.: Meet, discuss, and segregate! Complexity 7(3), 55–63 (2002)

13. Deffuant, G.: Comparing extremism propagation patterns in continuous opinion models. Journal of Artificial Societies and Social Simulation 9(3) (2006)
14. Moore, T., Finley, P., Hammer, R., Glass, R.: Opinion dynamics in gendered social networks: An examination of female engagement teams in Afghanistan. Social Computing, Behavioral-Cultural Modeling and Prediction, 69–77 (2012)
15. Ballerini, M., Cabibbo, N., Candelier, R., Cavagna, A., Cisbani, E., Giardina, I., Lecomte, V., Orlandi, A., Parisi, G., Procaccini, A.: Interaction ruling animal collective behavior depends on topological rather than metric distance: Evidence from a field study. Proceedings of the National Academy of Sciences 105(4), 1232–1237 (2008)
16. Ben-Jacob, E., Cohen, I., Gutnick, D.L.: Cooperative organization of bacterial colonies: From genotype to morphotype. Annual Reviews in Microbiology 52(1), 779–806 (1998)
17. Gautrais, J., Jost, C., Theraulaz, G.: Key behavioural factors in a self-organised fish school model. Annales Zoologici Fennici 45, 415–428 (2008)
18. Moussaïd, M., Perozo, N., Garnier, S., Helbing, D., Theraulaz, G.: The walking behaviour of pedestrian social groups and its impact on crowd dynamics. PloS One 5(4), e10047 (2010)
19. Salathé, M., Bonhoeffer, S.: The effect of opinion clustering on disease outbreaks. Journal of The Royal Society Interface 5(29), 1505–1508 (2008)
20. Schelling, T.C.: Dynamic models of segregation. Journal of Mathematical Sociology 1(2), 143–186 (1971)
21. Kozma, B., Barrat, A.: Consensus formation on adaptive networks. Phys. Rev. E 77(1) (January 2008)
22. Scott, J.: Social Network Analysis: A Handbook. Sage Publications Ltd., London (2000)
23. Christakis, N.A., Fowler, J.H.: The collective dynamics of smoking in a large social network. New England Journal of Medicine 358(21), 2249 (2008)
24. Ester, M., Kriegel, H.P., Sander, J., Xu, X.: A density-based algorithm for discovering clusters in large spatial databases with noise. In: Proceedings of the 2nd International Conference on Knowledge Discovery and Data Mining, vol. 1996, pp. 226–231 (1996)
25. Erdős, P., Rényi, A.: On the evolution of random graphs. Magyar Tud. Akad. Mat. Kutató Int. Közl 5, 17–61 (1960)
26. Newman, M.E.J., Girvan, M.: Finding and evaluating community structure in networks. Physical Review E 69(2), 026113 (2004)

Achieving Social Optimality
with Influencer Agents

Jianye Hao and Ho-fung Leung

Department of Computer Science and Engineering
The Chinese University of Hong Kong
jyhao,lhf@cse.cuhk.edu.hk

Abstract. In many multi-agent systems (MAS), it is desirable that the agents can coordinate with one another on achieving socially optimal outcomes to increase the system level performance, and the traditional way of attaining this goal is to endow the agents with social rationality [7] - agents act as system utility maximizers. However, this is difficult to implement when we are facing open MAS domains such as peer-to-peer network and mobile ad-hoc networks, since we do not have control on all agents' behaviors in such systems and each agent usually behaves individually rationally as an individual utility maximizer only. In this paper, we propose injecting a number of influencer agents to manipulate the behaviors of individually rational agents and investigate whether the individually rational agents can eventually be incentivized to coordinate on achieving socially optimal outcomes. We evaluate the effects of influencer agents in two common types of games: prisoner's dilemma games and anti-coordination games. Simulation results show that a small proportion of influencer agents can significantly increase the average percentage of socially optimal outcomes attained in the system and better performance can be achieved compared with that of previous work.

1 Introduction

In certain multi-agent systems (MASs), an agent needs to coordinate effectively with other agents in order to achieve desirable outcomes, since the outcome not only depends on the action it takes but also the actions taken by others. How to achieve effective coordination among agents in multi-agent systems is a significant and challenging research topic, especially when the interacting agents represent the interests of different parties and they have generally conflicting interests.

It is well-known in game theoretic analysis that, if each agent behaves as an individual utility maximizer in a game, and always makes its best response to the behaviors of others, then the system can result in a Nash equilibrium such that no agent will have the incentive to deviate from its current strategy [1]. The equilibrium solution has its merits considering its desirable property of stability, however, it can be extremely inefficient in terms of the overall utilities the agents receive. The most well-known example is the prisoner's dilemma (PD) game: the

K. Glass et al. (Eds.): COMPLEX 2012, LNICST 126, pp. 140–151, 2013.
© Institute for Computer Sciences, Social Informatics and Telecommunications Engineering 2013

agents will reach the unique Nash equilibrium of mutual defection if the agents act as individually rational entities, while the outcome of mutual cooperation is the only optimal outcome in terms of maximizing the sum of both agents' payoffs. In certain domains, nevertheless, a more desirable alternative is a social optimal solution [7], in which the utilitarian social welfare (i.e., the sum of all agents's payoffs) is maximized.

One approach of addressing this problem is to enforce the agents to behave in a socially rational way - aiming at maximizing the sum of all agents' utilities when making their own decisions. However, as mentioned in previous work [4], this line of approach suffers from a lot of drawbacks. It will greatly increase the computational burdens of individual agents, since each agent needs to consider all agents' interests into consideration when it makes decisions. Besides, it may become infeasible to enforce the agents to act in a socially rational manner if the system is open in which we have no control on the behaviors of all agents in the system. To solve these problems, one natural direction is considering how we can incentivize the individually rational agents to act towards coordinating on socially optimal solutions. A number of work [4,9,8,5] has been done in this direction by designing different interaction mechanisms of the system while the individual rationality of the interacting agents is maintained and respected at the same time. One common drawback of previous work is that certain amount of global information is required to be accessible to each individual agent in the system and also there still exists certain percentage of agents that are not able to learn to coordinate on socially optimal outcomes (SOs).

In this work, we propose inserting a number of influencer agents into the system to incentivize the rest of individually rational agents to behave in a socially rational way. The concept of influencer agent is first proposed by Franks et al [2] and has been shown to be effective in promoting the emergence of high-quality norms in the linguistic coordination domain [11]. In general, an influencer agent is an agent with desirable convention or prosocial behavior, which is usually inserted into the system by the system designer aiming at manipulating those individually rational agents into adopting desirable conventions or behaviors. To enable that the influencer agents exert effective influences on individually rational agents, we consider an interesting variation of sequential play by allowing entrusting decision to others similar to previous work [6]. During each interaction between each pair of agents, apart from choosing an action from its original action set, each agent is also given the option of choosing to entrust its interacting partner to make a joint decision for both agents. It should be noted that such a decision to entrust the opponent is completely voluntary, hence the autonomy and rationality of an agent are well respected and maintained. The influencer agents are socially rational in the sense that they will always select an action pair that corresponds to a socially optimal outcome should it becomes the joint decision-maker. Besides, each agent is allowed to choose to interact with an influencer agent or an individually rational agent, and then it will interact with a randomly chosen agent of that type. Each agent (both individually rational and influencer agents) uses a rational learning algorithm to make its decisions in terms of which type of agent to interact with and which action to choose, and improves its learning policy based on the reward it receives from the interaction. We evaluate the performance of the learning framework in two representative types of games: PD game and anti-coordination (AC) game. Simulation results show that a small proportion of influencer agents can efficiently incentivize most of the purely rational agents to coordinate on the socially optimal outcomes and better performance in terms of the average percentage of socially optimal outcome attained can be achieved compared with that of previous work [5].

The remainder of the paper is organized as follows. An overview of related work is described in Section 2. In Section 3, we give a description of the problem we investigate in this work. In Section 4, the learning framework using influencer agents we propose is introduced. In Section 5, we present our simulation results to compare the performance under the learning framework using influencer agents with previous work. Lastly conclusion and future work are given in Section 6.

2   Related Work

Hales and Edmonds [4] first introduce the tag mechanism originated in other fields (e.g., artificial life and biological science) into multi-agent systems research to design effective interaction mechanism for autonomous agents to achieve desirable outcomes in the system. They focus on the Prisoner's Dilemma game (called PD game hereafter), and each agent is represented by $1 + L$ bits. The first bit indicates the agent's strategy (i.e., playing C or D), and the remaining L bits are the tag bits, which are used for biasing the interaction among the agents and are assumed to be observable by all agents. In each generation each agent is allowed to play the PD game with another agent with the same tag string. The agents in the next generation are formed via fitness proportional reproduction scheme together with low level of mutations on both the agents' strategies and tags. This mechanism is demonstrated to be effective in promoting high level of cooperation among agents when the length of tag is large enough. However, there are some limitations of this tag mechanism. Since the agents mimic the strategy and tags of other more successful agents and the agents choose the interaction partner based on self-matching scheme, an agent can only play the same strategy as that of the interacting agent. Thus socially rational outcome can be obtained if and only if the agents need to coordinate on identical actions.

McDonald and Sen [8][9] propose three different tag mechanisms to tackle the limitations of the model of Hales and Edmonds. Due to space limitation, here we only focus on the third mechanism called paired reproduction mechanism since this is the only one that is both effective and practically feasible. It is a special reproduction mechanism which makes copies of matching pairs of individuals with mutation at corresponding place on the tag of one and the tag-matching string of the other at the same time. The purpose of this mechanism is to preserve the matching between this pair of agents after mutation in order to promote the survival rate of cooperators. Simulation results show that this mechanism can help sustaining the percentage of coordination of the agents at a high level in both PD games and Anti-Coordination games (AC games hereafter), and this mechanism can be applied in more general multi-agent settings where payoff sharing is not allowed. However, similar to Hales and Edmonds' model [4], these mechanisms all heavily depend on mutation to sustain the diversity of groups in the system. Accordingly this leads to the undesired result that the variation of the percentage of coordination is very high.

Considering the disadvantages of evolutionary learning (heavily depending on mutation), Hao and Leung [5] develop a tag-based learning framework in which each agent employs a reinforcement learning based strategy to make its decisions. Specifically, they propose a Q-learning based strategy in which each agent's learning process is augmented with an additional step of determining how to update its Q-values, i.e., update its Q-values based on its own information or information of others in the system. Each update scheme is associated with a weighting factor to determine which update scheme to use each time and these weighting factors are adjusted adaptively based on a greedy strategy. They evaluate their learning framework in both PD game and AC games and simulation results show that better performance can be achieved in terms of both average percentage of socially optimal outcomes attained and the stability of the system compared with the paired reproduction mechanism [8].

3 Problem Description

The general question we are interested in is how individually rational agents can learn to coordinate with one another on desirable outcomes through repeated pairwise interactions. In particular, we aim to achieve socially optimal outcomes, under which the utilitarian social welfare (i.e., the sum of all agents' payoffs) is maximized. At the same time, we desire that the rationality and autonomy of individual agents be maintained. In other words, the agents should act independently in a completely individually rational manner when they make decisions. This property is highly desirable particularly when the system is within an open, unpredictable environment, since the system implemented with this kind of property can largely withstand the exploitations of selfish agents designed by other parties.

Specifically, in this paper we consider studying the learning problem in the context of a population of agents as follows. In each round each agent chooses to interact with another agent (i.e., to play a game with that agent), which is constrained by the interaction protocol of the system. Each agent learns concurrently over repeated interactions with other agents in the system. The interaction between each pair of agents is formulated as a two-player normal-form game, which will be introduced later. We assume that the agents are located in a distributed environment and there is no central controller for determining the agents' behaviors. Each agent can only know its own payoff during each interaction and makes decisions autonomously.

Following previous work [4,8,5], we focus on two-player two-action symmetric games for modeling the agents' interactions, which can be classified into two different types. For the first type of games, the agents need to coordinate on the outcomes with identical actions to achieve socially rational outcomes. One representative game is the well-known PD game (see Fig. 1), in which the socially optimal outcome is (C, C), however, choosing action D is always the best strategy for any individually rational agent. The second type of games requires the agents to coordinate on outcomes with complementary actions to achieve socially optimal outcomes. Its representative game is AC game (see Fig. 2), in which either outcomes (C, D) or (D, C) is socially optimal. However, the row and column agents prefer different outcomes and thus it is highly likely for individually rational agents to fail to coordinate (achieving inefficient outcomes (C, C) or (D, D)). For both types of games, we are interested in investigating how the individually rational agents can be incentivized to learn to efficiently coordinate on the corresponding socially optimal outcomes.

A's payoff, B's payoff		Agent B's action	
		C	D
Agent A's action	C	R,R	S,T
	D	T,S	P,P

A's payoff, B's payoff		Agent B's action	
		C	D
Agent A's action	C	L,L	H,H
	D	H,H	L,L

Fig. 1. Prisoner's dilemma (PD) game satisfying the constraints of $T > R > P > S$ and $2R > T + S > 2P$

Fig. 2. Anti-coordination (AC) game satisfying the constraints of $H > L$

4   Learning Framework

We first give a background introduction on the concept of influencer agent and how it can be applied for solving our problem in Section 4.1. Then we describe the interaction protocol within the framework in Section 4.2. Finally the learning strategy the agents adopt to make decisions is introduced in Section 4.3.

4.1   Influencer Agent

The concept of influencer agent is firstly termed by Franks et al. [2], and there is also a number of previously work with similar ideas. In general, an influencer agent (IA) is an agent inserted into the system usually by the system designer in order to achieve certain desirable goals, e.g., emergence of efficient convention or norms [2]. Sen and Airiau [10] investigate and show that a small portion of agents with fixed convention can significantly influence the behavior of large group of selfish agents in the system in terms of which convention will be adopted in the system. Similarly Franks et al. [2] investigate the problem of how a small set of influencer agents adopting pre-fixed desirable convention can influence the rest of individually rational agents towards adopting the convention the system designer desires in the linguistic coordination domain [11].

Since we are interested in incentivizing individually rational agents to behave in the socially rational way, here we consider inserting a small number of influencer agents, which are socially rational, into the system. To enable the influencer agents to exert effective influence on individually rational agents' behaviors, we consider an interesting variation of sequential play by allowing entrusting decision to others similar to previous work [6]. During each interaction between each pair of agents, apart from choosing an action from its original action set, each agent is also given an additional option of asking its interacting partner to make the decision for both agents (denoted as choosing action F). If an agent A chooses action F while its interacting partner B does not, agent B will act as the leader to make the joint decision for them. If both agents choose action F simultaneously, then one of them will be randomly chosen as the joint decision-maker. The influencer agents are socially rational in that they will always select the socially optimal outcome as the joint action pair to execute whenever it becomes the joint decision-maker. If there exist multiple socially optimal outcomes, then these socially optimal outcomes will be selected with equal probability. For those individually rational agents, we simply assume that they will always choose the outcome under which their own payoffs are maximized as the joint action for execution, whenever they are entrusted to make joint decisions.

4.2 Interaction Protocol

From previous description, we know that there exist two different types of agents in the system: influencer agents (IA) and individually rational (or 'selfish') agents (SA). In each round, each agent is allowed to choose which type of agent to interact with, and then it will interact with an agent randomly chosen from the corresponding set of agents. This is similar to the commonly used interaction model that the agents are situated in a fully connected network in which each agent randomly interacts with another agent each round [10,12]. The only difference is that in our model the population of agents are divided into two groups and each agent is given the freedom to decide which group to interact with first but the specific agent to interact with within each group is still chosen randomly. Our interaction model can better reflect the realistic scenarios in human society, since human can be classified into different groups according to their personality traits and different persons may have different preferences regarding which group of people they are willing to interact with.

Each agent uses a rational learning algorithm to make its decisions in terms of which type of agent to interact with and which action to choose, and improves its policy based on the rewards it receives during the interactions. Besides, each agent chosen as the interacting partner also needs to choose an action to respond accordingly depending on the type of its interacting agent. We assume that during each interaction each agent only knows its own payoff and cannot have access to its interacting partner's payoff and action. The overall interaction protocol is shown in Algorithm 1.

Algorithm 1. Interaction Protocol

1: **for** a fixed number of rounds **do**
2:     **for** each agent i in the system **do**
3:         determine which type of agents to interact with
4:         interact with one agent randomly chosen from the corresponding set of agents

5:         update its policy based on the reward received from the interaction
6:     **end for**
7: **end for**

4.3 Learning Algorithm

For the individually rational agents, it is natural that they always choose the strategy which is a best response to its partner's current strategy in order to maximize its own payoff. If a learning algorithm has the property that it can converge to a policy that is a best response to the other players' policies when the other players' policies converge to stationary ones, then it is regarded as being rational [1]. A number of rational learning algorithms exist in the multi-agent learning literature and here we adopt the Q-learning algorithm [13], which is the most commonly used.[1] Specifically, each individually rational agent maintains two different set of Q-tables: one corresponding to the estimates of the payoffs for actions for interacting with influencer agents, Q_{IA}, and the other corresponding to the estimates of the payoffs for the set of actions for interacting with individually rational agents, Q_{SA}. In the following discussion, a_{IA} refers to an action when interacting with an influencer agent and a_{SA} refers to an action when interacting with an individually rational agent. In each round t, an individually rational agent i makes its decision (which type of agent to interact and which specific action to choose) based on the Boltzmann exploration mechanism as follows. Formally any action a_{IA} belonging to the set of actions available for interacting with influencer agents is selected with probability

$$\frac{e^{Q_{IA}(a_{IA})/T}}{\sum_{a_{IA}} e^{Q_{IA}(a_{IA})/T} + \sum_{a_{SA}} e^{Q_{SA}(a_{SA})/T}} \tag{1}$$

Any action a_{SA} belonging to the set of actions available for interacting with individually rational agents is selected with probability

$$\frac{e^{Q_{SA}(a_{SA})/T}}{\sum_{a_{IA}} e^{Q_{IA}(a_{IA})/T} + \sum_{a_{SA}} e^{Q_{SA}(a_{SA})/T}} \tag{2}$$

The temperature parameter T controls the exploration degree during learning, and initially it is given a high value and decreased over time. The reason is that initially the approximations of the Q-value functions are inaccurate and the

---

[1] Other rational learning algorithms such as WOLF-PHC [1] and Fictitious Play [3] will be investigated as future work.

agents have no idea of which action is optimal, thus the value of T is set to a relatively high value to allow the agents to explore potential optimal actions. After enough explorations, the exploration has to be stopped so that the agents can focus on exploiting the actions that has shown to be optimal before.

An individually rational agent selected as the interacting partner makes decisions depending on which type of agent it will interact with. If it interacts with another individually rational agent, then it will choose an action from the set of actions available for interacting with individually rational agents, and any action a_{SA} is chosen with probability

$$\frac{e^{Q_{SA}(a_{SA})/T}}{\sum_{a_{IA}} e^{Q_{SA}(a_{SA})/T}} \tag{3}$$

If it is chosen by an influencer agent, then it will pick an action from the set of actions available for interacting with influencer agents, and any action a_{IA} is selected with probability

$$\frac{e^{Q_{IA}(a_{IA})/T}}{\sum_{a_{IA}} e^{Q_{IA}(a_{IA})/T}} \tag{4}$$

After the interaction in each round t, each agent updates its corresponding Q-table depending on which type of agent it has interacted with. There are two different learning modalities available for performing update [12]: 1) multi learning approach: for each pair of interacting agents, both agents update their Q-tables based on the payoffs they receive during interaction, 2) mono learning approach: only the agent who initiates the interaction updates its Q-table and its interacting partner does not update. In mono learning approach, each agent updates its policy in the same speed, while in the multi learning approach, some agents may learn much faster than others due to the bias of partner selection. We investigate both updating approaches and the effects of both updating approaches on the system level performance will be shown in Section 5. Formally, each agent updates its Q-tables during each interaction as follows depending on which type of agent it has interacted with,

$$Q_{IA/SA}^{t+1}(a) = \begin{cases} Q_{IA/SA}^t(a) + \alpha_i(r_i^t - Q_{IA/SA}^t(a)) & \text{if } a \text{ is chosen in round } t \\ Q_{IA/SA}^t(a) & \text{otherwise} \end{cases}$$

$$\tag{5}$$

where r_i^t is the reward agent i obtains from the interaction in round t by taking action a. α_i is the learning rate of agent i, which determines how much weight we give to the newly acquired reward r_i^t, as opposed to the old Q-value. If $\alpha_i = 0$, agent i will learn nothing and the Q-value will be constant; if $\alpha_i = 1$, agent i will only consider the newly acquired information r_i^t.

For influencer agents, we do not elevate their learning abilities above the rest of the population. They make decisions in the same way as individually rational agents. The only difference is that the influencer agents behave in a socially rational way in that they will always select the socially optimal outcome(s) as the joint action pair(s) to execute whenever it is selected as the joint decision-maker as described in Section 4.1.

5   Simulation

In this section, we present the simulation results showing how the influencer agents can significant influence the population's behaviors towards socially optimality in two types of representative games: PD game and AC game. All the simulations are performed in a population of 1000 agents. Following previous work [2], we consider 5% of influencer agents in a population (i.e., 50 agents out of 1000) to be an appropriate upper bound of how many agents can be inserted into a system in practical application domains. However, for evaluation purpose, we perform simulations with the percentage of influencer agents up to 50 % in order to have a better understanding of the effects of influencer agents on the dynamics of the system. We first give an analysis of the effects of the number of influencer agents and different update modalities on the system level performance in each game in Section 5.1 and 5.2 respectively, and then compare the performance of our learning framework using influencer agents with that of previous work [5] in Section 5.3.

5.1   Prisoner's Dilemma Game

For the PD game, we use the following setting: $R = 3$, $S = 0$, $T = 5$, $P = 1$. Fig. 3 shows the average percentage of socially optimal outcome in the system when the number of influencer agents (IAs) inserted varies using mono-learning update. When no IAs are inserted into the system (i.e., a population of individually rational agents (SAs)), the percentage of socially optimal outcomes (SOs) achieved quickly reaches zero. This is obvious since choosing action D is always the best choice for each individually rational agent when it interacts with another individually rational entity in PD game. By inserting a small amount of IAs with proportion of 0.002 (2 IAs in the population of 1000 agents), we can see a significant gain in terms of the percentage of SOs attained up to 0.8. The underlying reason is that most of SAs are incentivized to voluntarily choose to interact with IAs and also select action F. Further increasing the number of IAs (to 10 IA agents) can significantly improve the speed of increase of percentage of SOs and also bring in small improvement of the percentage of SOs finally attained. We hypothesize that it is because some IAs' behaviors against SAs are not optimal when the number of IAs is small, and more IAs can successfully learn the optimal action against SAs when the number of IAs becomes larger. However, as the number of IAs is further increased, the increase in the final value of the proportion of SOs attained becomes less obvious. The reason is that with the number of IAs increasing, there is little additional benefit on the behaviors of IAs and the small amount of increase in the percentage of SOs is purely the result of the increase of the percentage of IAs itself.

Fig. 4 shows the differences between updating using multi-learning and mono-learning approach on the system level performance, i.e., the average percentage of SOs attained in the system, in PD game. From previous analysis, we have known that most SAs learn to interact with IAs and choose action F, thus IAs are given much more experience and opportunities to improve their policies against SAs under multi-learning approach. Accordingly it is expected that higher level of SOs can be achieved compared with that under mono-learning approach. It is easy to verify that the simulation results in Fig. 4 are in accordance with our predication.

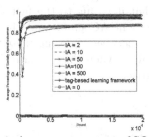

Fig. 3. Average percentage of SOs with different number of IAs under mono-learning update (PD game)

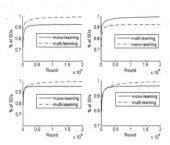

Fig. 4. Mono-learning approach v.s. multi-learning approach in PD game (IA = 2, 20, 100, 500)

5.2   Anti-coordination Game

In AC game, the following setting is used to evaluate the learning framework: $L = 1$, $H = 2$. Fig. 5 shows the average proportion of socially optimal outcomes attained in the system when the number of influencer agents varies. Different from the PD game, when there is no influencer agents inserted in the system, most of the selfish agents (up to 85%) can still learn to coordinate with each other on socially optimal outcomes. Initially the percentage of socially optimal outcomes is very high because most of the agents learn that choosing action F is the best choice which prevents the occurrence of mis-coordination. However, gradually the agents realize that they can benefit more by exploiting those peer agents choosing action F by choosing action C or D and thus this inevitably results in mis-coordination (i.e., achieving outcome (C, C) or (D, D)) when these exploiting agents interact with each other. Thus the average percentage of socially optimal outcomes gradually drops to around 85%. Besides, the mis-coordination rate converges to around 15 %, which can be understood as the dynamic equilibrium that the system of agents has converged to, i.e., the percentages of agents choosing action C, D, and F are stabilized.

Significant increase in the average percentage of socially optimal outcome attained (up to almost 100%) can be observed when a small amount of influencer agents (2 IAs out of 1000 agents) is inserted into the system. This can be explained by the fact that the SAs learn to entrust those IAs to make the joint decisions and also the IAs choose between the two socially optimal outcomes randomly. There is little incentive for the SAs to deviate since most of SAs have learned to interact with IAs and respond to SAs with action C or D, thus there is no benefits to exploit other SAs due to high probability of mis-coordination. When the number of IAs is further increased (the number of IAs is 50), there

is little performance increase in terms of the percentage of socially optimal outcomes achieved, and we only plot the case of $IA = 50$ for the purpose of clarity.

Fig. 6 shows the differences between updating using multi-learning and mono-
learning approach on the system level performance, i.e., the average percentage
of SOs attained in the system, in AC game with different number of IAs. Different
from PD game, we can observe that slightly lower percentage of SOs is attained
under multi-learning approach. We hypothesize that it is due to the fact that
there exist two different socially optimal outcomes in AC game and thus it
becomes easier for the agents to switch their policies between choosing these two
outcomes and increase the chances of mis-coordination when they update their
policies more frequently under multi-learning approach.

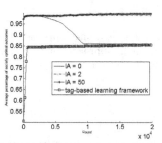

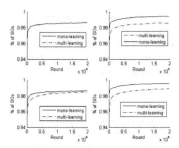

Fig. 5. Average percentage of SOs with different number of IAs under mono-learning update (AC game)

Fig. 6. Mono-learning approach v.s. multi-learning approach in AC game (IA = 2, 10, 100, 500)

5.3   Comparison with Previous Work

We compare the performance of our learning framework using influencer agents
with that of the tag-based learning framework [5] in both PD game and AC
game. The number of influencer agents are set to 50 (5 % of the total number of
agents) in our learning framework. The experimental setting for the tag-based
learning framework follows the setting given in [5].

Fig. 3 and 5 show the performance comparisons with the tag-based learning
framework in PD game and AC game respectively. For both cases, we can observe
that there is a significant increase in the average percentage of SOs under our
learning framework using influencer agents. Besides, the rate in which the agents
converge to SOs is higher than that using the tag-based learning framework. The
underlying reason is that under the tag-based learning framework, the agents
learn their policies in a periodical way, and the coordination towards socially
optimal outcomes requires at least two consecutive periods' adaptive learning
between individual learning and social learning. Also our learning framework
using IAs can better prevent the exploitations from SAs since the IAs act in the
same way as SAs if their interacting partners do not choose action F. Accordingly, higher percentage of SAs can be incentivized to cooperate with IAs and
thus higher percentage of socially optimal outcomes can be achieved.

6   Conclusion and Future Work

In this paper, we propose inserting influencer agents into the system to manipulate the behaviors of individually rational agents towards coordination on socially optimal outcomes. We show that a small percentage of influencer agents can successfully incentivize individually rational agents to cooperate and thus achieve socially optimally outcomes.

As future work, we are going to give detailed analysis of the learning dynamics of both types of agents (IAs and SAs) in order to better understand the effects of influencer agents. Another interesting direction is to apply this learning framework to other MAS domains (e.g., other types of games), and investigate the effects of influencer agents on the learning dynamics of individually rational agents as well.

References

1. Bowling, M., Veloso, M.: Multiagent learning using a variable learning rate. Artificial Intelligence 136, 215–250 (2002)
2. Franks, H., Griffiths, N., Jhumka, A.: Manipulating convention emergence using influencer agents. In: AAMAS (2012)
3. Fudenberg, D., Levine, D.K.: The Theory of Learning in Games. MIT Press (1998)
4. Hales, D., Edmonds, B.: Evolving social rationality for mas using "tags". In: AAMAS 2003, pp. 497–503. ACM Press (2003)
5. Hao, J.Y., Leung, H.F.: Learning to achieve social rationality using tag mechanism in repeated interactions. In: ICTAI 2011, pp. 148–155 (2011)
6. Hao, J., Leung, H.-F.: Learning to achieve socially optimal solutions in general-sum games. In: Anthony, P., Ishizuka, M., Lukose, D. (eds.) PRICAI 2012. LNCS, vol. 7458, pp. 88–99. Springer, Heidelberg (2012)
7. Hogg, L.M., Jennings, N.R.: Socially rational agents. In: Proceeding of AAAI Fall Symposium on Socially Intelligent Agents, pp. 61–63 (1997)
8. Matlock, M., Sen, S.: Effective tag mechanisms for evolving coordination. In: AAMAS 2007, pp. 1–8 (2007)
9. Matlock, M., Sen, S.: Effective tag mechanisms for evolving coperation. In: AAMAS 2009, pp. 489–496 (2009)
10. Sen, S., Airiau, S.: Emergence of norms through social learning. In: IJCAI 2007, pp. 1507–1512 (2007)
11. Steels, L.: A self-organizing spatial vocabulary. Artificial Life 2(3), 319–392 (1995)
12. Villatoro, D., Sen, S., Sabater-Mir, J.: Topology and memory effect on convention emergence. In: WI-IAT 2009, pp. 233–240 (2009)
13. Watkins, C.J.C.H., Dayan, P.D.: Q-learning. Machine Learning, 279–292 (1992)

The Dynamics of Propagation Fronts
on Sets with a Negative Fractal Dimension

Alfred Hubler and Josey Nance

Center for Complex Systems Research, University of Illinois at Urbana-Chamapaign,
1110 West Green Street, 61801 Urbana, IL, U.S.A.
hubler.alfred@gmail.com

Abstract. In sets with a fractal dimension greater than 1, the average number of neighbors increases with distance. For that reason spherical pulses propagate outward in systems with nearest neighbor interactions. In sets with a negative fractal dimension, such as the set of individual coordinates of a population of a small city, the average number of neighbors decreases with distance in a precise way relating the number of neighbor to the fractal dimension of the set. We study the propagation of diffusive pulses and waves on such sets. We find that on sets with negative fractal dimension, the velocity of pulse peak is negative (*i.e.* the median radius of circular pulses decreases as a function of time). Eventually the pulse broadens and disappears. We discuss applications in physical systems, such as the spreading of heat and sound, as well as applications in social systems, such as the spread of infectious diseases and the spread of rumors.

Keywords: Negative fractal dimension, diffusion, propagation front.

1   Introduction

Many interesting objects can be thought of as finite sets with a large number of elements. Social networks consist of a finite set of people. Solid objects, such as a copper wire, a snow crystal, or the branches of a tree are made of a finite number of molecules. The relative position of the elements with respect to their neighbors (or *distribution*) determines if the object is called one dimensional, two dimensional, or three dimensional. For instance, in a sheet of paper, the molecules are mostly located near a two dimensional plane. Therefore the sheet is called two dimensional. When the same molecules are arranged along a line, like in a thread, the resulting object is called one dimensional. And if the molecules agglomerate into a spherical lump, the object is called three dimensional.

There are several methods to quantify the dimension of a set. The Hausdorff dimension [1] and fractal dimension [2] use estimates of the volume of the set at various levels of course graining to determine the dimension of the set. The correlation dimension [3] uses a count of the number of neighboring elements in close proximity to determine the dimension of the set.

K. Glass et al. (Eds.): COMPLEX 2012, LNICST 126, pp. 152–158, 2013.
© Institute for Computer Sciences, Social Informatics and Telecommunications Engineering 2013

The algorithm for computing the correlation dimension is much simpler than the algorithm for computing the fractal dimension, but the correlation dimension sometimes produces unintuitive results. For instance, the correlation dimension of a sheet of paper is 3, because at distances less than the thickness of the sheet, the number of neighbors with radius R scales like in a three dimensional object.

There is another issue with the correlation dimension: the surface problem. That is, for elements which are close to the surface of a D dimensional object with a "lump" geometry and with a dimension D > 1, such as a sphere or a circle, the object appears D-1 dimensional near the surface. If the elements are sorted in terms of their distance from the center of mass, the largest group is the group near the surface of the object. Therefore for the largest group of elements the object appears to be D-1 dimensional.

In this paper we introduce a definition of dimension which overcomes these problems.

2   Objects with a Negative Fractal Dimension

We consider objects which are made of N elements. The positions of the elements are x_i, where i=1,..,N. We compute the distance of the elements from the center of mass $x_c = (1/N) \Sigma_{i=1,...,N} x_i$, and count the number of elements C(r), which are within a shell of radius r = k Δr and width Δr from x_c, where k=1,2,3,…, that is,

$$C(r)= \Sigma_{i=1,...,N} H(|x_i-x_c|- k \Delta r) - H(|x_i-x_c|- k \Delta r - \Delta r),$$

(1)

where r=$|x_i-x_c|$ is the distance of the element from the center. H is the Heaviside step function. For an object with a common shape, such as a sphere or a square, C(r) is a power law within a certain range of r-values, where the exponent is one less than the intuitive dimension of the object. Therefore the fractal dimension is defined as

$$D = 1 + d (\ln C) / d(\ln r).$$

(2)

Figure 1 and Fig. 2 show a circular object and a star-shaped object along with their corresponding functions C(r) and D(r). Both objects consist of about 3200 elements. In the star shaped object, the elements are on a square grid with side length 1. In the star-shaped object, the position of the elements (x,y) satisfies the condition: |x|<200 and |y|<200 and (|y|<=|x| and |y| <= 250/|x|$^\alpha$ or |x|<=|y| and |x| <= 250/|y|$^\alpha$), where $\alpha > 0$.

The fitting function for the number of elements at distance r from the center for the circular object in Fig. 1a is: ln C =   1.0467 ln r + 1.6791, for 0 < r < 32. The continuous line in Fig. 1b is a graph of the fitting function. With Eq. 1 we obtain a numerical value for the dimension D = 2.05, for 0 < r < 32. Fig 1c is a graph of the numerical value of the fractal dimension D versus the radius and the theoretical value. Fig. 1d shows that the numerical value for the fractal dimension is in good agreement with the intuitive value.

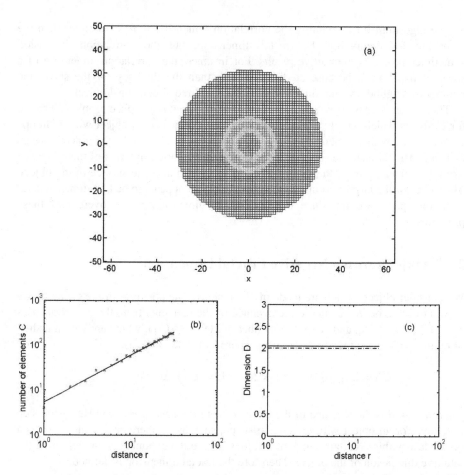

Fig. 1. Plot (a) shows a circular object. The colors indicate the amplitude of a spreading pulse (*red =very high, green=high, blue =low, black = very low*). The elements are located at the intersections and corners. The lines indicate connections between elements.   Plot (b) shows a histogram of the numbers of elements versus their distance from the center of mass for bin size $\Delta r=1$. The lines are least-square fits. Plot (c) shows the dimension D versus the distance from the center. The dashed line indicates the theoretical value, D=2.

The fitting function for the number of elements at distance r from the center for the star-shaped object (Fig. 2a, Fig. 2b) is: ln C = 1.0643 ln r + 1.6403, for $0 < r < 19$ and ln C = -1.3339 ln r + 8.1032, for $18 < r < 100$. The continuous line in Fig. 2c is a graph of the fitting functions. With Eq. 1, we obtain a numerical value for the dimension D = 2.06, for $0 < r < 32$ and D = -0.33, for $18 < r < 100$. Fig 2d is a graph of the numerical value of the dimension D versus the radius and the theoretical value.

We find that the numerical values for the dimension are in good agreement with the intuitive value. The star is expected to be 2-dimensional near the center, whereas for distances that exceed the solid area in the center, the dimension of the object is expected to be D = α+1, because the number of points inside the rays is approximately equal to the width of the rays W= 500/|y|$^{\alpha}$.

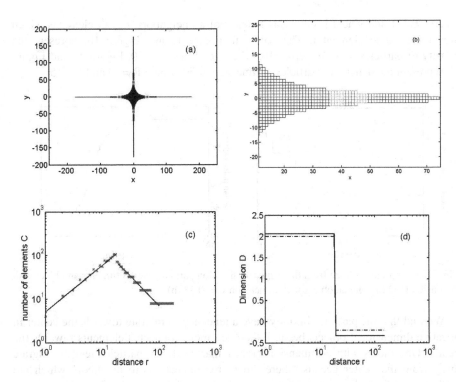

Fig. 2. Plot (a) shows a star-shaped object with $\alpha = 1.2$. The colors indicate the amplitude of a spreading pulse (*red =very high, green=high, blue =low, black = very low*). Plot (b) shows an enlargement of Plot (a). The elements are located at the intersections and corners. The lines indicate connections between elements. Plot (c) shows is a histogram of the numbers of elements versus their distance from the center of mass for bin size $\Delta r = 1$. The lines are least-square fits. Plot (d) shows the dimension D versus the distance. The dashed line indicates the theoretical value, D=2 for small distances and D= $-\alpha$ +1 = -0.2 for large distances from the center of mass.

3   Diffusion on Objects with a Negative Fractal Dimension

We assume that each element stores $q_i(n)$ particles at time step n, where n=1,2,3,...,T. The initial particle distribution is a square wave between R_1 and R_2, *i.e.* if $R_0 < r < R_1$ then $q_i(0) = 1$, where $r^2=x_i^2+y_i^2$, otherwise $q_i(0) = 0$. The elements and the connections form a grid.

We assume that the particles do a random walk on this grid. The dynamics of the density of random walkers is asymptotically equivalent to a diffusion process. We model the diffusion process with a relaxation dynamics

$$q_i(n+1) = q_i(n) + \lambda \Sigma_j (q_j(n) - q_i(n)), \tag{3}$$

where $\lambda=0.1$ and where j represents indexes of all neighbors, *i.e.* all elements which are connected to element i. The colors in Figure 1a and Figure 1b represent the density of particles at time step n=10, where $R_0=7$ and $R_1=10$. Figure 2a and Figure 2b represent the density of particles at time step n=50, where $R_0=40$ and $R_1=45$.

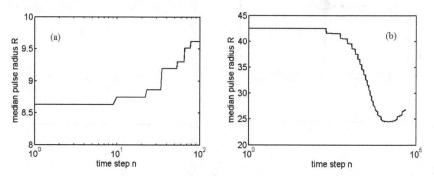

Fig. 3. The median distance from the center of diffusion particles versus time for an object with dimension D=2 (a), and an object with dimension D= -0.33 (b)

We find that the particles initially have a tendency to migrate towards the center in negative dimensional objects because there are more connected elements towards the center. On objects with a dimension greater than D=1, the particles tend to migrate away from the center because there are more connections to neighbors which are further away from the center. The median distance of the particles from the center is defined by the following inequality

$$\Sigma_{i=1,2,3,...N} \; q_i \; H(R^2 - x_i^2 - y_i^2) < Q/2 \tag{4}$$

where $Q = \Sigma_{i=1,2,3,...N} \; q_i$. Fig. 3 shows the dynamics of R.

For circular objects, R increases (Fig 3a), whereas for the negative dimensional object R initially decreases (Fig. 3b).

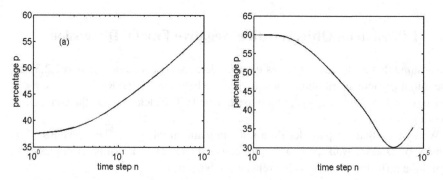

Fig. 4. The percentage of diffusion particles which are further away than the center of the initial square wave distribution versus time for an object with dimension D=2 (left) and an object with dimension D= -0.33 (right)

The quantity p tracks the percentage of particles which is further away than the center of the initial square wave

$$p = (1/Q) \sum_{i=1,2,3,..,N} H(x_i^2 + y_i^2 - R_w^2),$$ (5)

where $R_w^2 = (R_1 - R_2)^2/4$. R_w is the center of the initial square wave distribution. Figure 4 shows the dynamics of p for an object with dimension 2 and object with dimension D=-0.33. On the D=2 object, the number of p increases whereas on the negative dimensional object p initially decreases.

Another quantity which illustrates these tendencies to migrate towards or away from the center is the mean square deviation of the particles from the center

$$<r^2> = (1/N) \sum_{i=1,2,3,..,N} q_i (x_i^2 + y_i^2).$$ (6)

Figure 5 shows the dynamics of $<r^2>$ for an object with dimension 2 and object with dimension D=-0.33. On the D=2 object the number of $<r^2>$ increases whereas on the negative dimensional object $<r^2>$ initially decreases.

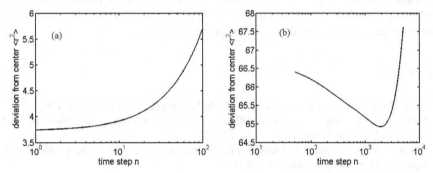

Fig. 5. The mean square deviation $<r^2>$ of diffusing particles from the center of the object versus time for a circular object (a) and for a star-shaped object with dimension D= -.33

4   Discussion

Diffusion on a two dimensional grid differs from diffusion on a three dimensional grid, because each node on a two dimensional grid has 4 neighbors and each node on a three dimensional grid has 6 neighbors. Similarly, diffusion particles appear to migrate away from the center of two and three dimensional objects because most neighboring elements are further away from the center than the element itself. This is different for objects with a negative dimension, since most neighboring elements are closer to the center than the element itself. Fig. 3 and Fig. 4 illustrate that in contrast to objects with dimension greater than 1, on negative dimensional objects, diffusing particles appear to migrate towards the center because the median distance from the center decreases due to an increasing number of particles is closer to the center than initially and because the mean square deviation of the particles from the center decreases.

Diffusing particles are equally likely to move to one of the neighboring elements. Many other systems behave in the same way. A sound wave spreads equally to all neighboring locations inside an object if the object is homogeneous. Similarly, heat spreads to all neighboring locations equally in solids, liquids and gases and charges in a resistor/capacitor network spread to all neighboring capacitors equally. Therefore we expect that acoustic pulses, heat pulses and charges spread like the pulse in Fig. 1 and Fig 2. We expect that in objects with a dimension greater than D=1, sound pulses, heat pulses, and charges distributions with radial symmetry tend to move outward, whereas in negative dimensional objects, these pulses move toward the center of the object.

It is conceivable that social systems can have a negative dimension. For example, in most cities, the density of people is highest in the center and decreases gradually as a function of the distance from the city center. If the density ρ decreases like a power law with an exponent less than -1 as a function of the distance from the city, *i.e.* $\rho = a\, r^{\alpha}$, where $\alpha < -1$, then the set of people in the city is an object with a negative dimension. If we assume that a rumor or infection starts at a certain distance from the city center and travels quickly from person to person, then the rumor or infection travels mostly towards the city center, *i.e.* more people close to the city center know about the rumor or are infected than those people further away.

References

1. Hausdorff, F.: Dimension und äußeres Maß. Mathematische Annalen 79, 157–179 (1919)
2. Mandelbrot, B.: How Long is the Coast of Britain? Statistical Self-Similarity and Fractional Dimension. Science 156, 636–638 (1967)
3. Grassberger, P., Procaccia, I.: Measuring the Strangeness of Strange Attractors. Physica D: Nonlinear Phenomena 9, 189–208 (1983)

Mathematical Model of Hit Phenomena as a Theory for Human Interaction in the Society

Akira Ishii[1], Hidehiko Koguchi[2,3], and Koki Uchiyama[4]

[1] Department of Applied Mathematics and PhysicsTottori University, Koyama,
Tottori 680-8554, Japan
[2] Perspective Media, 3-6-9 Kakinokizaka, Meguro-ku, Tokyo 152-0022 Japan
[3] M Data Co.Ltd., Toranomon, Minato-ku, Tokyo 105-0001 Japan
[4] Hottolink, Kanda-nishikicho, Chiyoda-ku, Tokyo 101-0054, Japan
ishii@damp.tottori-u.ac.jp

Abstract. A mathematical model for the hit phenomenon in entertainment within a society is presented as a stochastic process of interactions of human dynamics. The model uses only the time distribution of advertisement budget as an input, and word-of-mouth (WOM) represented by posts on social network systems is used as data to compare with the calculated results. The unit of time is a day. The calculations for the Japanese motion picture market based on to the mathematical model agree very well with the actual residue distribution in time.

Keywords: Hit phenomena, stochastic process, movie.

1   Introduction

Human interaction in real society can be considered in the sense of "many body" theory where each person can be treated as atoms or molecules in the ordinary many body theory. With the popularization of social network systems (SNS) like blogs, Twitter, Facebook, Google+, and other similar services around the world, interactions between accounts can be stocked as digital data. Though the SNS society is not the same as real society, we can assume that communication in the SNS society is very similar to that in real society. Thus, we can use the huge stock of digital data of human communication as observation data of real society [1-4]. Using this observed huge data (so -called "Big Data"), we can apply the method of statistical physics to social sciences. Since word-of-mouth (WOM) is very significant in marketing science [5-8], such analysis and prediction of the digital WOM in the sense of statistical physics become very important today.

Recently, we present a mathematical theory for hit phenomena where effect of advertisement and propagation of reputation and rumors due to human communications are included as the statistics physics of human dynamics [9]. This theory has also been applied to the analysis of the local events in Japan successfully [10]. Our model was originally designed to predict how word-of-mouth communication spread over social networks or in the real society, applying it to conversations about movies in particular, which was a success. Moreover, we also found that when they overlapped their predictions with the actual revenue of the films, they were very similar.

K. Glass et al. (Eds.): COMPLEX 2012, LNICST 126, pp. 159–164, 2013.

In the model [9], the key factors to affect the mind of the consumers are three: advertisement effects, the word-of-mouth (WOM) effects and the rumour effects. Recognising that WOM communication, as well as advertising, has a profound effect on whether a person goes to see a movie or not, whether this is talking about it to friends (direct communication or WOM) or overhearing a conversation about it in a café (indirect communication or the rumour), we accounted for this in our calculations. The difference between our theory and the previously presented researches [11-26] are discussed in ref.9.

We found that the effects of advertisements and WOM are included incompletely and the rumor effect is not included in the previous works [11-26]. Therefore, from the point of view of statistical physics, we present in our previous paper a model to include these three effects: the advertisement effect, the WOM effect, and the rumor effect. The presented model has been applied to the motion picture business in the Japanese market, and we have compared our calculation with the reported revenue and observed number of blog postings for each film.

In this paper, the responses in social media are observed using the social media listening platform presented by Hottolink. Using the data set presented by M Data Co.Ltd monitors the exposure of each films.

2   Model

We start the modeling from the viewpoint of the individual consumer. We define the purchase intention of the individual consumer, labeled i, at time t as $I_i(t)$. We assume that the number of products adopted until time t can be written as

$$Y(t) = p \int_{t_0}^{t} \sum_{i=1}^{N} I_i(t) dt,$$

(1)

where N is the maximum number of adopted persons, p is the price of the product and t_0 is the release date of the product. Thus, our problem is to define the equation of the purchase intention of each consumer $I_i(t)$.. We consider the modeling of the effects of advertisements, WOM, and rumor for the purchase intention in the following subsections.

The advertisement effect through mass media like TV, newspapers, magazines, the Web, Facebook, or Twitter is modeled as an external force for the equation of the purchase intention of the individual consumer:

$$\frac{dI_i(t)}{dt} = c_i A(t),$$

(2)

where A(t) is the time distribution of the effective advertisement effect per unit time and the coefficient   describes the impression of the advertisement for consumer i. The external force A(t) can be considered trends in the world or political pressure on the market.

Usually, a film's success spreads through WOM. Such WOM sometimes has a very significant effect on the success of the movie. Thus, the WOM effect should be included in our theory. The WOM effect should be distinguished into two types: WOM direct from friends, and indirect WOM as rumors. We name the WOM effect between friends *direct communication*, because customers obtain information directly from their friends. In the previous paper[9], we include also the communication between non-adopters. It is very significant for movie entertainment, especially before the opening of the movie. Let us consider that person i hears information from person j. The probability per unit of time for the information to affect the purchase intention of person i can be described as $D_{ij}I_j(t)$, where $I_j(t)$ is the purchase intention of person j and D_{ij} is the coefficient of the direct communication. Thus, we can write the effect of the direct communication as follows:

$$\sum_{j=1}^{N} D_{ij}I_j(t)$$
                                                                        (3)

where the summation is done without j = i.

In this paper, rumor effect is named *indirect communication*. In this form of communication, a person hears a rumor while chatting on the street, overhearing a conversation from the next table in a restaurant or on a train, or finds the rumor in blogs or on Twitter. To construct the theory using mathematics, we focus on one person who listens to a conversation happening around him/her. Let us consider that person I overhears the conversation between person j and person k. The strength of the effect of the conversation can be described as $D_{jk}I_j(t)I_k(t)$. The probability per unit time for the conversation to affect the purchase intention of person i is defined as $Q_{ijk}D_{jk}I_j(t)I_k(t)$, where Q_{ijk} is the coefficient. Thus, the indirect communication coefficient can be defined as $P_{ijk} = Q_{ijk}D_{jk}$.

Therefore, direct communication is two-body interaction and indirect communication is three-body interaction. Thus, our theory for the hit phenomenon like hit movies can be described as the equation of the purchase intention of person i with two-body interaction and three-body interaction terms.

According to the ref.9, we write down the equation of purchase intention at the individual level as

$$\frac{dI_i(t)}{dt} = -aI_i(t) + \sum_j d_{ij}I_j(t) + \sum_j \sum_k h_{ijk}d_{jk}I_j(t)I_k(t) + f_i(t)$$
                                                                        (4)

where d_{ij}, h_{ijk}, and $f_i(t)$ are the coefficient of the direct communication, the coefficient of the indirect communication, and the random effect for person i, respectively. We consider the above equation for every consumer so that i = 1, ..., N_p.

Taking the effect of direct communication, indirect communication, and the decline of audience into account, we obtain the above equation for the mathematical model for the hit phenomenon. The advertisement and publicity effect for each person can be described as the random effect $f_i(t)$.

Eq. (12) is the equation for all individual persons, but it is not convenient for analysis. Thus, we consider here the ensemble average of the purchase intention of individual persons as follows:

$$\langle I(t) \rangle = \frac{1}{N} \sum_i I_i(t)$$
                                                                        (5)

Taking the ensemble average of eq. (12), we obtain for the left-hand side:

$$\left\langle \frac{dI_i(t)}{dt} \right\rangle = \frac{1}{N}\sum_i \frac{dI_i(t)}{dt} = \frac{d}{dt}\left(\frac{1}{N}\sum_i I_i(t) \right) = \frac{d\langle I \rangle}{dt}$$

(6)

For the right-hand side, the ensemble average of the first, second, and third is as follows:

$$\langle -aI_i \rangle = -a\frac{1}{N}\sum_i I_i(t) = -a\langle I(t) \rangle$$

(7)

$$\left\langle \sum_j d_{ij}I_j(t) \right\rangle = \left\langle \sum_j dI_j(t) \right\rangle = \frac{1}{N}\sum_i\sum_j dI_j(t) = \sum_i d\frac{1}{N}\sum_j I_j(t) = Nd\langle I(t) \rangle$$

(8)

$$\left\langle \sum_j\sum_k p_{ijk}I_j(t)I_k(t) \right\rangle = \left\langle p\sum_j\sum_k I_j(t)I_k(t) \right\rangle$$

$$= \frac{1}{N}\sum_i p\sum_j\sum_k I_j(t)I_k(t)$$

$$= \sum_i p\frac{1}{N}\sum_j\sum_k I_j(t)I_k(t)$$

$$= Np\sum_i \frac{1}{N}\sum_j I_j(t)\frac{1}{N}\sum_k I_k(t)$$

$$= N^2 p\langle I(t) \rangle^2$$

(9)

where we assume that the coefficient of the direct and indirect communications can be approximated to be

$$d_{ij} \cong d$$

$$h_{ijk}d_{jk} = p_{ijk} \cong p$$

under the ensemble average.

For the fourth term, the random effect term, we consider that the random effect can be divided into two parts: the collective effect and the individual effect:

$$f_i(t) = \langle f(t) \rangle + \Delta f_i(t)$$

(10)

$$\langle f_i(t) \rangle = \frac{1}{N}\sum_i f_i(t) = \langle f(t) \rangle$$

(11)

where $\Delta f_i(t)$ means the deviation of the individual external effects from the collective effect, $\langle f_i(t) \rangle$. Thus, we consider here that the collective external effect term $\langle f_i(t) \rangle$ corresponds to advertisements and publicity to persons in the society. The deviation term $\Delta f_i(t)$ corresponds to the deviation effect from the collective advertisement and publicity effect for individual persons, which we can assume to be

$$\langle \Delta f_i(t) \rangle = \frac{1}{N} \sum_i \Delta f_i(t) = 0$$

(12)

Therefore, we obtain the equation for the ensemble-averaged purchase intention in the following manner as shown in ref.9:

$$\frac{d\langle I(t) \rangle}{dt} = -a\langle I(t) \rangle + D\langle I(t) \rangle + P\langle I(t) \rangle^2 + \langle f(t) \rangle$$

(13)

where

$$Nd = D$$
$$N^2 p = P$$

(14)

Eq. (13) can be applied to the purchase intention in the real market.

3   Calculation

Here, we show the results for three movies in figs.3 4 and 5. In fig.3, we show the observed daily number of blog postings and calculation for the Japanese movie "Unfear the answer". The calculation shows good agreement with the observed blog postings.

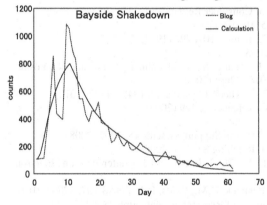

Fig. 3. Our calculation and the corresponding blog posting for the Japanese move, *Bayside Shakedown (Odoru Daisousasen)*. The dashed curve is the observed blog posting counts and the solid curve is our calculation.

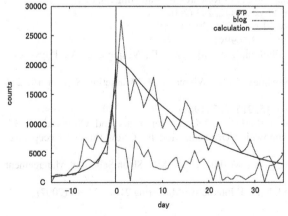

Fig. 4. Our calculation and the corresponding blog posting for the Japanese move, *Once in a Blue Moon*. The dotted curve is the observed blog posting counts, the gray curve is the gross rate point and the red curve is our calculaion.

In fig.4, we show the observed daily number of blog postings, the advertisement cost curve (gross rate point) and calculation for the Japanese movie "Once in a Blue Moon". The calculation shows good agreement with the observed blog postings.

We find that our calculations agree well with the observed data of the daily number of blog postings. It means that the exposure data from the data of M Data works as well as the daily advertisement cost used in our previous paper of ref.9. It means that the application of this model is very easy for many people, because everyone can buy the exposure data of each movie or something from M Data. The daily advertisement cost is very difficult to get even for scientific research in Japan and probably all over the world.

References

1. Allsop, D.T., Bassett, B.R., Hoskins, J.A.: J. Advertising Research 47, 398 (2007)
2. Kostka, J., Oswald, Y.A., Wattenhofer, R.: Word of Mouth: Rumor Dissemination in Social Networks. In: Shvartsman, A.A., Felber, P. (eds.) SIROCCO 2008. LNCS, vol. 5058, pp. 185–196. Springer, Heidelberg (2008)
3. Bakshy, E., Hofman, J.M., Mason, W.A., Watts, D.J.: Proceedings of the Fourth ACM International Conference on Web Search and Data Mining
4. Jansen, B.J., Zhang, M., Sobel, K., Chowdury, A.: J. Am. Soc. Inform. Sci. Tech. 60, 2169 (2009)
5. Brown, J.J., Reingen, P.H.: Journal of Consumer Research 14, 350 (1987)
6. Murray, K.: Journal of Marketing 55, 10 (1991)
7. Banerjee, A.: Quarterly Journal of Economics 107, 797 (1992)
8. Taylor, J.: Brandweek, 26 (June 2, 2003)
9. Ishii, A., Arakaki, H., Matsuda, N., Umemura, S., Urushidani, T., Yamagata, N., Yoshida, N.: New Journal of Physics 14, 063018 (22pp) (2012)
10. Ishii, A., Matsumoto, T., Miki, S.: Prog. Theor. Phys. (suppl. 194), 64–72 (2012)
11. Elberse, A., Eliashberg, J.: Marketing Science 22, 329 (2003)
12. Liu, Y.: Journal of Marketing 70, 74 (2006)
13. Duan, W., Gu, B., Whinston, A.B.: Decision Support Systems 45, 1007 (2008)
14. Duan, W., Gu, B., Whinston, A.B.: J. Retailing 84, 233 (2008)
15. Zhu, M., Lai, S.: Proceeding of the 2009 International Conference on Electronic Commerce and Business Intelligence
16. Goel, S., Hofman, J.M., Lahaie, S., Pennock, D.M., Watts, D.J.: PNAS 107, 1786 (2010)
17. Karniouchina, E.V.: International Journal of Research in Marketing 28, 62 (2011)
18. Sinha, S., Raghavendra, S.: Eur. Phys. J. B42, 293 (2004)
19. Pan, R.K., Sinha, S.: New J. Phys. 12, 115004 (2010)
20. Asur, S., Huberman, R.A.: aiXiv:1003.5699v1
21. Ratkiewicz, J., Fortunato, S., Flammini, A., Menczer, F., Vespignani, A.: Phys. Rev. Lett. 105, 158701 (2010)
22. Eliashberg, J., Jonker, J.-J., Sawhney, M.S., Wierenga, B.: Marketing Science 19, 226 (2000)
23. Bass, F.M.: Management Science 15, 215–227 (1969)
24. Bass, F.M.: The Adoption of a Marketing Model: Comments and Observations. In: Mahajan, V., Wind, Y. (eds.) Innovation Diffusion Models of New Product Acceptance. Ballinger (1986)
25. Dellarocas, C., Awad, N.F., Zhang, X.: Working paper. MIT Sloan School of Management (2004)
26. Dellarocas, C., Zhang, X., Awad, N.F.: J. Interactive Marketing 21(4), 23–45 (2007)

Hospitals as Complex Social Systems: Agent-Based Simulations of Hospital-Acquired Infections

José M. Jiménez, Bryan Lewis, and Stephen Eubank

Network Dynamics and Simulation Science Laboratory, Virginia Bioinformatics Institute
1800 Pratt Drive, Research Building XV, Blacksburg, Virginia 24061
jjimenez@vbi.vt.edu

Abstract. The objective of this study was to develop a highly-detailed, agent-based simulation to compare medical treatments against healthcare-acquired infections (HAIs). A complex hospital model was built using patient information and healthcare worker data from two regional hospitals in Southwest Virginia. A specific HAI, *Clostridium difficile*, was chosen among other HAIs as the pathogen for the study due to its increased prevalence in the United States. The complex hospital simulation was created using the first principles of agent-based simulation. The simulation was then tested using a disease model with two different scenarios: a baseline with no medical treatment antimicrobials, and the use of an antimicrobial (fidaxomicin). The model successfully simulated over 164,000 personal contacts between patients and healthcare workers. Each medical treatment was evaluated one hundred times using one month of real hospital data. The mean case count was 2.66 for scenario 1 and 2.33 for scenario 2. The highest case count for scenario 1 was 21 cases whereas scenario 2 had a maximum of 11 cases. Understanding complex interactions between patients and hospital personnel could help hospitals understand transmission of infections while simultaneously reducing healthcare costs.

Keywords: Complex systems, healthcare systems, healthcare-acquired infections, *Clostridium difficile*, agent-based simulation.

1   Introduction

Hospitals are by definition complex systems containing multiple subsystems. The environment inside a hospital is a collection of interactions between numerous subsystems. Hospital subsystems include the internal and external environment, personnel, and technology. Although sometimes the subsystems may seem to work autonomously, there is an overarching objective that governs all of them. This goal aims to improve health of the patients by performing tasks unique to that subsystem. Take for example the interactions between the environmental services or janitorial department and the nursing staff within an intensive care unit (ICU). The objective of the environmental services department is to ensure that all surfaces in a patient room are cleaned and sanitized. Members of this department have very specific standard work that includes directives on

K. Glass et al. (Eds.): COMPLEX 2012, LNICST 126, pp. 165–178, 2013.
© Institute for Computer Sciences, Social Informatics and Telecommunications Engineering 2013

how to sanitize, what surfaces to clean, and cleaning times for each room and major area. Nurses, on the other hand, have a very different mission. They are tasked with patient care by following physician orders for treatment. If a patient's room needs to be cleaned, but at the same time the patient needs a particular treatment, there might be a conflict of objectives between the two subsystems. While at times it may appear that both subsystems are competing for the same space to conduct different activities, they share one goal. Their end goal is to guarantee that the patient regains health and that he is not harmed further during his stay in the hospital.

Healthcare-acquired infections (HAIs) occur within this complex system. Interactions between multiple subsystems such as hospital personnel, patients, and technology can create a difficult environment to identify and treat HAIs. In order to investigate this type of environment without reducing it to a very simplistic model, one can utilize highly-resolved simulations. Highly-resolved simulation refers to agent-based simulation models that can evaluate infection exposure, interventions and individual behavior change for populations in the hundreds of millions, while maintaining the resolution of the individual agent. This study is unique in its kind because it combines very specific factors. First, it makes use of multiple populations inside a hospital. These populations include patients, physicians, nurses, respiratory therapists, occupational therapists, speech therapists, physical therapists, and environmental services associates. Second, it utilizes actual patient data in the form of electronic medical records. Third, the study uses the technique of shadowing hospital workers to develop activity schedules from multiple disciplines to include in the simulation. Lastly, due to the amount of data being processed, it is clear that less advanced simulation software would not be able to calculate the interactions of thousands of agents in an efficient manner. The use of highly-resolved simulation was essential for the epidemiological study. For this reason it is necessary to combine the resolution of highly-resolved simulation with the power of high performance computing.

1.1   Objective of the Study

The objective of the study was to develop a highly-detailed, agent-based simulation to compare medical treatments against healthcare-acquired infections (HAIs). In order to achieve this objective the number of daily exposures and contacts between patients and healthcare workers was obtained through a hospital simulation. By determining these contacts, their durations, and locations it was possible to suggest to hospital staff better treatment measures against HAIs. This paper represents the initial stages of the study and therefore looks at the practicability of conducting highly-resolved simulations for hospitals.

1.2   Healthcare-Acquired Infections

Healthcare-acquired infections are those infections defined as being transmitted and acquired inside a healthcare facility. Healthcare facilities include hospitals, clinics, outpatient facilities, and nursing homes. There are multiple types of pathogens that can infect a patient or be transmitted within a healthcare facility. The transmission, control and prevention, and treatment are very different for each type of infection.

Clostridium difficile is a normal occurring bacteria in the intestinal flora, however certain strains of the bacteria can cause *Clostridium difficile*-associated disease (CDAD). CDAD can produce watery diarrhea with the mildest of cases, but can also produce severe colitis requiring surgery in the harshest cases. The severe form of the infection can also lead to death. CDAD has been linked to risk factors such as being 65 years-old or older, suffering a severe underlying illness, going through a nasogastric intubation, taking antiulcer medications, having a prolonged hospital stay, and receiving treatment with certain antibiotics [1, 2]. New treatments are currently being developed to fight CDAD to include new antibiotics, fecal transplants, and a new vaccine. *Clostridium difficile* was chosen as a model infection for this study due to its prevalence in the hospital environment and chain of infection.

A recent review of data from the Healthcare Cost and Utilization Project (HCUP) showed that the average cost of *Clostridium difficile* treatment could be as high as $24,400 and the aggregate costs of all *Clostridium difficile* treatments in the United States was $8,238,458,700 in 2009. Of the total $8 billion in *Clostridium difficile* infections costs 67.9 percent was covered by Medicare, 9.1 percent by Medicaid, and 18.8 percent was covered by private insurance [3]. Several studies have estimated the cost of individual treatment between $3,000 and $32,000 [4-7].

1.3   A Hospital as a Complex System

Hospitals are by virtue of their structure complex systems. Each hospital is a set balanced connections of multiple subsystems and thousands of individual agents. The personnel subsystem and the technological subsystem are two of the most important subsystems in the hospital. The personnel subsystem is composed of all the individuals that interact within the hospital. This subsystem includes patients, hospital workers and hospital visitors. The technological subsystem is sometimes seen as synonymous with machines or computers, but it also includes the knowledge shared by the hospital personnel, the level of automation of its internal processes, and as mentioned before the equipment that is utilized. Furthermore, the organizational structure of a hospital can be daunting. A complex organizational structure can follow a large number of procedures, rules, and guidelines. Nurses, as an example, can have supervisors from different reporting structures. For example, a nurse can report to a nurse manager, a resident doctor, and a chief of staff all at the same time. With nested structures such as these, standard operating procedures can become complex and with additional stress added to a person, errors are more prevalent [8].

In addition to the multiple subsystems there are thousands of individuals interacting constantly within the walls of a hospital. For this study, one of the regional hospitals had over 37,000 patients (non-recurring admissions) and over 4,000 hospital workers. The same hospital had over 1,000 locations. The combination of personnel and locations can produce millions of activities during a year inside a hospital.

Complex systems with so many individual parts interacting together are difficult to study without reducing to very simplistic models. Most of the regularly used studies do not have the capacity to analyze the number of agents that have been described above. It is also impractical and costly to conduct complex experiments on interventions in a hospital. The size of the overall population and the interactions of the multiple subsystems make this task very difficult to achieve. In the past, researchers have looked at different types of study designs to test their hypotheses such as case-control studies or retrospective studies. One of the weaknesses of a case-control or a retrospective study is that experiments of this fashion would not be able to encompass the detailed dynamics within a hospital. The use of highly-resolved, agent-based simulation can help overcome these weaknesses. Highly-resolved simulation uses data obtained from the population data sources to create a realistic synthetic (*in-silico*) population. The *in-silico* representation of the population includes information such as location, daily activities, age, and other important demographics [9]. Even though all this information is incorporated into the simulation, the identity of the patients and hospital workers are protected through anonymity. Highly-resolved simulation has been used in the past to analyze possible interventions in response to hypothetical disease outbreaks such as influenza [10, 11].

1.4   Highly-Resolved Simulations

In the past, simulations have been used on multiple occasions to help explain the complexities of health systems. Simulations are representations of complex systems, but they can never be a perfect representation of reality. Additionally, one cannot say that any kind of simulation is better than the other. Simulations that are more graphical and user-friendly have the advantage of a short learning curve and ease of use. People that unfamiliar with simulation can learn to use the software in a matter of hours. However, user-friendly simulations lack the resolution needed for analyzing large number of individual agents. On the other hand, one can find simulations that are computer code-intensive and that are not intuitively user-friendly. Written directly in language such as C, C++, or Java language, these type of simulations can provide better resolution for large amounts of data, but require training and users that are computer-code savy [12]. The selection of the right simulation software depends on the type of problem that a researcher seeks to answer.

EpiSimdemics is a novel simulation software that looks at simulation in large social networks. EpiSimdemics is a parallel scalable algorithm used to simulate the spread of infectious diseases over extremely large contact networks. This algorithm can also simulate other population factors such as fear and behaviors. The Network Dynamics and Simulation Science Laboratory (NDSSL) at Virginia Tech developed EpiSimdemics with the objective of studying the effects of pharmaceutical and non-pharmaceutical interventions. EpiSimdemics is an agent-based simulation in which every person is an individual agent. EpiSimdemics is based on social network theory, but can overcome its limitations of use for only small groups of people. Prior studies have utilized this tool for analysis of large populations at a scale of 100,000,000 people. In the simulation, agents and locations are identified as nodes in a network graph (or a sociogram) and the edges between the nodes are considered contacts between agents or visits of an agent to a specific location. Every activity for a particular agent is defined in an activity schedule. The schedule determines the location of the activity, the activity performed, and the duration of the activity. Interactions between people are calculated using a stochastic model. If one or more individuals are in the same location as an infected individual then they might or might not become infected based on probability calculations. These probabilities are defined in the simulation in the disease model, which we explain later on. Other simulation models do not have the same level of resolution that EpiSimdemics has due to the ability to utilize extremely large populations through social networks. Other advantages of EpiSimdemics are the ability to combine policy interventions as well as individual behaviors in the same model. NDSSL has used EpiSimdemics for several studies of large populations for multiple government agencies [13-15].

Furthermore, "first principles" or guidelines have also been developed by NDSSL[16]. The first step in developing the simulation is the creation of a synthetic population or proto-population. In other models that describe epidemics or pandemics in large areas, the use of massive databases was needed to replicate the population of a city [10], a region, or an entire country [16, 17]. In the case of the hospital simulation, the study uses data collected directly from patient records and shadowing of hospital workers to develop the proto-population. The second step in developing highly-resolved simulations is to create activity schedules for the entire proto-population. The third step is to create a disease model that will be used to "infect" agents as they come in contact with other agents. All these steps will be explained further in the next section.

EpiSimdemics has the distinct ability to quantify not only the number of agents that become infected with a specific disease in the simulation but it can also identify the contacts between individual agents in the simulation. These contacts occur for different times depending on the activity and location of the person-agent. The software can count the number of contacts and the time that each contact lasts. Figure 1 below, shows an example representation of contacts in an ICU's social network. Every time that there is an agent-to-agent contact the software keeps a record of its location, time, and of the agents who interacted. This record is used to create the sociogram or network diagram at each time step.

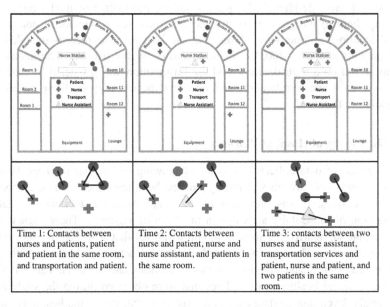

| Time 1: Contacts between nurses and patients, patient and patient in the same room, and transportation and patient. | Time 2: Contacts between nurse and patient, nurse and nurse assistant, and patients in the same room. | Time 3: contacts between two nurses and nurse assistant, transportation services and patient, nurse and patient, and two patients in the same room. |

Fig. 1. Contacts at different times in an ICU social network. In this representation of the simulation different agents come into contact with each other causing a potential transmission of the infection. With more contacts there is a higher probability of transmission. The *red circles* represent patients, the *blue crosses* represent nurses, the *green hexagons* represent transportation services, and the *yellow triangle* represents the nurse assistant.

2   Methodology

The methodology of the study consisted of four phases: electronic data collection, shadowing of hospital workers, disease modeling and simulation of the hospital. In order to create a more realistic simulation the study gathered data that represented reality as close as possible. For this reason, electronic medical records were requested directly from the hospital. Additionally, the study included shadowing of different hospital disciplines while hospital workers performed their daily activities. It is difficult to obtain 100 percent validation on simulation models, however when the agents in the model are based on accurate data directly from patient records and activities from direct observation, the fidelity of a large portion of the simulation can be verified.

2.1   Data Collection

As explained before, the first step in creating a highly-resolved simulation is to develop a synthetic or proto-population that resembles the real population with high resolution and high fidelity. In order to achieve this detail of resolution it was necessary to obtain actual data of the multiple populations that interact within a hospital. Two regional hospitals from Southwest Virginia provided data for the study. The first type of data obtained was in the form of electronic medical records. The patient records were de-identified to protect patient information by the hospital before submitting them to the research team. The electronic records included one year of patients' locations and activities throughout the hospital. The records included over 37,000 patients and over 400,000 single activities by the patients. The anonymous data was stored on a SQL database and the information was protected by numerous measures such as deidentification and restricted access only to those researchers participating in the study.

2.2   Shadowing of Hospital Workers

The next phase of data collection was the shadowing of hospital workers. Hospital workers are not typically tracked throughout the healthcare facility as they conduct their daily routines. It was important for the study however, to obtain a realistic representation of those daily activities in the form of vignettes. These vignettes are schedules that summarize the most common activities that a worker would perform throughout the day. The vignettes include locations, activities, contacts with other people or technology, and durations of those activities and contacts. Table 1 below shows the different healthcare disciplines that were observed during the study at the two hospitals. Even though data exists on most departments for the hospitals, not all the disciplines have been included currently in the simulation. Further work is needed to develop specialty departments within the hospitals such as the emergency department, the operating rooms, the neonatal unit, and the cafeterias.

Table 1. List of hospital disciplines shadowed during the study

Intensive Care Unit (ICU) Nurses	Infection Preventionists	Dieticians
Vascular Intensive Care Unit Nurses	Infection Control Physicians	Social Workers
Progressive Care Unit Nurses	Resident Physicians	Case Managers
Ostomy Nurses	Respiratory Therapists	Imaging Specialists (radiography/ultrasound)
Neonatal Intensive Care Unit (NICU) Nurses	NICU Respiratory Therapists	Facility Managers
Dialysis Nurses	Physical Therapists	Phlebotomists
Emergency Department Nurses	Occupational Therapists	Laboratory Technicians (General)
Nurse Assistants	Speech Therapists	Laboratory Technicians (specializing in *Clostridium difficile*)
	Environmental (Janitorial) Services	

Figure 2 below shows an example of a vignette. As shown on the diagram a day shift Intensive Care Unit (ICU) nurse arrives to the hospital in the morning and participates in a shift meeting. Then, the nurse conducts a preliminary assessment of his two patients. After that the nurse could continue to monitor patients, relieve another nurse, or goes to the cafeteria to eat lunch. This information was coded into a computer program that create multiple schedules for a large number of nurses that had similar jobs. Stochasticity was also added into the program so that all nurses would not be conducting the same activity at the same time and location.

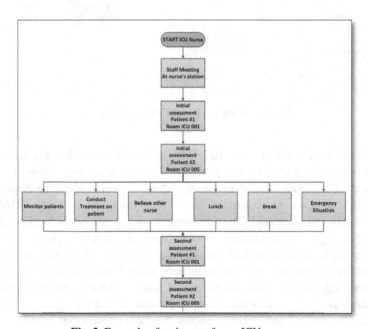

Fig. 2. Example of a vignette for an ICU nurse

2.3   Disease Models

The next step in highly-resolved simulation is designing a disease model based on the medical literature from the infection that will be used to "infect" the proto-population. Information gathered from the literature or from the facility's infection prevention professionals to identify the problem served as a base to build the disease model. The disease model is a probabilistic finite-state machine (FSM), a state-transition model that is used on the entire proto-population. Probabilities and distributions are fed into the FSM in order to develop a realistic model of how the disease spreads inside the hospital. Figure 3 shows an example of a disease model for *Clostridium difficile*.

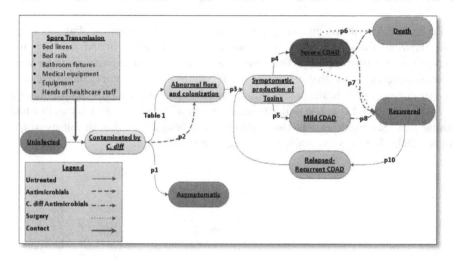

Fig. 3. Disease model for *Clostridium difficile*

The disease model represents the different health states that any agent in the simulation can be in at any particular time. For this study two different scenarios were used to compare several treatments. Both scenarios had nine health states: uninfected, colonized, not colonized, infected, asymptomatic, severe CDAD, mild CDAD, death, and recovered. The first scenario was a control or baseline therefore No medical or preventive treatment was utilized. The second scenario incorporated *Clostridium difficile* antimicrobials for mild and severe CDAD as well as a vaccine to avoid infection. Specific antimicrobials have been identified as a risk factor for *Clostridium difficile*. For this reason, use of antimicrobials is the first health state (uninfected) as a link towards the next step, colonization, if the agent was exposed to the pathogen. Table 2 and 3 below shows the parameters utilized for the different disease models.

Table 2. Scenarios used in the simulation

	Disease Model 1	Disease Model 2
First Disease Link	Exposure to *Clostridium difficile*, use of antimicrobials to trigger infection	Exposure to *Clostridium difficile*, use of antimicrobials to trigger infection
Treatments Tested	No preventive or medical treatments	*Clostridium difficile* antimicrobials for mild CDAD *Clostridium difficile* vaccine *Clostridium difficile* antimicrobials for severe CDAD

Table 3. Parameters for the two scenarios

Disease State	Disease Link	Probability	Next Disease State
Uninfected	Untreated	0.95	Colonized
	Untreated	0.05	Not Colonized
	Antimicrobials	0.96	Colonized
	Antimicrobials	0.04	Not Colonized
	Vaccine (Scenario 2)	0.25	Colonized
	Vaccine (Scenario 2)	0.75	Recovered
Colonized	Untreated	0.50	Infected
	Untreated	0.50	Asymptomatic
Infected	Untreated	0.1	Severe CDAD
	Untreated	0.9	Mild CDAD
Not Colonized	None	None	None
Asymptomatic	None	None	None
Severe CDAD	Untreated	1.0	Death
	Clostridium difficile antimicrobials (Scenario 2)	0.68	Recovered
	Clostridium difficile antimicrobials (Scenario 2)	0.32	Death
Mild CDAD	*Clostridium difficile* antimicrobials (Scenario 2)	0.85	Recovered
	Clostridium difficile antimicrobials (Scenario 2)	0.15	Severe CDAD
	Untreated	1.0	Severe CDAD
Death	None	None	None
Recovered	None	None	None

3   Results

3.1   Simulation Runs and Parameters

This study simulated the entire patient population of the hospital for an entire year. Particular importance was given to the ninth floor of the simulated hospital for the purposes of studying the interactions of the multiple populations of hospital workers. Only hospital workers of the ninth floor and their activities were simulated due to the complexity of creating additional departments for the hospital such as the emergency department and the operating room. These departments will be included in further simulations. The study was conducted for thirty days of hospital activities and it included agents acting as patients, physicians, nurses, respiratory therapists, occupational therapists, speech therapists, physical therapists, and environmental services associates. Each of the agents had individual activity schedules that identified each location, activity type, and duration. Each simulation of thirty days ran through 100 iterations for each scenario.

3.2   Assumptions and Simplifications

In order to run the simulation in a fast and efficient manner some assumptions were made regarding the model. The model does not distinguish between the ages of the agents. In reality, the age of the patient is an important risk factor as people over 65 years old are more likely to be infected with HAIs than younger people. Additionally, the medical condition of the agents did not play a factor in the simulation with the exception of those individuals that are under an antimicrobial regiment. Infections other than CDAD were not taken into account. The next assumption was that every agent in the *in-silico* hospital population could be infected. Similarly, all agents could progress through the disease models equally with no difference on whether they were patients or hospital workers. Finally, the study assumed a "barrier" around the hospital for agents. This means that once a patient-agent left the premises of the hospital, the infection and the disease model did no longer affect him.

3.3   Exposures and Contacts

A very interesting point of the simulation is to be able to observe the different contacts that occur between *in-silico* agents. The average number of activities for the entire month was 8,234 activities. The total number of agent-to-agent contacts during the 30 days was 164,176. These contacts are represented on figure 5 as a sociogram. Every node in the sociogram represents an agent-person. When two agent-persons are in the same location at the same time they are considered in contact with each other. An edge or line represents their contact.

3.4   Scenario Summary

Two graphs were obtained from the multiple iterations of the simulation. Each graph includes the cumulative case count of infected agents for each of the two scenarios as explained in section 3.1. Each scenario was run 100 times with different probability seeds for each run. The stochasticity of the models based on the seeds explains the differences in the number of cases from one curve to another. During one of the iterations for scenario 1 the number of cases grew to 21, whereas in several other iterations there were no cases at all. It is important to understand that the purpose of the simulation is not to replicate the exact results from the two regional hospitals but to evaluate the possible contacts between the different populations. These contacts and exposures can later be analyzed for improving systemic interventions such as hand washing, education programs, or better disinfection in rooms. The cumulative case curves are presented to show the disease transmission in a manner that is consistent, but not exactly the same as an actual hospital. Scenario 1 had an average number of cases of 2.66, compared to 2.33 from scenario 2. The highest number of cases for scenario 2 was only 11 compared to 21 on scenario 1. Figures 5 and 6 show the cumulative cases for scenarios 1 and 2.

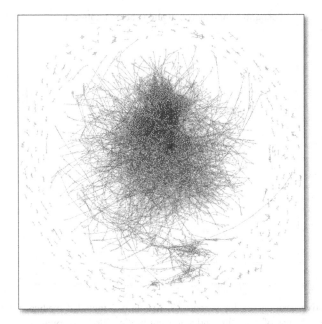

Fig. 4. Network diagram of the *in-silico* hospital population during one month

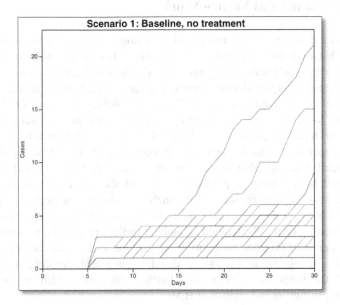

Fig. 5. Cumulative case curves for scenario 1. Scenario 1 is the baseline scenario and did not include any medical or preventive treatment against the infection.

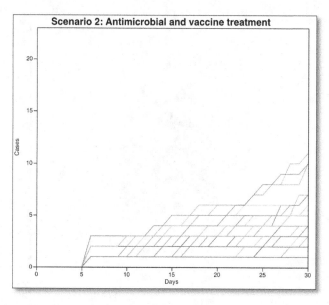

Fig. 6. Cumulative case curves for scenario 2. Scenario 2 included medical treatment of the infection through the use of a vaccine and a new antibiotic called fidaxomicin.

4   Conclusions and Future Work

A hospital is a complex system composed of multiple subsystems and thousands of individuals that work for the health of the patients. It is difficult and impracticable to perform complex experiments taking into consideration multiple populations within a hospital. The use of highly-detailed simulation for the study of healthcare-acquired infections can be of great use not only for researchers but also for healthcare professionals. In an era where "every penny counts" for healthcare system managers a simulation can represent an easy and affordable tool to understand infection control. This initial paper represents the first steps in our study of simulations of HAIs. The study provided evidence that simulation of a large complex system with high resolution and high fidelity is practicable. One of the advantages of this type of simulation is that real data from hospitals, such as electronic medical records, can be modified to create *in-silico* representations of people. The simulation was able to produce interesting results that should be studied further. One of the important results is that for over 164,000 contacts between agents, the highest number of cases was only 21. This is significant because it could explain that the current prevention practices in the hospitals are effective. Further study is needed in the area of simulation of prevention and control of HAIs.

Additional work in the study of HAIs in hospitals is already being performed at NDSSL in diverse areas to improve the simulation performance. First, to ensure the completeness of the hospital system it is necessary to add additional *in-silico* populations. These populations should include the entirety of the hospital workers and the visitors to the hospital. Visitors can be a very important factor in the transmission of infections. Visitors are susceptible to infections, especially if their demographics carry any of the risk factors for a specific disease. Hospital visitors are also more mobile than patients and could potentially transmit infections to different locations inside and outside of the hospital. The demographics of the population should also be included to future iterations of the study. Risk factors linked to the demographics of the agents could give further insight into effective treatment. The current simulation model did not include all the floors of the hospital due to the complexity of certain specialty departments. Further models should include every floor and department of the hospital to include the operating room and the emergency department.

In order to increase the realism of the simulation multiple hospitals could be added to the simulation. This network of hospitals could work similar to a state or regional health department evaluating an outbreak. Finally, the realism of the simulation could increase if other infections, not necessarily other HAIs, would be added into the activity files.

References

[1] Bignardi, G.E.: Risk factors for Clostridium difficile infection. J. Hosp. Infect. 40, 1–15 (1998)

[2] Sunenshine, R.H., McDonald, L.C.: Clostridium difficile-associated disease: new challenges from an established pathogen. Cleve. Clin. J. Med. 73, 187–197 (2006)

[3] Lucado, J., Gould, C.V., Elixhauser, A.: Clostridium difficile infections (CDI) in hospital stays, 2009: statistical brief #124 (2012)

[4] Scott, D.: The direct medical costs of healthcare-associated infections in US hospitals and the benefits of prevention. In: Centers for Disease Control and Prevention, N. C. f. P. Division of Healthcare Quality Promotion, Detection, and Control of Infectious Dieseases (2009)

[5] Dubberke, E.R., Reske, K.A., Olsen, M.A., McDonald, L.C., Fraser, V.J.: Short- and long-term attributable costs of Clostridium difficile-associated disease in nonsurgical inpatients. Clin. Infect. Dis. 46, 497–504 (2008)

[6] Kyne, L., Hamel, M.B., Polavaram, R., Kelly, C.P.: Health care costs and mortality associated with nosocomial diarrhea due to Clostridium difficile. Clin. Infect. Dis. 34, 346–353 (2002)

[7] Dubberke, E.R., Wertheimer, A.I.: Review of current literature on the economic burden of Clostridium difficile infection. Infect. Control Hosp. Epidemiol. 30, 57–66 (2009)

[8] Karsh, B.T., Brown, R.: Macroergonomics and patient safety: The impact of levels on theory, measurement, analysis and intervention in patient safety research. Appl. Ergon. 41, 674–681 (2010)

[9] Lewis, B.: In silico Public Health: The Essential Role of Highly Detailed Simulations in Support of Public Health Decision-Making. Doctor of Philosophy, Genetics, Bioinformatics, and Computational Biology, Virginia Tech, Blacksburg, VA (2011)

[10] Eubank, S., Guclu, H., Kumar, V.S., Marathe, M.V., Srinivasan, A., Toroczkai, Z., Wang, N.: Modelling disease outbreaks in realistic urban social networks. Nature 429, 180–184 (2004)

[11] Halloran, M.E., Ferguson, N.M., Eubank, S., Longini Jr., I.M., Cummings, D.A., Lewis, B., Xu, S., Fraser, C., Vullikanti, A., Germann, T.C., Wagener, D., Beckman, R., Kadau, K., Barrett, C., Macken, C.A., Burke, D.S., Cooley, P.: Modeling targeted layered containment of an influenza pandemic in the United States. Proc. Natl. Acad. Sci. U S A 105, 4639–4644 (2008)

[12] Keeling, M.J., Rohani, P.: Modeling infectious diseases in humans and animals. Princeton University Press, Princeton (2008)

[13] Barrett, C.L., Bisset, K.R., Eubank, S.G., Feng, X., Marathe, M.V.: EpiSimdemics: An efficient algorithm for simulating the spread of infectious disease over large realistic social networks. Presented at the Proceedings of the 2008 ACM/IEEE Conference on Supercomputing, Austin, Texas (2008)

[14] Bisset, K.R., Aji, A., Bohm, E., Kale, L.V., Kamal, T., Marathe, M., Yeom, J.-S.: Simulating the Spread of Infectious Disease over Large Realistic Social Networks using Charm++. Presented at the to appear at 17th International Workshop on High-Level Parallel Programming Models and Supportive Environments, HIPS (2012)

[15] Bisset, K.R., Feng, X., Marathe, M., Yardi, S.: Modeling interaction between individuals, social networks and public policy to support public health epidemiology. Presented at the Winter Simulation Conference, Austin, Texas (2009)

[16] Barrett, C., Beckman, R., Khan, M., Kumar, A., Marathe, M., Stretz, P., Dutta, T., Lewis, B.: Generation and Analysis of Large Synthetic Social Contact Networks. In: Winter Simulation Conference (2009)

[17] Chao, D.L., Halloran, M.E., Obenchain, V.J., Longini Jr., I.M.: FluTE, a publicly available stochastic influenza epidemic simulation model. PLoS Comput. Biol. 6, e1000656 (2010)

Hypernetworks for Policy Design
in Systems of Systems of Systems

Jeffrey Johnson

Faculty of Mathematics, Computing and Technology
The Open University, MK7 6AA, UK
j.h.johnson@open.ac.uk

Abstract. Hypernetworks generalise networks and hypergraphs, allowing relations between many things to be modelled by *hypersimplices* with richer structure than hypergraph edges. They provide a way of integrating bottom-up and top-down micro, meso and macrolevel dynamics in multilevel systems. They provide a natural way of representing social structures, enabling polices to be tested by computation and big data before they are implemented.

Keywords: networks, hypergraphs, hypernetworks, modelling, Q-analysis, Galois lattice, policy, social informatics, policy informatics, gangs.

1   Introduction

Social networks have been intensely studied since the nineteen sixties when computers enabled increasingly large social systems to be studied [1]. However most social systems involve networks of networks, and the interaction of many agencies and recent work has begun to explore the properties of coupled networks. To date the focus has been on *binary relations* between pairs of things analysed using network theory. Surprisingly the possibility of *n*-ary relations between any *n* things has received less attention, even though they are ubiquitous in all systems.

Hypergraphs [2, 3] provided an early attempt to model relations between more than two things. They mark a big step forward in the study of higher relations but they are set-theoretic and lack the representational power needed for complex systems.

A further step forward is to use *simplices* rather than hypergraph edges to represent related entities [4, 5, 6, 7]. The vertices of simplices are *ordered*, and they have a multidimensional connectivity structure. Algebraic topology provides much useful theory for representing coupled dynamical subsystems. Even so, the orientation of simplices and related algebraic operations in simplicial complexes have limited representational power for social systems.

In contrast, hypernetworks [8, 9] have much richer structure to represent social dynamics. Hypernetworks are formed from *relational simplices*, otherwise called *hypersimplices*, in which the relational structure is explicit. Like the formation or

K. Glass et al. (Eds.): COMPLEX 2012, LNICST 126, pp. 179–189, 2013.
© Institute for Computer Sciences, Social Informatics and Telecommunications Engineering 2013

breaking of links in networks, the formation and disintegration of hypersimplices represent the discrete dynamics of systems. They generalise the dynamics of network formation, when vertices join or leave the system, and links are formed or cease to exist. The formation or disintegration of a hypersimplex is a structural *event*. Such events mark time in systems, where this is related to but different from clock time. For example time series are usually (more or less) continuous mappings of simplices to numbers in clock time, but the formation or disintegration of the simplices are discrete events as the structure of the system changes.

Hypersimplices are *wholes* formed their vertices as *parts*. This part-whole structure allows simplices at one level to become vertices at higher levels in multilevel systems, and the mappings defined on the simplices aggregate or disaggregate accordingly. This gives a way of representing and integrating the bottom-middle-top up-down-diagonal dynamics of complex multilevel systems of systems of systems.

The *state* of a system at any instance in time is represented by its multilevel simplices and patterns of numbers defined on them. Policy *goals* can be defined to be desirable system states, and policy can be defined as the decisions and actions intended to move the system onto trajectories that will achieve the goals. This paper will present hypernetworks as a generalisation of networks and show how they can be used for modelling complex systems in the context of policy and designing the future.

2   Graphs, Networks, Hypergraphs, and Simplicial Complexes

A *graph* is a set of points called *vertices* or *nodes* and a set of pairs of vertices called *edges* or *links*. Let V be a set of vertices, $E = \{ (v, v') \mid v \text{ and } v' \text{ belong to } V \}$. Then $G = (V, E)$ is a graph. An edge is *directed* or *oriented* if $(v, v') \neq (v', v)$. A graph with directed edges is called a *digraph*. A *network* is a digraph with mappings assigning numbers to the vertices and the edges, $f : V \rightarrow R$ and $f : E \rightarrow R$ where R is a number system such as the integers, the rationals or the reals. In many networks $f(v, v')$ represents a *weighting* or a *flow* from v to v'. A graph or network is *bipartite* if its vertex V set can be partitioned into two sets A and B, $V = A \cup B$ and $A \cap B = \varnothing$, such that every edge can be written as (a, b) where a belongs to A and B belongs to B. The literature often uses the terms graph and network interchangeably.

Graphs and networks are used to represent relationships between things. Let R be a relation between the sets A and B. Let $V = A \cup B$ and $E = \{ (a, b) \mid a \text{ is } R\text{-related to } b \}$. Then (V, E) is a graph. When numbers are assigned to its vertices and edges it is a network. For example, an airline network has vertices airports and edges the pairs of airports related by having direct flights between them. Then f. (a) can be the number of people leaving a and $f_+(b)$ can be the number of people arriving at b with $f(a, b)$ being the number of people flying from a to b in a given time.

The *degree* of a vertex a in a graph is the number of edges (a, b) in the graph. The *out-degree* of a vertex a in a directed graph is the number of edges (a, b) in the graph and the *in-degree* of a vertex b is the number of edges (a, b). For example Figure 1(a) shows twenty six girls with links to their favourite two dining partners. The degree of Adele is 5, and her in-degree is 3. The out-degree of all the girls is 2.

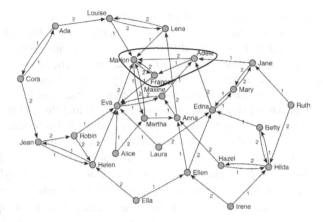

(a) Network of girls' preferred dining partners. (Source: de Noy *et al* [10]).

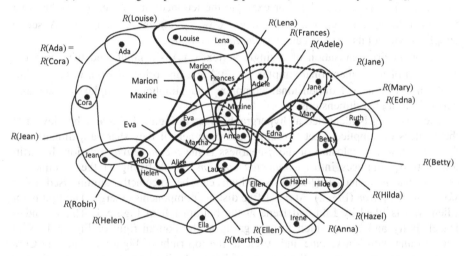

(b) Hypergraph of girls' preferred dining partners

Fig. 1. Networks and hypergraphs

The edges of graphs are restricted to having two vertices. Hypergraphs remove this restriction. Let R be a relation between set A and B. In principle, given a set B, any class of subsets of B is a *hypergraph*. Let $R(a_i)$ be the set of all members of B that are R-related to a_i, where a_i is in A, $R(a_i) = \{ b_j \mid$ for all bj with $a_i R b_j \}$. Then $H_A(B, R) = \{ R(a_i) \mid$ for all a_i in $A\}$ is a hypergraph. Figure 1(b) shows the hypergraph of the relation $H_{Girls}(Girls, R_{dining\_prepference})$. The hypergraph edge $R(g_i)$ is the set of all girls whose first or second choice of dinner companion is girl g_i.

In the network representation of the dinner partners data, the in-degree is a measure of the popularity of a girl. For example, Eva and Marion each have six girls who like their company, and Edna and Hilda have four. By comparison none of the girls has

Ella, Irene, Laura, or Alice as their first two preferences. The hypergraph representation adds information to this by explicitly listing the girls. Also the hypergraph representation has a different kind of connectivity through the intersections of the edges, which may have more than two vertices.

Hypergraphs provide a way of modelling n-ary relation when $n > 2$. However, hypergraphs lack the structure needs to make some fundamental distinctions. For example, consider the relation between the set of words and the alphabet. Then, for example $R(\text{dog}) = \{d, o, g\}$ and $R(\text{god}) = \{g, o, d\}$. But set-theoretically $\{d, o, g\} = \{g, o, d\}$, so $R(\text{dog}) = R(\text{god})$. This can be overcome by using simplices.

Let V be a set of vertices. The sequence of vertices $\langle v_0, v_1, ..., v_p \rangle$ is defined to be an abstract p-simplex. A p-simplex has a *geometric realisation* as a polyhedron in an n-dimensional space, $n \geq p$. For example, the geometric representation for $\langle v_0, v_1, v_2 \rangle$ is a triangle in 2-D space, that for $\langle v_0, v_1, v_2, v_3 \rangle$ is a tetrahedron in 3-D space, and so on. The simplex $\langle v_0, ..., v_p \rangle$ is a p-*dimensional face* of the simplex $\langle v'_0, ..., v'_q \rangle$ if $\{v_0, ..., v_p\} \subset \{v'_0, ..., v'_q\}$. For example the tetrahedron $\langle v_0, v_1, v_2, v_3 \rangle$ has four triangular faces, $\langle v_0, v_1, v_2 \rangle$, $\langle v_0, v_1, v_3 \rangle$, $\langle v_0, v_2, v_3 \rangle$ and $\langle v_1, v_2, v_3 \rangle$. A set of simplices with all its faces is called a *simplicial complex*.

The geometric realisation of a single vertex, $\langle v_1 \rangle$ is a point, and the geometric realisation of a 1-dimensional simplex $\langle v_1, v_2 \rangle$ is a line oriented from v_1 to v_2. Thus every network is a 1-dimensional complex, and simplicial complexes generalise networks to higher dimensions.

The sets of girls identified as hypergraph edges in Figure 1 can also be viewed as the vertices of simplices. Let two simplices be q-near if they share a q-dimensional face. The transitive closure of the q-nearness is the q-*connectivity* relation. It partitions sets of simplices into q-*connected components*. A listing of the q-connected components is called a *Q-analysis*, which can be succinctly summarised by a *skyscraper diagram* (Fig. 2). At $q = 1$ five distinct components emerge. The first is for Ellen who is liked by Ella and Irene at the bottom left of Figure 1. The second is Hazel, Betty and Hilda which form a group at the bottom right of Figure 1. Next comes Edna with Mary, Jane and Adele at the top right of Figure 1. Ada and Cora form a small components at the top left of Figure 1. The largest component clusters around the popular Eva and Marion at the top centre of Figure 1. *Q*-analysis reveals structure that is not obvious in the network representation, and also provides a more tractable way of computing Galois pairs of simplices for bipartite relations [2, 9].

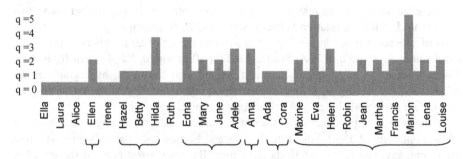

Fig. 2. The Q-analysis skyscraper diagram for the girls' dining preference relation

Figure 3 shows five faces, f_1, f_2, f_3, f_4, and f_5 made of the shapes s_1, s_2, s_3, s_4, s_5 and s_6. These face shapes can be represented by the simplices $\sigma(f_1) = \langle s_1, s_3, s_5, s_6 \rangle$, $\sigma(f_2) = \langle s_1, s_4, s_5, s_6 \rangle$, $\sigma(f_3) = \langle s_2, s_3, s_5, s_6 \rangle$, and $\sigma(f_4) = \langle s_2, s_4, s_5, s_6 \rangle$, $\sigma(f_5) = \langle s_2, s_3, s_5, s_6 \rangle$.

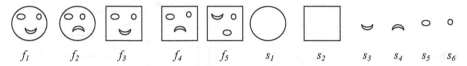

| f_1 | f_2 | f_3 | f_4 | f_5 | s_1 | s_2 | s_3 | s_4 | s_5 | s_6 |

Fig. 3. Faces made up from shapes

Using simplices also has problems. For example, $\sigma(f_3) = \langle s_2, s_3, s_5, s_6 \rangle = \sigma(f_5)$, and the simplex representation cannot discriminate them. Let R be a relation on an ordered set of four vertices $\langle x_1, x_2, x_3, x_4 \rangle$, defined as: place x_2 under the centre of x_1; place x_3 above and to the left of the centre of x_1; place x_4 above and to the right of the centre of x_1. Let $\langle x_1, x_2, x_3, x_4; R \rangle$ be defined to be a *relational simplex*, or *hypersimplex*. Now $\sigma(f_3) = \langle s_2, s_3, s_5, s_6; R \rangle$ while $\sigma(f_5) = \langle s_2, s_5, s_3, s_6; R \rangle$, so that $\sigma(f_3) \neq \sigma(f_5)$.

A hypernetwork is defined to be a collection of hypersimplices. Generally hypernetworks support patterns of numbers that represent properties and flows through the system being modelled. The relationship between the structures defined here is given in Figure 4.

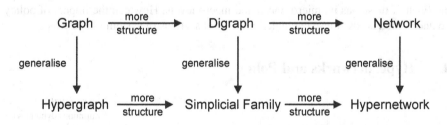

Fig. 4. The relationship between graphs, network, hypergraph and hypernetworks

3   Multilevel Structure

Hypersimplices provide a method of representing multilevel structure. Imposing an n-ary relations on sets of elements creates objects at higher levels. *E.g.* the three blocks a, b, and c are assembled by R into a structure, $R: \{a, b, c\} \rightarrowtail \langle a, b, c; R \rangle$ and given the *name* arch. If the elements exist at, say, *Level N* then the structured object exists at a higher level, say *Level N+1*. Here the higher level structure has an *emergent property* not possessed by its elements, namely there is a 'gap' between the assembled blocks.

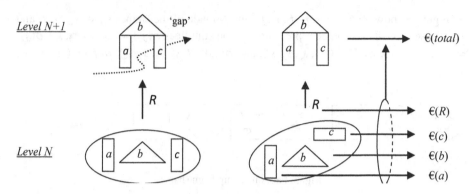

Fig. 5. Aggregating *Level N* parts to *Level N+1* wholes, and aggregating numbers

One of the challenges of modelling complex systems is to integrate micro and macro theories of behaviour. From a policy perspective, the target systems almost always include people as social and economic systems. Social systems have structures at many levels, from the individual at micro levels through organisations and institutions at meso levels to nations and international structures at macro levels. Alongside this are economic considerations with individual's costs and benefits at the microlevel aggregated through social groups at meso levels to the costs and benefits perceived by the Finance Ministry, European Bank or the World Bank at macro level. Alongside this there are considerations of the particular system being managed, such as 'health', 'welfare', 'transport', 'environment, 'food', 'education', 'housing', and so on. Each of these has its micro, meso and macro levels. How can the impact of policy be understood across these entangled multilevel systems of system of systems?

4   Hypernetworks and Policy

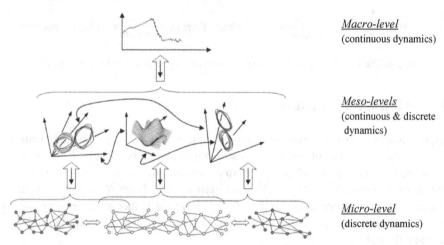

Macro-level
(continuous dynamics)

Meso-levels
(continuous & discrete dynamics)

Micro-level
(discrete dynamics)

Fig. 6. The challenge of modelling multilevel systems of systems of systems

Figure 6 illustrates the challenge of creating coherent multilevel theories to support policy. At the microlevel the dynamics emerge from the interactions of individuals. Currently a favoured scientific way of investigating the microlevel behaviour is agent-based simulation, which generates two-level dynamics at the level of the individual and the emergent behaviour of the community. Simulations do not give 'predictions' of future system behaviour, but give insights in to possible system behaviour and, at best, estimate the likelihood for a policy to have its desired outcome at that level.

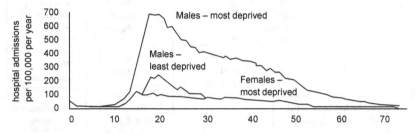

Fig. 7. Admissions to English NHS hospitals for assault involving 13 to 14 year olds. [11]

Figure 7 illustrates macro-level statistics that provide a macro-level snapshot in the terms of Figure 6. They come from a report entitled 'Dying to Belong' published in 2009 by the Centre for Social Justice (CSJ) which gives an in-depth analysis of street gangs in Britain [12].

Between 6th and 10th August in 2011 London and other British cities experienced violent riots, looting, arson, assault and robbery. Many thousands of lawless people took to the streets including rival street gangs acting together.

The report 'Ending Gang & Youth Violence', published in November 2011 by the Secretary of State for the Home Department was the basis for a policy response to the riots. "One thing that the riots in August did do was to bring home to the entire country just how serious a problem gang and youth violence has now become." [13] This report set out detailed policy plans for the agencies to work together, including providing support to local areas; preventing young people becoming involved in serious violence; pathways out of violence and gang culture; punishment and enforcement to suppress the violence of those refusing to exit violent lifestyles, and partnership working to join up local area responses to gangs and youth violence.

The 2012 CSJ report 'Time to Wake Up' questions the effectiveness of the police practice of identifying and removing gang 'elders': "it seems that an unintended consequence of the arrest of senior gang members has been to heighten tensions and violence. ... There was a consensus that the current gangs neither have [no] cohesive leadership, which is resulting in increased chaos, violence and anarchy. [14]

These reports discuss the problem of gangs at the macro level of state policy, at meso level of local authority policy, and at the microlevel of dealing with individuals. For example, 'Ending Gang and Youth Violence' documents the life of "Boy X" from his birth to his seventeen year old crack cocaine addicted mother to life imprisonment for murder at age twenty one, and his many contact points with the social and emergency services (Fig. 8). [13]

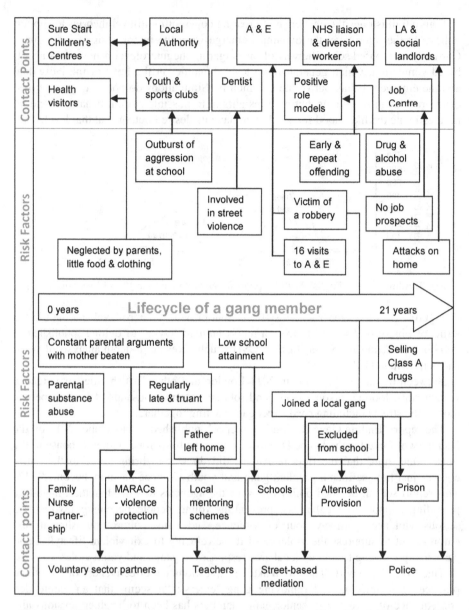

Fig. 8. The lifecycle of a gang member (Source: HMG: Ending Gang Youth Violence [13])

If this were the story of just one person it would be regrettable but not an issue for policy. Figure 8 is a kind of model of the life of this individual at the microlevel and his interactions with the mesolevel social and emergency services. Implicitly it is intended to generalise to *classes* of individuals at higher mesolevels of aggregation. Policy cannot target individuals at the microlevel but requires *models* of the behaviour of individuals within classes or higher level aggregate entities, *i.e.* hypernetworks.

The various reports on gangs cited above contain propositions about the behaviour of gang members. These include postulating classes of 'Elders' and 'Youngers' and their modes of interaction at the microlevel, and classes of rival gangs and their modes of interaction at higher mesolevels. Much of this is empirically based and forms a theory of gang behaviour on which to base policy. What can formal modelling with hypernetworks add to this?

Social policy is inevitably expressed in natural language within a legal framework for implementation. In comparison, technical or formal models of systems are stated in their own language which may include mathematics and computation, but they are always embedded in a metalanguage such as vernacular English. This is as true in engineering as it is in social administration.

The problem with the theories in the reports cited above is that their vernacular models are *untestable* before implementation. For example, the failure of the 2011 police policy to remove Elders from gang was *predicted* in 2009 [12] but this was ignored or overlooked by the police. Why? Perhaps due to predictions presented in natural language carrying little more weight than opinions because the outcome of their premises and logic cannot be demonstrated before empirical testing on the street.

Technical models can translate vernacular models into formal theories that can be implemented in computers to generate their logical and empirical consequences. Given the many possible initial conditions such models must be run many times to characterise the space of possible policy outcomes. For policy purposes the output of the computational model will again be evaluated in a vernacular metalanguage, but the intermediate step of *computation* can add a lot in terms of understanding the real social system and its dynamics.

Many hypersimplices can be abstracted from Figure 8, *e.g.* ⟨ neglected by parents, parental substance abuse, parental violence; $R_{\text{experienced\_0-5}}$⟩ suggests a child at high risk. The hypersimplex ⟨ outbursts of aggression as school, involved in street violence, many visits to A&E; $R_{\text{experienced\_10-15}}$⟩ indicates a child becoming increasingly violent and dangerous, while ⟨ regularly late and truant, school low attainment, excluded from school;; $R_{\text{experienced\_10-15}}$⟩ is a likely precursor to ⟨ joined a local gang, selling Class A drugs, early and repeat offending, drug and alcohol abuse; $R_{\text{experienced\_16-21}}$⟩.

In many cases subsequent enquires discover that neighbours, welfare agencies, schools, the police and even the postman had evidence of tragic events to come, but this evidence was not 'joined up' as a hypersimplex. This is an easy conclusion but it is difficult to rectify the lack of structure to prevent further incidents. How can micro-level individuals join up the information spread across their mesolevel organisations? How can institutional structures such as ⟨ welfare agencies, schools, police, postman, milkman; $R_{\text{share\_and\_synthesie\_information}}$⟩ be formed and run at politically bearable costs?

5   Hypernetworks, Big Data and Policy Informatics

Recently it has become clear that the dynamics of society are carried by an unprecedented flux of data. *Big Data* includes billions of phone calls, text messages, emails, financial transactions, personal data, and so on. Data mining has well established techniques for abstracting useful information from these data, and many involve discovering hypersimplices and relationships between them in hypernetworks.

Policy informatics is the process of building computer models of social systems with the explicit intention of using them as policy tools. Increasingly these models have the structure shown in Figure 6, with massive microlevel simulations creating more aggregate information at higher levels. For example, the TRANSIMS system developed at Los Alamos National Laboratory in the nineteen nineties simulates the trip making of millions of travellers in US cities, including explicit representations of each individual traveller, their family structure, their activity patterns and so on.

Policy informatics is built on *social informatics*, the process of building computer models of social systems to investigate their dynamics. Social informatics is not necessarily policy-driven and can be pure research. Today big data plays an important part in social informatics and the development of new models of social processes.

Policy informatics may answer the question of forming institutional structures able to synthesise atomic social data for policy purposes. In future it is likely that computers will shift from having their constructs supplied by humans to them abstracting 'relevant' constructs for themselves, where a *construct* is a hypersimplex, of subordinate constructs assembled by a relation. To illustrate this consider the faces shown in Figure 3. What is the correct way of grouping these faces? For example, the faces could be aggregated on the basis of round versus square, or they could be aggregated on the basis of smiles versus frowns. Of course there is no 'correct' way of aggregating them, only more or less *useful* way of aggregating them.

Hypernetworks provide essential structure for computer systems to build vocabularies automatically for searching big data in policy informatics. [9]

6   Conclusions

Network theory is fundamental in both social informatics and policy informatics. Hypernetwork generalise networks and hypergraphs, allowing relations between many things to be modelled by hypersimplices which have much richer structure than edges in hypergraphs. Hypersimplices provide a coherent way of representing multilevel systems to integrate their bottom-up and top-down micro-, meso- and macrolevel dynamics. They provide a natural way of representing structures in a policy context, and this enables polices to be tested by computation before they are implemented with possibly unexpected consequences. Hypernetworks have immediate interpretation as data structures and can bridge the gap between vernacular models and computation of social processes. Hypernetworks will play an important role in the use of big data in social informatics and policy informatics, and will play a central role in future policy formulation and implementation.

References

1. Caldarelli, G., Catanzaro, M.: Networks: a Very Short Introduction. OUP, Oxford (2012)
2. Berge, C.: Graphs and Hypergraphs. North-Holland, Amsterdam (1973)
3. Freeman, L.C., White, D.R.: Using Galois Lattices to represent Network Data. Social Methodology 23, 127–146 (1993)

4. Atkin, R.H.: From Cohomology in Physics to Q-connectivity in Social Science. I. J. Man-Machine Studies 4, 139–167 (1972)
5. Atkin, R.H.: Mathematical Structure in Human Affairs. Heinemann, London (1974)
6. Atkin, R.H.: Combinatorial Connectivity in Social Systems. Birkäuser, Basel (1977)
7. Gould, P., Johnson, J., Chapman, G.: The Structure of Television, Pion, London (1984)
8. Johnson, J.: Hypernetworks of Complex Systems. In: Zhou, J. (ed.) Complex 2009. LNICST, vol. 4, pp. 364–375. Springer, Heidelberg (2009)
9. Johnson, J.H.: Hypernetworks in the Science of Complex Systems. Imperial College Press, London (2013)
10. de Nooy, W., Mrvar, A., Batagely, V.: Exploratory Social Network Analysis with Pajek. Cambridge University Press, Cambridge (2005)
11. Bellis, M.A., Hughes, K., Wood, S., Wyke, S., Perkins, C.: National five-year examination of inequalities and trends in emergency hospital admission for violence across England. Injury Prevention 17, 319–325 (2011)
12. Centre for Social Justice: Dying to belong, London (February 2009), http://www.centreforsocialjustice.org.uk/client/downloads/DyingtoBelong FullReport.pdf
13. H M Government: Ending Gang and Youth Violence. A cross-Government Report, Government Command Paper 8211, H. M. Stationary Office, London (2011), http://www.homeoffice.gov.uk/crime/knife-gungang-youth-violence/
14. CSJ: Time to wake up. Tackling gangs one year after the riot. CSJ, London (October 2012), http://www.centreforsocialjustice.org.uk/client/images/Gangs%20Report.pdf

To Trade or Not to Trade: Analyzing How Perturbations Travel in Sparsely Connected Networks

Marshall A. Kuypers[1,2], Walter E. Beyeler[1], Matthew Antognoli[1,3],
Michael D. Mitchell[1], and Robert J. Glass[1]

[1] Complex Adaptive System of Systems (CASoS) Engineering
Sandia National Laboratories, Albuquerque, New Mexico, USA
{mkuyper,webeyel,mantogn,micmitc,rjglass}@sandia.gov
http://www.sandia.gov/CasosEngineering/
[2] Management Science and Engineering Department, Stanford University
Stanford, California, USA
mkuypers@stanford.edu
[3] School of Engineering, University of New Mexico
Albuquerque, New Mexico, USA
mantogno@unm.edu

Abstract. In global economics, nations are often faced with the opportunity to open or close new avenues of trade or to join new markets. These actions can be beneficial or harmful for a nation because entering a market exposes that nation to the perturbations caused by others in the market. However, joining a new market offers the benefit of lower prices and increased security against domestic perturbations because shocks are spread across all trading partners. This risk/benefit tradeoff is relatively straightforward for one market, but the effects are more complicated when multiple markets are introduced. We use an agent-based model to analyze how the connection pattern of markets affects perturbations that travel across networks. We find that shocks are not easily transmitted across networks unless the perturbed resource is directly traded and we discuss the tradeoffs associated with opening new international market connections.

Keywords: Complex adaptive systems, Agent-based model, Trade, International markets, Perturbations, Industry protection.

1 Introduction

Although international transactions are a significant portion of nearly every nation's economy, the dynamics of trade are still only partially understood. The incentives and politics of trade have been studied without uncovering a complete picture of the optimal strategy for choosing a trading network. Furthermore, the complicated relationships and feedbacks among trading entities often lead to surprises for policy makers. Understanding international transactions is important because trade opens up avenues for improved efficiency but also makes the nation vulnerable to supply shocks.

K. Glass et al. (Eds.): COMPLEX 2012, LNICST 126, pp. 190–200, 2013.

In this paper, we describe an agent-based model (ABM) designed to study resource exchanges between economic agents. We initialize a simple system consisting of two nations and observe how perturbations travel across sectors and between nations. We find that shocks are not transmitted easily between nations unless the resource produced by the perturbed sector is internationally traded. If the perturbed resource is internationally traded, the consuming sectors experience gains and losses that are determined by the global connection pattern of markets.

1.1 International Relations and Trade as Complex Systems

While many researchers have modeled supply chains and organizations as complex adaptive systems (CAS) there has been very little work on international relations modeled as CAS [1, 2]. Axelrod's work in the 1990s proved complex systems are a useful perspective for analyzing international relations (IR), but almost 20 years later the IR community has yet to embrace CAS techniques even though IR problems contain many of the CAS criteria outlined in Choi et al. 2001 such as interacting agents, nonlinear relationships, and adaptive components [3, 4, 5].

Contagion, or the spread of economic shocks between countries, is one of the few processes that has been understood using CAS theory [6]. However, most of the literature on contagion is limited to finance, with little attention paid to commodities or goods. The interest in contagion started in the late 1990s when a financial crisis in East Asia spread to South America but it has gained more prominence in the last five years after the financial collapse of 2008 and the current Eurozone crisis [7, 8]. Others have studied how shocks in industries such as airlines travel between international economies but this work is very sparse [9]. Our work broadly studies the tradeoffs with opening new avenues of trade and shows another example of applying CAS modeling to IR problems.

2 Model Formulation

We use an agent-based model developed at Sandia National Laboratories to analyze this system [10]. The agents are defined by a set of coupled nonlinear first-order differential equations and interact through trading events, resulting in continuous behavior that triggers discrete exchanges.

Each agent is called an entity. Each entity must consume, produce, and store resources to remain competitive in its environment. A 'health' state variable controls the agent's production processes and is a function of the agent's resource consumption rates. The nominal level of health is 1 and higher health levels are caused by above-nominal consumption rates. Low health levels are caused by agents reducing consumption due to low demand for its product, high input prices, or low money levels. Protracted reduction in consumption can cause an agent to die although this did not occur in the scenarios examined in this study.

Agents interact in discrete exchange events that are facilitated by markets. Each agent buys and sells resources in a market and is matched for trading using a double auction. A collection of agents and markets makes up a nation. The specific representation of this hierarchical structure is irrelevant and can represent a variety of institutions, such as companies, industries or nations.

Each agent processes resources to remain competitive in its environment. Agents obtain resources at a market and store them. Then a production process occurs, converting the input resources to an output resource. The output resources are stored in an output tank before being sold at a different market.

Perturbations are introduced into the model by removing a defined amount of resource from an agent's output storage tank which causes the agent to readjust its production and consumption as it balances the competing interests of filling its storage tank back up and selling its produced resource to obtain money. The resulting behavior affects other agents and the shock propagates through the system. For these simulations we chose a set of parameter values where the perturbation propagated through the production network but did not cause any agent to die.

Nation State

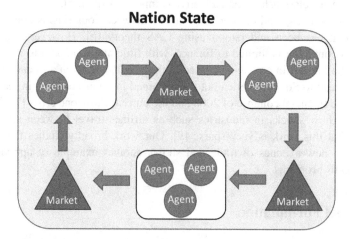

Fig. 1. A nation's structure. Agents are circles, rectangles are sectors, and triangles are markets. A collection of agents, sectors, and markets makes up a nation.

For this study, we initialize two identical nations, A and B. Each nation has six sectors made of one agent each. The sectors are fully connected within the nation so that each sector consumes resources from every other sector and produces a resource that every other sector needs (figure 2). This model can be interpreted as representing a closed system of several resources such as food, water, goods, energy, and labor. Each sector requires a non-zero amount of every other resource although some substitution of resources is possible. Our model assumes the resource requirements are symmetric for each sector but we have the flexibility to represent realistic asymmetric trade flows, which we would like to study in the future.

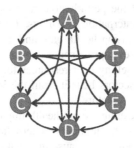

Fig. 2. The structure of the network for each nation. Each lettered node is a sector made up of a single agent. The network is fully connected. A study on how network structure affects the dynamics of the systems can be found in Kuypers et al. 2011 [11].

We study the effect of international markets that allow trade between two nations. There is no additional cost imposed on international markets, resulting from geographic proximity although this feature will be implemented in the future. Geographic constrains might create more realistic behavior by imposing a transportation cost which would incentivize domestic trading over international.

We introduce perturbations by removing the entire product inventory of a resource (F) in nation A and observe the changes in health in all sectors in both nations. The total impact of a perturbation is measured by integrating the difference in health between a perturbed run and an unperturbed run where the system maintains an equilibrium health level equal to one for the total duration of the simulation.

3   Perturbations in Non-internationally Traded Markets

First, we consider perturbations that are only indirectly spread through nations, meaning the perturbed resource is not internationally traded. If the number of internationally traded markets is increased, thereby increasing the number of connections between the two nations, then perturbations travel more easily from nation A to nation B. In figure 3, we expect to see a larger response in case 2, since more connections are available.

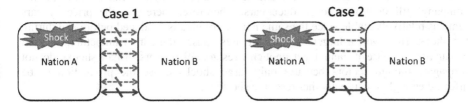

Fig. 3. Dashed arrows denote markets in non-perturbed resources and solid arrows denote markets in the perturbed resource. An uninterrupted line means that resource is internationally traded while a break-line means that the resource is not internationally traded. Case 1 shows no resource internationally traded, while case 2 shows 5 resources (all except the resource produced in the perturbed sector) internationally traded. Shocks are introduced in nation A.

Indeed, Table 1 shows that the perturbation travels to the second nation state more easily when there are more international markets. Furthermore, as more markets are opened nation B reaps an increasing health gain and nation A sees an increasing health loss. When a perturbation causes a price spike in a closed nation, the perturbed sectors recover part of their losses by selling their resources for high prices. When international markets are open, nation B's sectors benefit from the increased price which delays the recovery of the perturbed nation's sectors.

The total system impact, meaning the net health impact for nations A and B summed together, is constant regardless of the number of international markets. The consistency of the perturbation impact is not obvious because the total health loss could rise as more of the perturbation leaks into the new nation which is initially in equilibrium. Although more of the perturbation travels from nation A to nation B, the total system impact does not change; only the distribution of gains and losses changes.

Table 1. Health values as a function of the number of international markets in non-perturbed resources. The error term is the estimated standard deviation calculated from three runs.

Number of International Markets in Non-Perturbed Resources	Net Health Impact of Perturbed Nation (Nation A)	Net Health Impact of Un-Perturbed Nation (Nation B)	Total Health Impact (nation A + Nation B)
0	-12.27 ± 0.06	-0.02 ± 0.00	-12.29 ± 0.06
1	-12.29 ± 0.10	-0.08 ± 0.06	-12.37 ± 0.08
2	-12.94 ± 0.26	0.36 ± 0.09	-12.58 ± 0.29
3	-13.23 ± 0.05	0.84 ± 0.02	-12.38 ± 0.05
4	-13.83 ± 0.12	1.31 ± 0.08	-12.52 ± 0.07
5	-14.11 ± 0.17	1.70 ± 0.05	-12.41 ± 0.22

Nation B only experiences a net impact that is 15% the magnitude of nation A's response when all the non-perturbed markets are internationally traded, showing that shocks cannot travel efficiently through markets that are not directly perturbed. Small shocks are dampened by the relationship between inputs and outputs. For example, a small shortage of tires causes a car manufacturer to pay a higher price for tires, but the company still continues to produce cars. Therefore, therefore, the price of cars remains relatively stable since the same number of cars is produced before and during the shock. However, a larger shortage of tires causes the car manufacturer to limit production because it cannot get enough tires for all its cars. Small shocks do not propagate though sectors because only large shocks cause the production to be affected enough to cause another resource shortage.

4   Perturbations in Internationally Traded Markets

Shifting attention to cases where the perturbed resource is internationally traded provides new insights. First, we can look at the net health impact as we increase the number of international markets. The connection pattern is illustrated in figure 4.

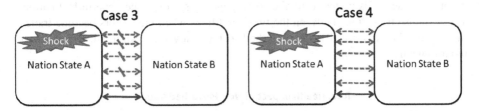

Fig. 4. Case 3 and case 4. The resource produced by the perturbed sector (the solid line) is internationally traded. Other international markets are opened (dashed lines). Case 3 has one international market, and Case 4 has all resources traded internationally.

Case 3 becomes case 4 by introducing additional markets. The total system health impact is shown in figure 5. Allow the perturbed resource to be traded internationally causes a significantly worse total system impact than an isolationist strategy. However, the total system impact becomes less severe as more international markets are added. Trading internationally becomes the optimal choice once 2/3rds or more of the markets are traded internationally.

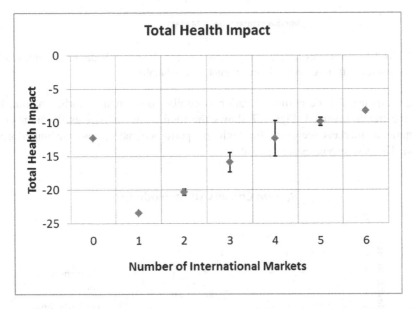

Fig. 5. Total health impact for nation A and B as a function of the number of international markets. The first market to be opened to international trade is the resource produced by the perturbed sector. Error bars show the estimated standard deviation based on three runs.

At the scale of the individual nations, we see a different story. For the non-perturbed nation, the non-perturbed sectors (A-E) follow two trend lines depending on whether they are internationally traded. An internationally traded project always causes a sector's health to decrease in comparison to not trading internationally because the sector is underbid by the corresponding sector in the unperturbed nation. The response transitions between the two trend lines when the sectors become traded internationally. Additionally, the perturbed sector follows its own trend, which grows monotonically.

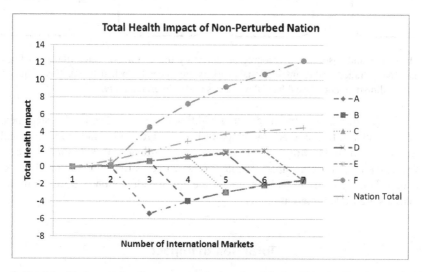

Fig. 6. Total health impact for the non-perturbed nation. Market F is the first internationally traded market, and then A through E are brought in sequentially.

The response for the perturbed nation is similar to the non-perturbed nation, but with the trends reversed. Figure 7 shows the total health impact as the number of international markets increases for both the perturbed nation and the unperturbed nation. The two responses are opposite.

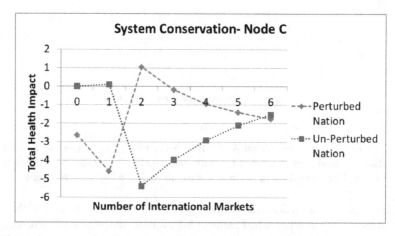

Fig. 7. System conservation for node C. Note that node C behaves inversely between the two nations.

4.1 Perturbation Flows

Many of the model's dynamics are characterized by universal trends, but significant qualitative changes occur as well. The model responds qualitatively differently depending on whether the perturbed sector is internationally traded or the perturbed sector plus one non-perturbed sector is traded internationally.

The model output is similar to two closed nations when only the perturbed sector's output is traded internationally,. The perturbation does not travel between nations very effectively through one sector alone. Once another sector is added the perturbation travels through the nations much more. The requirement to have two international markets to transmit shocks can be understood by considering water flow. When two tanks of water are connected by one pipe which is pressurized, nothing too interesting happens. Some water flows from the first tank to the second tank, but the pressure in the second tank will quickly build, preventing more flow through the pipe. Adding another pipe causes a return flow of water, which allows more water to flow through the first pipe. This creates more interesting dynamics.

Similarly, in the model, a single connection is not enough to transmit an economic shock. A return mechanism is required for the two systems to affect each other. The return mechanism is a second international market.

4.2 Time Series Analysis

Examining a single perturbation and how it travels from one nation to another reveals some important dynamical processes of the model. Figure 8 shows the response when only the perturbed sector is internationally traded. First, the resource is removed. Immediately, the perturbed sector's health falls because it has no resource to sell, which causes its money level to drop and it scales back consumption. Additionally, all the sectors in the perturbed nation experience a health loss. Although the perturbed nation's sectors can still obtain input resources in the international market, the perturbed sector F is not consuming resources they produce. Therefore, the health loss of the non-perturbed sectors results from the reduction in demand for their output, not because of a shortage of any input resource. The perturbed sector in the non-perturbed nation is able to benefit slightly from the additional demand for its product causing it to gain money and increase consumption, which gives more money flow to the other sectors in its nation, resulting in a slight health gain.

These processes continue until the perturbed sector resumes selling its resource. After the perturbation, the perturbed agents have competing interests to build up their storage tank and accept the high price being offered for its goods. Once the price is high enough and the input tank level is high enough, the agent starts to sell and begins to make profits resulting in higher money levels, which spurs additional consumption. Once the health values have reached their extrema the trends reverse. The perturbed sector starts producing again and benefits from the high price of its good. It increases consumption which allows the other sectors in its nation to recover also. The increased consumption and production rates provide some momentum so that the perturbed sector reaches the maximum health of any sector before the perturbation is damped out.

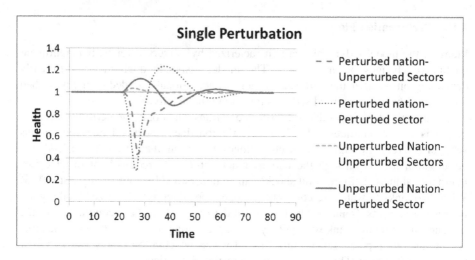

Fig. 8. Time series for a single perturbation on nation A, where the perturbed sector is the only internationally traded market. The time series are classified by their nation and sector.

These processes lead to an interesting result. When a perturbation occurs, there is not a net flow of resources into the hole that is created. There is a flow of resources away from the hole created by the perturbation. The agents find the best market, which is far away from the shock. For example, if you sell apples in a town where a tornado hits, it makes sense to leave and go to a town where there is still demand for apples.

5   Discussion

The applications of this study are far reaching. Our model points to interesting tradeoffs due to connections in international markets. Many developing countries practice industry protection where they shield a developing sector to allow it to become strong enough before entering the international market [12]. For example, it is unlikely an Indian car manufacturer could compete with the Japanese, European, and American auto makers. Therefore, the Indian government might create incentives for consumers to buy the Indian product and discourage foreign suppliers by capping imports or setting tariffs. Of course, this can also stifle innovation and turn out worse if the protected company decides it does not need to compete or innovate. Industry protection is still a hotly debated topic, with numerous studies offering data supporting its adoption and denouncing it [13, 14, 15]. This paper is not an argument for or against industry protection but instead points out some interesting features that result from different trading structures.

This work shows that it is not obvious who will win and who will lose from a given perturbation. Nations should consider feedback effects carefully when protecting industries.

5.1 Pushing the System to Complexity

The system may not currently be displaying complexity, since we have a low number of homogenous agents. We would like to scale these simulations up to include more realistic trading patterns, more realistic network structures, and heterogeneous agents. New emergent behaviors might become evident under these conditions.

6 Conclusion

In this paper, we present an ABM of two idealized nations that trade resources using several exchange patterns. We observe how the pattern of international connections affects a nation's response to a shock. We find that perturbations do not travel across nations efficiently if the perturbed sector is not internationally connected since the unperturbed sectors absorb a significant amount of the shock and only transmit it if the price shock is large enough to disrupt their production.

We also find that when the perturbed sector is not internationally traded the total system impact is unaffected by additional international markets. However, if the perturbed sector is internationally traded, then additional international markets reduce the net system impact. In both cases the distribution of winners and losers changes based on the connection pattern.

When a perturbation occurs the goods flow toward stronger economies instead of rushing in to fill the void left by the shock. Agents seek to preserve their markets and move to where their goods can demand higher prices which is away from the perturbation.

Our work shows how nations might evaluate their trading network and offers some insights into the industry protection argument. Complicated feedbacks lead to gains and losses across sectors and nations that are not intuitive.

References

1. Choi, T.Y., Dooley, K.J., Rungtesanantham, M.: Supply Networks and Complex Adaptive Systems: Control Versus Emergence. J. Oper. Manag. 19, 351–366 (2001)
2. Schneider, M., Somers, M.: Organizations as complex adaptive systems: Implications of complexity theory for leadership research. Leadership Quart. 17, 351–356 (2006)
3. Axelrod, R.: The Complexity of Cooperation: Agent-Based Models of Competition and Collaboration. Princeton University Press, Princeton (1997)
4. Tezcan, M.Y.: The EU Foreign Policy Governance as a Complex Adaptive System. In: Minai, A., Braha, D., Bar-Yam, Y. (eds.) Unifying Themes in Complex Systems Volume VI: Proceedings of the Eighth International Conference on Complex Systems. New England Complex Systems Institute Series on Complexity, Springer, NECSI Knowledge Press, Cambridge (2006)
5. Favero, C.A., Giavazzi, F.: Is the International Propagation of Financial Shocks Non-linear? Evidence from the ERM. J. Int. Econ. 57, 231–246 (2002)

6. Markose, S., Giansante, S., Gatkowski, M., Shaghaghi, A.R.: Too interconnected to fail: Financial contagion and systemic risk in network model of cds and other credit enhancement obligations of us banks. University of Essex Discussion Paper Series, No. 683 (2010)
7. Claessens, S., Dornbusch, R., Park, Y.C.: Contagion: why crises spread and how this can be stopped. In: Claessens, S., Forbes, K. (eds.) International Financial Contagion, pp. 19–42. Kluwer Academic Publishers, Boston (2001)
8. Kolb, R.W.: Financial Contagion: The Viral Threat to the Wealth of Nations. John Wiley & Sons, Inc. (2011)
9. Gillen, D., Lall, A.: International transmission of shocks in the airline industry. J. Air Trans. Manag. 9, 37–49 (2003)
10. Beyeler, W.E., Glass, R.J., Finley, P.D., Brown, T.J., Norton, M.D., Bauer, M., Mitchell, M., Hobbs, J.A.: Modeling systems of interacting specialists. In: Sayama, H., Minai, A., Braha, D., Bar-Yam, Y. (eds.) Unifying Themes in Complex Systems Volume VIII: Proceedings of the Eighth International Conference on Complex Systems. New England Complex Systems Institute Series on Complexity, pp. 1043–1057. NECSI Knowledge Press, Cambridge (2011)
11. Kuypers, M.A., Beyeler, W.E., Glass, R.J., Antognoli, M., Mitchell, M.: The impact of network structure on the perturbation dynamics of a multi-agent economic model. In: Yang, S.J., Greenberg, A.M., Endsley, M. (eds.) SBP 2012. LNCS, vol. 7227, pp. 331–338. Springer, Heidelberg (2012)
12. Botelho, A.J.J., Dedrick, J., Kraemer, K.L., Tigre, P.B.: To Industry Promotion: IT Policy in Brazil, Center for Research on Information Technology and Organizations, University of California Irvine (1999)
13. Baldwin, R.E.: The Case Against Infant-Industry Tariff Protection. J. Pol. Econ. 77, 295–305 (1969)
14. Hinton, M.N.A.: Infant Industry Protection and the Growth of Canada's Cotton Mills: A Test of the Chang Hypothesis. The Rimini Centre for Economic Analysis. Working paper (2012)
15. Chang, H.J.: Kicking Away the Ladder: Development Strategy in Historical Perspective. Anthem, London (2002)

Help from Hoarders: How Storage Can Dampen Perturbations in Critical Markets

Marshall A. Kuypers[1,2], Walter E. Beyeler[1], Matthew Antognoli[1,3],
Michael D. Mitchell[1], and Robert J. Glass[1]

[1] Complex Adaptive System of Systems (CASoS) Engineering
Sandia National Laboratories, Albuquerque, New Mexico, USA
{mkuyper,webeyel,mantogn,micmitc,rjglass}@sandia.gov
http://www.sandia.gov/CasosEngineering/
[2] Management Science and Engineering Department, Stanford University
Stanford, California, USA
mkuypers@stanford.edu
[3] School of Engineering, University of New Mexico
Albuquerque, New Mexico, USA
mantogno@unm.edu

Abstract. Critical resource supply chains are vulnerable to manipulation because of the un-substitutability of their goods. When a monopoly controls all or part of a market, it has the ability to profit from a reduction in supply of a critical resource. We model this complex adaptive system (CAS) using an agent-based model (ABM) and investigate a strategy to mitigate the potential for exploitation of a market by a monopoly. We find that when entities increase their input resource buffer, they decrease the reactivity of resource prices to supply disruptions, which limits the amount by which monopolies benefit from price fixing. This storage strategy also reduces total system losses due to a perturbation.

Keywords: Complex adaptive systems, Agent-based model, Supply networks, Buffer capacity, Monopoly, Perturbations, Storage, Price fixing.

1 Introduction

Critical resources and supply networks are important to model and understand because they have the potential to be crippling and disruptive to an economy. Often consisting of un-substitutable goods, these resource supply chains become even more vulnerable when operated by unfriendly stakeholders or controlled by a monopoly capable of exploitation by supply reduction and price fixing. In this paper, we examine methods by which a monopoly is able to exploit supply shocks and potential mitigation strategies to prevent this behavior.

Economic interactions have been recognized as complex adaptive systems (CAS) for over twenty years [1]. Recent research has confirmed that supply networks exhibit CAS behaviors and structures as well [2]. We use an agent-based model (ABM)

K. Glass et al. (Eds.): COMPLEX 2012, LNICST 126, pp. 201–212, 2013.
© Institute for Computer Sciences, Social Informatics and Telecommunications Engineering 2013

developed at Sandia National Laboratories to initialize an environment of interacting agents that consume, produce, and store resources. We use this model to study the ways a monopolized supply chain is vulnerable to price fixing through a scenario where a monopoly controls a critical resource in a fully connected network of interacting agents. We show how the monopoly can profit by reducing its production, thereby increasing the price of its good. Then, we examine how the other agents can reduce the monopoly's ability to profit by enlarging their input resource buffer tanks. Finally, we consider the system effects and policy implications of this work. We find that increasing the buffer capacity reduces the total system loss from a perturbation. The severe price reactivity caused by low storage generates more system damage, while larger buffers benefit the whole system.

1.1   Relevant Literature

This paper draws on multiple disciplines. Relevant work on supply networks as CASs, buffer capacities in supply networks, monopolies, and price fixing is summarized here. Engineers have been studying optimal buffer capacities for several decades [3]. Since then others have developed algorithms and methodologies for investigating supply chain operations in various environments [4, 5, 6]. However, most of these papers consider supply networks as deterministic processes, with a stochastic component occasionally represented to introduce uncertainty. Additionally, the analyzed supply networks have static structures and no adaptability.

Other researchers have argued that supply networks should be characterized as CAS [2], and other studies have rigorously tested the validity of treating supply networks as CAS, verifying this designation [7]. We believe that supply chains and economic networks can be studied more effectively by considering their complex adaptive characteristics. In our model, we treat supply networks as CAS and analyze the buffer capacity problem, which to our knowledge has not been done before.

Additionally, we consider a network with monopolistic players and investigate policies that restrict the monopoly's ability to profit from the system through price fixing. A monopoly's ability to price fix has also been well studied, under a variety of constraints [8, 9]. The practice of price fixing has been repeatedly demonstrated in the real world. The Organization of Petroleum Exporting Countries (OPEC) has been shown to manipulate prices through production levels [10, 11]. Other companies, such as De Beers, Enron, and Archer Daniels Midland have also been investigated for price fixing [12, 13, 14, 15]. While much of the literature hints at ways to prevent monopolies from reaping undue profits, little literature exists that explicitly investigates policies that could protect consumers. Our hope is that this paper will encourage more discussion on mitigating disruptions to critical supply networks.

2   Model Formulation

We use an ABM developed at Sandia National Laboratories to analyze this system [16]. Each agent or entity must consume resources to maintain viability in its environment. To obtain resources for consumption each agent produces and sells a unique resource: each also stores resources as a buffer against disruptions. Agents interact with each other through resource markets in which buyers and sellers are matched using a double auction. These exchanges are executed with the aid of a 'money' resource. Agents respond to their environment by adjusting production and consumption rates. We use a state variable 'health' to control the production capacity of an agent, which is a function of recent consumption rates. Consumption, production, and health dynamics are described by a set of coupled nonlinear first-order differential equations.

Each agent processes resources the same way. An agent obtains a resource from the market and stores it in an input tank. The size of this tank is designated by an inventory coverage time parameter. A translation process converts input resources to an output resource. The output resource is stored in another tank before being sent to the market for sale. Figure 1 illustrates this process.

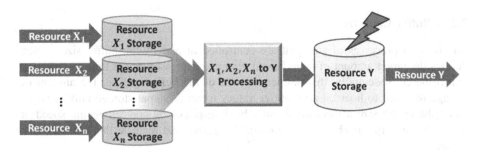

Fig. 1. An agent's resource process. The lightning bolt shows where the perturbation removes resources from the agent's output resource storage tank.

For this model, we use a fully connected network of resource interdependencies, as shown in Figure 2. An exploration of the effect of other types of network structure on the dynamics of the system has been previously explored [17]. For simplicity, we use the fully connected configuration.

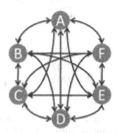

Fig. 2. A fully-connected network. Each node is an agent. Arrows denote resource flows; each agent consumes the resources produced by every other agent.

2.1   Perturbations and Reductions in Supply

Perturbations are introduced in the model by removing a defined amount of a resource from an agent's production storage tank. This creates a shock as the perturbed agent balances the competing interests of refilling its tank with selling its goods, which have become scarce in the market. The resource loss caused by this perturbation travels through the rest of the system by affecting other agents' input resources, which in turn affects their production.

We should also note that a perturbation that takes away some of an entity's produced resource is very similar to a reduction in production. The major difference is that perturbations remove the resource from the perturbed agent, while a reduction in production does not cause the unproduced resource to be lost. However, this distinction does not have an effect on health or money level as we are measuring them. Therefore, we can use perturbations to simulate how a reduction in supply will affect the system.

2.2   Buffer Capacity

The internal processes of an agent are controlled in part by the buffer size, which defines the target amount of stored input resource expressed as the time to consume that resource. Essentially, this defines the size of the buffer tank, or the amount of storage resource to hoard. Large buffer values mean the input storage tank is large. Also, the buffer size affects the agent's bidding prices by controlling the speed at which the bid price is changed. The casual processes that regulate this are shown in Figure 3.

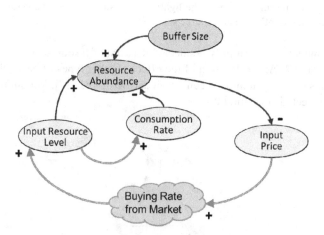

Fig. 3. A casual diagram of an entity's internal processes. Buffer size affects resource abundance, which is negatively correlated with the input price. Larger buffers reduce the premium an agent is willing to pay for a resource.

A small buffer size means the agent runs out of its input resource quickly and it will be willing to pay a premium to replenish its supply. Therefore, the size of the buffer determines how quickly the entity will react to a higher price: if the buffer is small, the agent will end up paying a premium quickly after the initial price spike. A larger buffer allows the entity to maintain production without chasing the price up, which alleviates the effects of the perturbation.

2.3   Substitution

The translation process that converts input resources to an output resource is not constrained to operate on one resource. The translation process can take in several inputs to produce one output. There is limited substitution between resources so that a shortage of one resource can be made up by increasing consumption of other resources. This tradeoff is reflected in Equation 1, which describes the influence of consumption rates on the evolution of health.

$$\frac{d}{dt}h(t) = \frac{1}{T_h}\left[\frac{h_0}{\left(\frac{1}{N_{Ch}}\sum_i \frac{1}{c_{h_i}^*(t)}\right)^{P_{Ch}}} - h(t)\right] \tag{1}$$

where $h(t)$ is the health level, T_h defines the decay time of health, h_0 is the nominal health level, N_{Ch} is the number of resources consumed to sustain health, and P_{Ch} is the power of the dependence of health on consumption.

This equation leads to diminishing returns as resources are substituted. For example, as the consumption of resource C_a^* decreases, more and more additional consumption of resource C_b^* is needed to maintain nominal health. A graph of this relationship is shown below, in Figure 4.

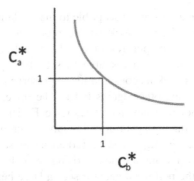

Fig. 4. Substitutability between two resources. As the consumption of resource C_a^* decreases, the amount of resource C_b^* that is needed to make up the difference increases.

This substitutability is generalizable to n different resources: each agent can consume multiple resources which are all substitutable as defined in equation 1. Note that this relationship requires that some nonzero amount of each input resource is consumed, regardless of the abundance of other resources.

2.4   Monopoly Strategy

The fact that each resource in this model is produced by a single agent means that each is controlled by a monopoly. Agents in this model do not have a cognitive strategy; they do not decide to price fix. Instead, price fixing emerges from the pricing mechanisms initialized in the model.

2.5   Clarification of Model

We would like to be precise about how we define and classify this model. The system we are modeling is undoubtedly a CAS. Our agent-based model has system dynamic components that govern the evolution of agents' states. The model may or may not represent CAS behavior for several reasons. We are observing a small number of agents, which may not be sufficient to demonstrate complexity under a formal definition. Also, the system may not have enough throughput to trigger complexity, just as a road with one car does not have enough throughput to trigger complex traffic patterns. Some behavior in this model could be classified as emergent, such as the price levels: our hope is to push this system to more interesting displays of emergence. Although the behavior analyzed in this study may not be complex, the ingredients for complexity exist. In this paper, we characterize some important dynamics of the system which is an important part of validating and testing the model. In future studies, we plan to observe complexity that conforms to a more rigorous definition.

3   Market Manipulation

Fixed supply and demand markets are susceptible to manipulation by the suppliers. In our model, we introduce a simple shock on the production tank of Agent F that removes 100% of its stored output resource. The health of the shocked agent decreases initially because it has no resource to sell. Its input flow of money is reduced and the agent scales back its consumption to preserve its money levels. While the resource is unavailable, the other agents bid up the price, since each agent still needs to consume some nonzero amount of resource F. Therefore, as soon as the perturbed agent F starts producing again, it can obtain a high price for its good since the demand for its resource is high. The perturbed agent sells its resources for a premium, which leads to its money level rising, which results in increasing consumption. As a result, the perturbed agent F sees a large health gain as it benefits from providing a scare resource that others need. This process reaches a maximum point, at which the health of unperturbed agents has decreased from paying such a high price that they can no longer afford the perturbed resource. At this point, F must bid down its selling price because agents stop buying its resource and its health begins to fall. This process effectively dampens the perturbation, but the health values experience some overshoot as all the values return to the nominal health level of one. These dynamics are shown in Figure 5.

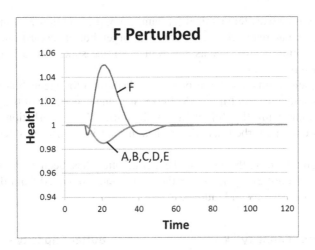

Fig. 5. A perturbation response. Node F is perturbed and experiences a health gain while all other nodes experience a health loss.

The final result of this simulation shows that the perturbed node experiences a net health gain. This result is robust to perturbation intensity, perturbation duration, and changes across a wide degree of parameter values. Some structures, such as the hub and spoke network, remove the incentive for price fixing because feedback patterns couple agents' health values. There are certainly more structures that exhibit this property as well.

However, for the fully-connected network structure the response pattern suggests that certain agents can benefit from inflicting a perturbation on themselves. Such a scenario has, in fact, happened. In 2000, middle-men like Enron convinced California power plants to shut down production, introducing an artificial scarcity into the markets [14]. California residential electricity is not tied to demand, so there is no incentive to reduce power consumption during peak hours. They continued to consume power at high rates, which resulted in numerous blackouts now known as the California Energy Crisis. Enron profited enormously by reducing the supply of its product.

OPEC has frequently generated similar market responses through reduction in production levels during times of high prices [10]. This situation can occur in any market where a supplier or suppliers can cooperate to control a significant portion of the market. By controlling the amount of resource available through a market, suppliers can fix the price.

We are interested in exploring ways to mitigate the consequences of disruptions. Are there policies that people could enact to dampen the perturbation caused by scarcity?

3.1   Dampening Perturbations

Buffer capacity is an excellent method for dampening perturbations. By increasing the level of stored input resource agents create a larger buffer against supply shocks. When a perturbation occurs, agents can wait out price spikes for a longer time by living off their input stores.

The graphs below illustrate the response to a disruption for agents holding different levels of input resources. By tripling the buffer size, the magnitude of the perturbation is reduced, along with the following oscillations. The perturbation duration is lengthened, but the perturbed agent's net health gain is significantly less when the buffer is larger.

For these simulations, the size of the perturbation has been kept consistent in absolute terms and not as a percentage: the total amount removed from the system is the same for both scenarios in Figure 6.

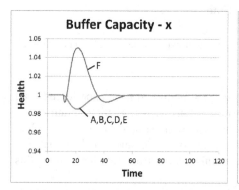
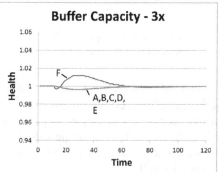

Fig. 6. Illustrating the effect of tripling the buffer size in every agent. The perturbation magnitude is greatly reduced.

What are the implications of this result? Our model suggests that perturbations can be dampened with larger input stores, allowing consumers to be less reactive to price spikes. It is unclear if this would work on a psychological level, since consumers might still behave irrationally and panic during price spikes. For example, before predicted natural disasters, long lines of cars often form at gas stations in anticipation of scarcity in the future.

If cars were made with gas tanks that are double the current size, gas prices might be less volatile. Creating more buffers on the scale of the national Strategic Petroleum Reserve could mitigate perturbations in the US domestic oil market. Of course, expanded resource storage works most effectively when storage costs are negligible and there is no product decay, as for oil and grains. Otherwise, consumers would be paying a premium to maintain above nominal stores, offsetting the savings from a reduced price spike. .

The manufacturing sector has largely decided that the cost of storage isn't worth the protection it offers. The popularity of just-in-time (JIT) manufacturing, lean six sigma, and the Toyota manufacturing system illustrate a sector-wide movement towards less inventory. The challenge is to accurately predict disruptions and risk in order to use these inventory-reducing strategies. Despite the carefully crafted predictions, many companies structured for JIT-like operations experienced huge losses following the September 11 2001 terrorist attacks, coming within hours of shutting down their production lines due to supply disruptions because they considered the probability of certain events to be unlikely [18]. Although JIT focuses on many aspects of production, September 11th prompted many manufactures to double their in-house inventory, despite the increased cost [19, 20].

3.2   Quantifying Returns

We are interested in understanding how effective a given increase in buffer capacity is at dampening the perturbation response. At some point, the benefit of additional inventory will be offset by the cost of adding additional storage.

We measure overall benefit by integrating the change in health from the nominal level of one over the simulation. The graphs in Figure 7 illustrate that the largest benefit comes from the first increase, with diminishing returns after that. It would be important to analyze where on this curve a product falls before implementing a policy.

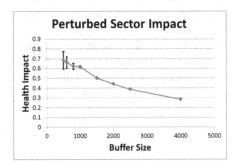
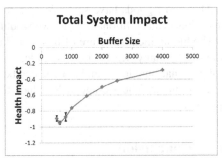

Fig. 7. Perturbed sector impacts. As the buffer size is increased, the perturbed sector's total health gain goes down, while the system total impact goes up. Error bars are the estimated standard deviation derived from 3 model runs.

The net system impact graph is particularly interesting to consider for policy applications. We find that the reduction in the net system impact is larger than the perturbed sector's reduction in profit. This is not a zero sum game. The system as a whole does better when the price spike is reduced.

The system improvement results from consumption processes becoming less efficient when the model is not at equilibrium. If a large price spike occurs, a large shortage of resources causes agents to substitute goods. Diminishing returns due to the substitution causes agents to operate less efficiently the more the price spikes.

In the real world, the inefficiency can be understood as agents paying premiums for inefficient service. For example, when you pay for a package's overnight delivery, the additional cost is spent on the airplane that rushes the box to your city. If the more efficient ground delivery system delivered your package, it would cost less but take more time. The consumer is paying a premium for time, which could add value if the package is holding up the production line, but if the overnight delivery is unnecessarily due to panic then the premium for express delivery is lost. In the model, we see something similar. The price spike causes consumption and production values to get pushed into inefficient regimes, resulting in value leaving the system.

The system level impact is greatly reduced when storage is implemented. Since the benefit is global, or to the entire system, it makes sense that the cost could be global as well. The policy maker would have an incentive to subsidize storage, since it decreases the net loss of the system. Regulating the model can create a more robust network.

3.3   Limitations

In the real world, additional storage comes at some cost. For some resources most of this cost comes in at the beginning as fixed capital costs. If a 100 barrel tank is built, there is an enormous cost to store the first barrel, but the next 99 are essentially free. Therefore, the cost of storage could be implemented as a step function with rising costs for the capacity of storage. Some resources incur additional marginal storage costs such as refrigeration or security. This could be implemented as a constant cost.

Also, most goods have some decay rate. Milk, grain, and gasoline all go bad over time. This is an additional feature we would like to implement into the model.

The hierarchical structure of this model allows us to generalize the buffer capacity solution to many scales. For example, the Strategic Petroleum Reserve is a buffer tank on a national level, and storage can be just as effective for a mid-level distributor or a consumer. The storage will reduce the price reactivity on whatever scale it is applied to.

4   Conclusion

In this paper, we modeled a group of interacting economic agents using an agent-based model. We showed how a perturbation or a reduction in supply can cause a monopoly to have a net profit due to price increases resulting from resource scarcity. Then, we showed how increasing the input buffer of the agents reduces the monopoly's ability to profit from a perturbation or reduction in supply.

We find that, in the case of a fully-connected network, the total system benefits from increasing the buffer capacity of the agents. Increasing the buffer size decreases the reactivity of prices and prevents the agents from entering inefficient regimes of consumption.

Our model does not consider storage costs or decay, but we would like to study these aspects in the future. There is a strong incentive for agents to price fix in the real world, because large profits can be made. This work suggests actions that can be taken by agents, sectors, or governments to mitigate the manipulation of markets by monopolies.

References

1. Holland, J.H., Miller, J.H.: Artificial Adaptive Agents in Economic Theory. Am. Econ. Rev.: Papers and Proceedings 81, 365–370 (1991)
2. Choi, T.Y., Dooley, K.J., Rungtesanantham, M.: Supply Networks and Complex Adaptive Systems: Control Versus Emergence. J. Oper. Manag. 19, 351–366 (2001)
3. Buzacott, J.A.: Automatic Transfer Lines with Buffer Stocks. Int. J. Prod. Res. 5, 183–200 (1967)
4. Shi, C., Gershqin, S.B.: An Efficient Buffer Design Algorithm for Production Line Profit Maximization. Int. J. Prod. Econ. 122, 725–740 (2009)
5. Cochran, J.K., Kokangul, A., Khaniyev, T.A.: Stochastic approximations for optimal Buffer Capacity of Many-Station Production Lines. Int. J. Math. Oper. Res. 1, 211–227 (2009)
6. Zequeira, R.I., Valdes, J.E., Berenguer, C.: Optimal buffer inventory and opportunistic preventive maintenance under random production capacity availability. Int. J. Prod. Econ. 111, 686–696 (2008)
7. Wycisk, C., McKelvey, B., Hulsmann, M.: "Smart Parts" Supply Networks as Complex Adaptive Systems: Analysis and Implications. Int. J. Phys. Distrib. 38, 108–125 (2008)
8. Cabral, L.M.B., Salant, D.J., Woroch, G.A.: Monopoly Pricing with Network Externalities. Int. J. Ind. Organ. 17, 199–214 (1999)
9. Harrington, J.E.: Optimal Cartel Pricing in the Presence of an Antitrust Authority. Int. Econ. Rev. 46, 145–169 (2005)
10. Kaufmann, R.K., Dees, S., Karadeloglou, P., Sanchez, M.: Does OPEC Matter? An Econometric Analysis of Oil Prices. Energy J. 25, 67–90 (2004)
11. MacFadyen, A.J.: OPEC and cheating: Revisiting the kinked demand curve. Energ. Policy 21, 858–867 (1993)
12. Labaton, S.: De Beers Agrees to Guilty Plea to Re-enter the U.S. Market. The New York Times (July 10, 2004)
13. Bergenstock, D.J., Maskulka, J.M.: The De Beers story. Are diamonds forever? Bus. Horizons 44, 37–44 (2001)
14. Weaver, J.L.: Can Energy Markets be Trusted? The Effect of the Rise and Fall of Enron on Energy Markets. Houston Bus. Tax Law J. 4, 69–72 (2004)
15. White, L.J.: Lysine and Price Fixing: How Long? How Severe? Rev. Ind. Organ. 18, 23–31 (2001)
16. Beyeler, W.E., Glass, R.J., Finley, P.D., Brown, T.J., Norton, M.D., Bauer, M., Mitchell, M., Hobbs, J.A.: Modeling systems of interacting specialists. In: Sayama, H., Minai, A., Braha, D., Bar-Yam, Y. (eds.) Unifying Themes in Complex Systems Volume VIII: Proceedings of the Eighth International Conference on Complex Systems. New England Complex Systems Institute Series on Complexity, pp. 1043–1057. NECSI Knowledge Press, Cambridge (2011)

17. Kuypers, M.A., Beyeler, W.E., Glass, R.J., Antognoli, M., Mitchell, M.D.: The impact of network structure on the perturbation dynamics of a multi-agent economic model. In: Yang, S.J., Greenberg, A.M., Endsley, M. (eds.) SBP 2012. LNCS, vol. 7227, pp. 331–338. Springer, Heidelberg (2012)
18. Terror Attacks Stall Industry; Long-term Impact Unknown, http://wardsautoworld.com/ar/auto_terror_attacks_stall/
19. Martha, J.: Just-in-case operations. Warehousing Forum 17, 1–2 (2002)
20. Lee, S.M., Hancock, M.E.: Disruption in Supply Chain Due to to September 11, 2001. Decision Line 36, 8–11 (2005)

Paid Sick-Leave: Is It a Good Way to Control Epidemics?

Shaojuan Liao[2], Yifei Ma[1,3], Jiangzhuo Chen[1], and Achla Marathe[1,4]

[1] Virginia Bioinformatics Institute, Virginia Tech, Blacksburg VA 24061, USA
[2] Department of Economics, Virginia Tech, Blacksburg VA 24061, USA
[3] Department of Computer Science, Virginia Tech, Blacksburg VA 24061, USA
[4] Department of Agricultural and Applied Economics, Virginia Tech,
Blacksburg VA 24061, USA

Abstract. This research considers an economic intervention i.e. a paid sick leave policy to control an Influenza epidemic. Research has shown that "presenteeism" i.e. sick workers coming to work, costs employers more than "absenteeism" because sick workers put their coworkers at risk and are less productive.

We examined the costs and benefits of a paid sick leave policy through its effect on productivity, medical costs and attack rate. We considered two kinds of workers' behavior: honest and rational. Honest workers take sick leave for days they are sick; but rational workers take all available sick leave. We ran agent-based epidemic simulations on large scale social contact networks with individual behavior modeling to study the coevolution of policy, behavior, and epidemics, as well as their impact on social welfare.

Our experimental results indicate that if the workers behave honestly, the society's economic benefits increase monotonically with the number of paid sick days, however if the workers behave dishonestly but rationally, the society's welfare is maximized when the number of paid sick days is equal to the number of mean days of sickness. This research shows that paid sick leave can be used as an effective policy instrument for controlling epidemics.

Keywords: epidemics, simulation, influenza, public health, economic analysis, social welfare, sensitivity analysis.

1 Introduction

Global disease outbreaks, such as H1N1 and H5N1, have severe morbidity, mortality, and economic consequences. For instance, the World Bank estimated in 2008 that a flu pandemic could cost $3 trillion, affect 70 million people worldwide and decrease the world gross domestic product by 5% [9]. Small and timely interventions can sometimes prevent isolated outbreaks from becoming epidemics or they may hold back the epidemics enough to deploy vaccines to the masses. In the absence of vaccines or antivirals, social distancing may be the only viable measure available in the early period of the epidemics. Previous researchers have studied a variety of intervention strategies, both pharmaceutical and non-pharmaceutical, to control the spread of an infectious disease. These include

K. Glass et al. (Eds.): COMPLEX 2012, LNICST 126, pp. 213–227, 2013.

social distancing strategies such as school closure, work place closure and quarantine [5,13]; and pharmaceutical strategies such as distribution of vaccines and antivirals [8,18,19], as well as herd immunity [1,2].

This paper focuses on an economically driven intervention, i.e. a paid sick leave policy that allows the workers to stay home from work without loss of income. The authors believe that the sick leave policy as a tool to contain epidemics has not been studied in detail in the health care literature, but noteworthy exceptions are [11–13,17]. Work by Gilleskie [11] explores the endogenous decision making of medical care consumption and absenteeism by sick employees in order to understand the behavior that contributes to increasing health care costs. The study of [17] goes one step further and incorporates an epidemiological model to the labor market and its consequent impact on absenteeism. It examines the endogenous determination of the optimal sick leave ratio. In particular, it looks at how sick leave serves to decrease the transmission rates, reduce the spread of disease and increase the social welfare, using a theoretical model. Authors in [12] show that sick leave results in an abrupt decrease in the magnitude of the epidemics. Work by [13] simulates the effectiveness of a set of potentially feasible intervention strategies including the liberal leave policy using three different simulation models developed separately. Simulation results show that the liberal leave policy along with increasing community and workplace social distancing can reduce the disease prevalence significantly. This paper uniqueness lies in the fact that it uses an individual based detailed simulation model to do a parameterized study and considers both epidemiological and economic factors in detail.

Previous researchers have pointed out that presenteeism may be more damaging than absenteeism [14]. From public health viewpoint, it is desirable to reduce the disease *attack rate*, defined as the fraction of the population being infected. A liberal sick leave policy will discourage sick employees from coming to work which will help contain the disease; but it will affect the productivity of the society. To see whether a sick leave policy is indeed an effective tool for controlling the epidemics and whether it is economically efficient, our paper considers a variety of sick leave policies and worker behavior. The analysis takes a social welfare point of view to study the cost effectiveness by comparing the productivity loss of the sick workers with the socio-economic gain caused by a lower attack rate in the population.

A detailed experimental design considers a variety of scenarios based on the number of paid sick days allowed, disease type, the behavior of the workers and the compliance level of the employers with the sick leave policy. The worker behavior is assumed to be of two types: *rational* and *honest*. In case of rational behavior, the workers take the maximum number of sick days available regardless of how long they are sick for, but in case of honest behavior, a sick worker takes off only the number of days s/he is sick for. Honest behavior can also be thought of as a proxy for full information sharing, or symmetric information between the employer and employees; and the rational behavior can be interpreted as partial information sharing or asymmetric information between the employer and employees [17]. Our simulation results show that if employees behave honestly, a liberal sick leave policy would maximize the social benefit but if they behave rationally, the social benefit is maximized when the paid sick days are equal to the mean number of sick days. In the rational case, a liberal sick leave policy reduces the overall social welfare.

The paper is organized as follows: Section 2 explains the disease model, experiment parameters and the methodology. Section 3 describes the simulation results and provides a discussion from both epidemiology and economics point of view. In Section 4 shows results of the sensitivity analysis and the final section concludes the paper.

2 Methodology

2.1 Disease Model

This study assumes that an "Influenza-like-illness" is spreading across a synthetic population representing the city of Miami, Florida, via people-to-people contacts. The simulation is run using EpiFast, a fast agent-based epidemic simulation tool [6]. The disease model, the synthetic population modeling, and the people-to-people contact network model are described in detail in [3,4,7,10]. Our disease model assumes that the probability of transmission depends upon the health states of the individuals and the duration of their simultaneous presence in a small area.

The progression of disease within the host is based on the usual SEIR model and the duration of each state [13, 15]: at any given time, each individual in the population is in one of four health states: *susceptible, exposed, infectious,* or *removed* (SEIR). For each individual, the incubation period duration is sampled from a discrete distribution with mean 1.9 days and standard deviation 0.49 day; the infectious period duration is sampled from a discrete distribution with mean 4.1 days and standard deviation 0.89 day.

We assume that only 66.7% are symptomatically sick and of those only two-thirds are correctly diagnosed [13]. Only symptomatic people go to see the doctor and encounter medical costs. The asymptomatic individuals behave as healthy individuals but they can transmit the disease, although they are only half as infectious as the symptomatic ones. The epidemic is seeded with five randomly chosen individuals. Every day five new infections from external sources occur within the population in addition to those generated by transmission. The simulation is run for 300 days. Reported results are based on an average of 25 simulation replicates.

2.2 Factorial Experiment Design

We consider the following 5 factors in our experimental design: disease type, compliance level, maximum number of sick days allowed, workers' behavior and the productivity level of workers who work while they are sick. The disease types are moderate flu or catastrophic flu. The compliance levels of the employers/workplace are set at 50% and 100%, which refers to the fraction of work

Table 1. Factorial Design

Factor	Description	Values
Dis	disease types: catastrophic and moderate flu.	Cat, Moderate
Comp	compliance: probability each workplace complies with the sick leave policy	50%, 100%
$D_{\max}$	sick leave policy: max number of sick days allowed to diagnosed workers	3, 4, 5, 6
Beh	workers' behavior towards the sick leave policy: take the exact sick days off (honest) or take the maximum possible days off (rational)	rational, honest
e	productivity level of those working while sick	20%, 50%, 80%

locations that comply with the sick leave policy. In the 50% compliance case, only 50% of the work locations choose to comply and provide paid sick leave to its employees. In the 100% case, all work locations and hence all workers are given paid sick leave. However note that only diagnosed workers are allowed to take sick leave.

Regarding the number of maximum paid sick days, workers can take sick leave up to $D_{\max} = 3$, 4, 5, or 6 days without any income loss. Workers' behavior is considered to be of 2 types, rational and honest; if rational, workers take all available sick leave so the actual number of sick leave days taken is $D_{sl} = D_{\max}$. In the honest case, eligible workers take off only when they are sick: $D_{sl} = \min(D_{sick}, D_{\max})$, where D_{sick} is the actual number of sick days. All of the experimental factors are summarized in Table 1.

2.3 Interventions

2.4 Cost and Benefit Estimates

The procedure below describes the methodology used in estimating the economic costs and benefits of a paid sick leave policy. The costs include medical costs and loss in productivity of the sick workers. Benefits include lower attack rate and hence gain in productivity. Information used to calculate the costs and benefits include workers' income, medical costs for treating sick workers, health status, number of sick days, number of paid sick leave days used, and the age of workers. Workers' productivity is calculated based on their income in the following manner: $y_i = 1.154 I_i$ where y_i is the daily productivity of worker i and I_i is the income generated by worker i. We use a multiplier of 1.154 which reflects the ratio of productivity to income in the US i.e. Gross National Product (GNP) to Gross National Income (GNI) is $GNP/GNI = 1.154$.[1]

Let's assume h_i represents the health status of worker $i, i \in [1..n]$ and h_i is a binary variable that takes value 1 if i is symptomatic and 0 if i is healthy

---

[1] The ratio is calculated based on the US indices for year 2010. Data source: The World Bank.

or asymptomatic. Let $\epsilon_i = 1$ represent that i is diagnosed and 0 otherwise. We assume that the healthy and asymptomatic workers work at their full productivity level but if a worker is symptomatic and working, the productivity is $e < 1$. In our experiment design different parameter values are considered for e: $e = 0.2, 0.5, 0.8$.

Let med represent the medical costs of treatment per person. For ages between 0-19, the average medical cost for treating flu is \$249 per person; for 20-64, \$400.7 per person; and for 65 and above, it is \$415 per person. These numbers are based on estimates given in [16]. We assume that the asymptomatic people do not have medical expenditure.

$$med = (249 \times N_{0-19}) + (400.7 \times N_{20-64}) + (415 \times N_{65+}) \tag{1}$$

where N_x is the number of symptomatic people within x age bin.

The simulation experiments are done for a total of $T = 300$ days for the city of Miami, Florida. Equation 2 shows the for worker i for the honest case i.e. $prod^H$, when the sick leave policy is in effect, while equation 3 shows the same for the rational case ($prod^R$). $D_{sick,i}$ is the number of sick days for worker i and $D_{\max}$ is the maximum number of sick-leave days allowed by the policy. The first line of both equations shows the productivity of the healthy and asymptomatic workers as represented by $h_i = 0$. The second line shows the productivity of the workers who are symptomatic ($h_i = 1$) but undiagnosed ($\epsilon_i = 0$). The third line represents the productivity of workers who are symptomatic ($h_i = 1$) and diagnosed ($\epsilon_i = 1$).

$$prod_i^H = (1 - h_i) \sum_{t=1}^{T} y_{it} \tag{2}$$

$$+ h_i(1 - \epsilon_i) \left[\sum_{t=1}^{T-D_{sick,i}} (y_{it}) + \sum_{t=1}^{D_{sick,i}} (y_{it})e \right]$$

$$+ h_i \epsilon_i \left[\sum_{t=1}^{T-D_{sick,i}} (y_{it}) + \sum_{t=1}^{\max(D_{sick,i}-D_{\max},0)} (y_{it})e \right]$$

$$prod_i^R = (1 - h_i) \sum_{t=1}^{T} y_{it} \tag{3}$$

$$+ h_i(1 - \epsilon_i) \left[\sum_{t=1}^{T-D_{sick,i}} (y_{it}) + \sum_{t=1}^{D_{sick,i}} (y_{it})e \right]$$

$$+ h_i \epsilon_i \left[\sum_{t=1}^{T-D_{\max}} (y_{it}) - \sum_{t=1}^{\max(D_{sick,i}-D_{\max},0)} (y_{it})(1 - e) \right]$$

The loss in productivity for the above scenario is calculated by taking the difference between the productivity when there is no sickness in the society and the productivity as calculated in equation 2 or 3 for the honest or rational case respectively. Finally,

$$\text{Net Social Benefit} = \text{Productivity} - \text{Medical Costs} \qquad (4)$$

3   Results

3.1   Effect of the Sick Leave Policy on Epidemics

The epidemic curves derived from our simulation results are displayed in Figure 1 for the catastrophic flu and Figure 2 for moderate flu. In both figures, we show the epidemic curve for the base case (i.e. no sick leave at all) and the curves for D_{max} values set at 3 days and 6 days only. The epidemic curves for $D_{max} = 4$ and $D_{max} = 5$ are very close to the 3-day and 6-day cases and are hence omitted to avoid clutter. Table 2 shows the total attack rate, peak day and peak size for all values of D_{max} and compliance levels.

It is important to note that the epidemics do not change across honest and rational cases, because in the SEIR model a recovered individual does not transmit the disease to others and cannot be infected by others either. The honest case and the rational case differ only when a sick worker recovers before D_{max} is reached: an honest worker will then go back to work while a rational worker will remain off. But it does not matter any more with respect to the disease spread because this worker is already recovered. Therefore the epidemic curves are the same under these two cases and we show the curves only once.

Figure 1 and Table 2 clearly show that the sick leave policy has a significant effect on the epidemic dynamics in catastrophic case. At 50% workplace compliance, the sick leave policy of 5 days can reduce the peak size from 19,000 to 16,000, i.e. by 15%. It can be reduced by another 15% to 13,600 if the compliance goes up to 100%. The effect is more prominent than it appears because

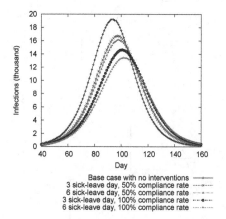

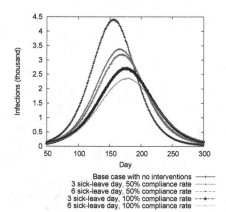

Fig. 1. Epidemic curves under the catastrophic flu case

Fig. 2. Epidemic curves under the moderate flu case

Table 2. This table shows the percentage of people infected (attack rate), peak infection day (peak day), and the maximum number of infections on one day (peak size) in the catastrophic and moderate flu cases along with the base case (no intervention)

scenario	catastrophic flu			moderate flu		
	attack rate	peak day	peak size	attack rate	peak day	peak size
base case	40.0%	93	19,000	20.0%	155	4,400
Comp=50%, D_{max}=3	37.3%	98	16,700	17.4%	165	3,400
Comp=50%, D_{max}=4	36.5%	98	16,200	16.9%	169	3,200
Comp=50%, D_{max}=5	36.2%	99	16,000	16.7%	168	3,200
Comp=50%, D_{max}=6	36.1%	98	16,000	16.7%	170	3,200
Comp=100%, D_{max}=3	34.5%	101	14,600	15.1%	177	2,700
Comp=100%, D_{max}=4	33.1%	101	13,800	14.3%	177	2,500
Comp=100%, D_{max}=5	32.5%	101	13,600	13.9%	180	2,400
Comp=100%, D_{max}=6	32.4%	102	13,400	13.8%	179	2,400

the targeted intervened people only account for at most 6% (in case of 100% compliance rate) or 3% (in case of 50% compliance rate) of the total population.[2] The results show that the higher the compliance, the lower is the overall attack rate and changes to the maximum number of sick days by even a single day can cause statistically significant change to the attack rate and the medical costs. See Section 4.1 for details. A longer sick leave results in lower attack rates which is not surprising but the marginal effect due to an extra day of sick leave is nonlinear. The sick leave policy can help postpone the peak day by 5-8 days depending upon the compliance rate.

Similar pattern is found in Figure 2 and Table 2 for the moderate case. The policy has even more significant effects. It reduces the peak size by 27% compared to the base case (from 4400 to 3200) when compliance is at 50%. At 100% compliance the peak size is reduced by another 25%. The peak day can be delayed by 15 days at 50% compliance and another 10 days at 100% compliance.

For both catastrophic flu and moderate flu, the marginal effect of additional one day sick leave is decreasing. When D_{max} changes from 3 to 4, the effect on attack rate reduction is fairly large but as we increase D_{max} to 5 and 6 days, the marginal effect of additional sick days on attack rate starts to decrease. This can be explained by the fact that the mean infectious period is 4.1 days (see Section 2.1) and hence most people stay infectious for 4 days; there are relatively fewer people who stay sick for 5 or 6 days. Therefore the greater number of sick leave days are not needed for a large proportion of the population and the marginal improvement in the attack rate from the additional days drops.

3.2 Economic Benefit and Loss

In this section, we compare the economic gains and losses from the sick leave policy. Our earlier analysis shows that the sick leave policy can significantly

---

[2] This is because only one third of the population is workers. In addition, the attack rate is at most 40%, the symptomatic rate is 2/3, and diagnosed rate is 2/3 which makes the targeted people only a small fraction of the society.

reduce the epidemic, but it is important to understand the cost as a result of the policy. Sick leave policy gives opportunity to not only the sick workers to stay home but also to recovered workers if they behave dishonestly. This results in big loss in productivity which may not necessarily be offset by the benefit of smaller attack rate and lower medical bills. We take a societal point of view to see if there is an overall net benefit to the society from the sick leave policy.

We split the scenarios in 4 cases and describe the results of each case in the following subsections. Case 1 assumes people's behavior to be "honest" and flu to be of type "catastrophic". Case 2 assumes the behavior to be "honest" and flu to be of type "moderate". Similarly, case 3 and case 4 assume behavior to be "rational" and flu to be "catastrophic" and "moderate" respectively.

Case 1: Honest and Catastrophic. This section calculates the net benefit to the society when the workers' behavior is honest and the flu is of catastrophic type. We measure change in attack rate, change in productivity (at different levels of e) and change in the medical costs and compare it with the base case of no sick leave. The results are shown in Table 3.

Results in Table 3 show that higher compliance rate and more sick leave lead to lower attack rate and higher net social benefit to the society. As $D_{\max}$ increases, the diagnosed workers take more days off which changes the social contact network and the probabilities for disease transmissions. The disease spread slows down which is reflected in the lower attack rate compared to the base case.

However, the marginal effect of each additional sick day is decreasing since most people are sick for 4 days according to the disease model. When $D_{\max} = 3$, not enough sick leave is being given to cover the period of sickness so when it increases from 3 to 4, there is a much higher drop in attack rate than when $D_{\max}$ increases from 4 to 5 or from 5 to 6. The change in productivity is greatly affected by the parameter e which represents the level of productivity of sick

Table 3. Case 1: behavior is "honest" and flu type is "catastrophic". Comp shows the compliance rate, D_{max} represents the maximum number of sick days allowed, ΔAttack rate, ΔProductivity, ΔMedical and ΔNet Benefit represent the change in attack rate, change in productivity, change in the medical costs and change in net social benefits respectively as compared to the base case. The variable e represents the level of productivity of people who are working while sick.

Comp	D_{max}	ΔAttack Rate	ΔProductivity(millions) $e=0.2$	$e=0.5$	$e=0.8$	ΔMedical (millions)	ΔNet Benefit(millions) $e=0.2$	$e=0.5$	$e=0.8$
0.5	3	-0.029	10.007	-0.362	-10.731	-15.450	25.457	15.088	4.719
	4	-0.036	12.928	0.650	-11.628	-19.179	32.107	19.829	7.551
	5	-0.039	14.025	1.072	-11.881	-20.675	34.700	21.747	8.794
	6	-0.040	14.369	1.188	-11.993	-21.140	35.509	22.328	9.147
1	3	-0.056	19.865	1.677	-16.511	-29.344	49.209	31.021	12.833
	4	-0.069	25.298	4.241	-16.815	-36.189	61.487	40.431	19.374
	5	-0.075	27.543	5.415	-16.714	-39.147	66.690	44.562	22.434
	6	-0.076	28.310	5.863	-16.584	-39.902	68.212	45.765	23.318

workers. If e is small, there are fewer gains to be had from keeping sick workers at work who are likely to spread the disease by just being present.

Hence at low levels of e, the gain in productivity caused by the lower attack rate outweighs the loss in productivity caused by the sick leave policy. However, when the productivity of sick workers jumps to $e = 0.8$, the loss in productivity outweighs the gain, resulting in a net drop in productivity. Intuitively, this makes sense since at $e = 0.8$ the sick workers are 80% as productive as healthy workers so giving them sick time off will result in a big productivity loss. The change in medical bills column shows that as the attack rate goes down, the medical costs go down too.

The net benefit column in Table 3 accounts for both the productivity and the medical costs. As D_{max} increases the productivity increases, the medical costs go down, and the net benefits go up. The last three columns show that the optimal policy is $D_{\mathrm{max}} = 6$ because no matter what compliance rate is and what e is, the net benefit is the highest. Such a result should be expected given the assumption that workers behave honestly and take exactly the required number of sick days. Giving more sick days off will only result in the sick workers taking the extra sick days which will keep them out of the social network and hence keep them from transmitting the disease. This policy will lower the attack rate and yet not sacrifice productivity (because of no dishonest workers).

Case 2: Honest and Moderate. The results under this scenario, as shown in Table 4, are similar to case 1 in terms of the trend of the variables but the magnitude of the numbers is smaller for moderate flu as compared to the catastrophic flu. We still observe that $D_{\mathrm{max}} = 6$ is the optimal policy. The net benefit increases with the increase in the number of sick days. Compared to the catastrophic case, the policies are even more efficient when the productivity parameter $e = 0.8$ i.e. the loss in productivity is much lower. From both these cases, we can conclude that as long as workers are honest, a liberal sick leave policy is the best, no matter what the parameter settings are.

Table 4. Case 2: behavior is "honest" and flu type is "moderate"

Comp	D_{max}	ΔAttack Rate	ΔProductivity(millions) e=0.2 e=0.5	e=0.8	ΔMedical (millions)	ΔNet Benefit(millions) e=0.2 e=0.5	e=0.8
0.5	3	-0.027	8.657 2.891	-2.874	-13.348	22.005 16.239	10.474
	4	-0.032	10.221 3.624	-2.974	-15.738	25.959 19.361	12.764
	5	-0.033	10.797 3.902	-2.993	-16.599	27.397 20.502	13.607
	6	-0.034	10.984 3.979	-3.026	-16.855	27.839 20.834	13.830
1	3	-0.049	15.507 5.984	-3.540	-24.108	39.616 30.092	20.568
	4	-0.057	17.853 7.264	-3.326	-27.849	45.703 35.113	24.523
	5	-0.060	18.811 7.849	-3.114	-29.493	48.304 37.342	26.379
	6	-0.061	19.068 7.992	-3.084	-29.904	48.972 37.896	26.820

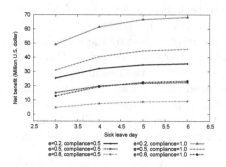
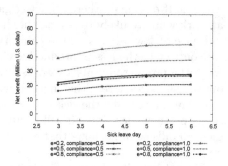

Fig. 3. Change in net benefit with honest behavior and catastrophic flu

Fig. 4. Change in net benefit with honest behavior and moderate flu

Figure 3 and 4 show the net benefits for different policies for case 1 and case 2 respectively. The net benefit is monotonically increasing as D_{max} increases and all cases result in positive net benefit, suggesting that all intervention strategies are economically efficient.

Case 3: Rational and Catastrophic. Next we consider the scenario where the workers behave rationally and the flu type is catastrophic. In the rational case, the eligible workers use up all the available sick leave days.

Comparing the productivity in Table 5 with the productivity in the honest case in Table 3 shows that rational behavior results in lower productivity at all levels of e compared to the honest case. If workers stay home longer than they are sick for, it causes a pure loss to the society because there is no gain due to less disease transmissions. As a result the change in net benefit is smaller too. In this case, the optimal policy changes to $D_{max} = 4$. The effect of D_{max} on net social benefit becomes non monotonic. As D_{max} increases from 3 to 4, the net benefit improves but when D_{max} increases from 4 to 5, the net benefit decreases. This is again because the mean sick days in our model is 4.1 so most of the people are sick for 4 days. If more than 4 days of sick leave is given, the loss in productivity outweighs any gains from lower attack rate. The optimal policy, $D_{max} = 6$, under the honest case now becomes the least favorable especially when e is high.

Case 4: Rational and Moderate. Finally, results for case 4 where rational behavior is combined with moderate flu are presented in Table 6. Here the optimal policy is $D_{max} = 5$ when $e = 0.2$, and $D_{max} = 4$ when $e = 0.8$. Note that when $e = 0.2$, the productivity loss from the extra day off (i.e. $D_{max} = 4$ vs. 5) by the sick workers is very little but their presence in the workforce increases the attack rate and the medical costs, which makes it socially optimal to have D_{max} set at 5. Increasing it further i.e. to $D_{max} = 6$ does not help as much because the mean sick days are 4.1. However when $e = 0.8$ the productivity loss is high from the extra day off and hence the optimal sick leave policy changes to $D_{max} = 4$.

Table 5. Case 3: behavior is "rational" and flu type is "catastrophic"

Comp	D_{max}	ΔAttack Rate	ΔProductivity(millions) e=0.2	e=0.5	e=0.8	ΔMedical (millions)	ΔNet Benefit(millions) e=0.2	e=0.5	e=0.8
0.5	3	-0.029	10.007	-0.362	-10.731	-15.450	25.457	15.088	4.719
	4	-0.036	11.323	-0.955	-13.233	-19.179	30.502	18.224	5.946
	5	-0.039	8.900	-4.053	-17.006	-20.675	29.576	16.623	3.670
	6	-0.040	4.742	-8.439	-21.620	-21.140	25.882	12.701	-0.480
1	3	-0.056	19.865	1.677	-16.511	-29.344	49.209	31.021	12.833
	4	-0.069	22.794	1.737	-19.319	-36.189	58.984	37.927	16.870
	5	-0.075	19.664	-2.465	-24.593	-39.147	58.811	36.683	14.555
	6	-0.076	13.713	-8.734	-31.182	-39.902	53.615	31.168	8.720

Figures 5 and 6 show the change in net social benefit when the behavior is rational and flu is catastrophic and moderate respectively. The net benefit is no longer monotonically increasing with the sick leave days.

4 Sensitivity Analysis

4.1 Epidemiological Variables

This section performs a detailed sensitivity analysis of the compliance rate, disease type and the maximum number of sick leave available on the two response variables i.e. the attack rate and medical costs using *analysis of variance* (ANOVA). Given that the human behavior and productivity level (e) can only affect the economic variables but not the epidemiological variables, we consider them separately. There are three factors which affect the attack rate and the medical costs: maximum sick leave days ($D_{\max}$), compliance rate (C) and disease type (Di). The 3-factor ANOVA model below shows how the factors are related to the response variable:

$$y_{ijsk} = \alpha + \beta_j + \gamma_s + \delta_k + \beta\gamma_{js} + \gamma\delta_{sk} + \beta\delta_{jk} + \beta\gamma\delta_{jsk} + \epsilon_{ijsk} \qquad (5)$$

Table 6. Case 4: behavior is "rational" and flu type is "moderate"

Comp	D_{max}	ΔAttack Rate	ΔProductivity(millions) e=0.2	e=0.5	e=0.8	ΔMedical (millions)	ΔNet Benefit(millions) e=0.2	e=0.5	e=0.8
0.5	3	-0.027	8.207	2.598	-2.874	-13.348	22.005	16.239	10.474
	4	-0.032	9.172	2.731	-3.573	-15.738	24.910	18.469	12.165
	5	-0.033	8.452	1.713	-4.889	-16.599	25.051	18.312	11.710
	6	-0.034	6.984	0.136	-6.576	-16.855	23.840	16.991	10.279
1	3	-0.049	15.058	5.691	-3.540	-24.108	39.616	30.092	20.568
	4	-0.057	16.561	6.127	-4.169	-27.849	44.411	33.977	23.680
	5	-0.060	15.760	4.954	-5.716	-29.493	45.253	34.447	23.777
	6	-0.061	13.779	2.859	-7.923	-29.904	43.684	32.764	21.981

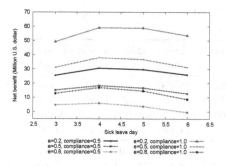

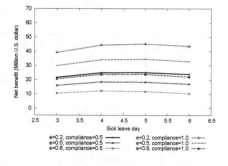

Fig. 5. Change in net benefit with rational behavior and catastrophic flu

Fig. 6. Change in net benefit with rational behavior and moderate flu

where y represents the response variable i.e. attack rate and medical costs; α is the constant term, the main effects or factors are $\beta = D_{\max}$, $\gamma = C$, $\delta = Di$; ϵ is the error term and the rest are interaction terms. Subscript i represents replicates and takes values $1 \ldots 25$; $j = 3,4,5,6$; $k = 50\%, 100\%$; and $s =$ catastrophic, moderate flu.

The sensitivity results as measured by ANOVA show that all three factors significantly affect the attack rate and the medical costs (p-value $< 1\%$). Both response variables are sensitive to the choice of policy days, disease type and the compliance rate. All interaction terms are significant too (p-value $< 1\%$).

In particular, we are interested in understanding if different number of sick leave days make a significant difference in the outcome variables. For this, we conduct *Tukey's honestly significant difference* (HSD) test to do a pairwise comparison. It considers all possible pairs of means and finds the ones that are significantly different. It is often used in conjunction with ANOVA. Suppose μ_i and μ_j are the means of two different treatments, and $\mu_i > \mu_j$, then Tukey's test statistic q_s is:

$$q_s = \frac{\mu_i - \mu_j}{SE} \tag{6}$$

where SE is the standard error of the data in question. The results of Tukey's test of different sick days ($D_{\max}$) are shown in Table 7 in the Appendix. All pairwise comparisons are significant, so 3, 4, 5, 6 days are all significantly different from each other in terms of their effects on the attack rate and the medical cost.

4.2   Economic Variables

Next, we analyze the sensitivity of the economic variables such as productivity and net social benefit. Besides the epidemiological factors, the economic variables are also affected by workers' behavior and their productivity levels during sick days. Hence we have a total of five factors i.e. $D_{\max}$, compliance (C), disease type (Di), behavior (B), productivity level of sick workers (e). Now the ANOVA model represents 5 factors:

$$y_{ijsklm} = \alpha + \beta_j + \gamma_s + \delta_k + \zeta_l + \lambda_m + I + \epsilon_{ijsklm} \tag{7}$$

where y represents the response variable i.e. productivity and net benefit. Note that this productivity is the overall productivity of the society as opposed to e which represents the productivity level of sick workers. $\beta = D_{max}$, $\gamma = C$, $\delta = Di$, $\zeta = B$, $\lambda = e$, I represents all the interaction terms and ϵ is the error term. The sensitivity results as measured by ANOVA show that all five factors are significant (p-value $< 1\%$) in explaining productivity and the net benefit. Most interaction terms are significant too (p-value $< 1\%$). Due to space limit, in Table 8 we only show the interaction terms that are not significant at 1% level.

The results of Tukey's test (omitted due to space limit) show that $D_{max} = 3, 4, 5$, or 6 days are all significantly different from each other in terms of their effects on the productivity and net benefit (p-value $< 10\%$). Specifically, although $D_{max} = 5$ and $D_{max} = 6$ yield very similar net benefit in the honest case, and $D_{max} = 4$ and $D_{max} = 5$ have similar net benefit in the rational case, they are still statistically significantly different.

5   Summary and Conclusions

This research aims to study the role of an economic intervention i.e. a paid sick leave policy, as an instrument, to effectively control an influenza epidemic. A liberal paid sick leave policy discourages sick workers from coming to work which reduces the transmission of the disease among the workers and hence the society but affects the productivity of the society. The analysis takes a social planner's point of view to study the cost effectiveness of such a policy. We consider productivity loss due to the sick leave policy and compare it with the benefits from reduced attack rate and lower medical bills. Our experiments test a variety of scenarios based on the number of paid sick days and behavior of the workers. The number of maximum sick days considered are 3, 4, 5 and 6. The worker behavior is assumed to be of two types: rational or honest. In case of rational behavior, the workers take the maximum number of sick days available but in case of honest behavior, only the necessary number of sick days are taken. The simulation results show that if workers behave honestly, a liberal sick leave policy is the most optimal since the social gains increase with the increase in the number of paid sick days. If the workers behave rationally and take the maximum available sick days, however, it is optimal to have paid sick days to be equal to the mean number of sick days.

Acknowledgement. We thank members of the Network Dynamics and Simulation Science Laboratory for their suggestions and comments. This work has been partially supported by DTRA Grant HDTRA1-11-1-0016, DTRA CN-IMS Contract HDTRA1-11-D-0016-0001, NIH MIDAS Grant 2U01GM070694-09, NIH MIDAS Grant 3U01FM070694-09S1, NSF ICES Grant CCF-1216000, NSF NetSE Grant CNS-1011769. The content is solely the responsibility of the authors and does not necessarily represent the official views of the NIH, NSF and DoD DTRA.

References

1. Anderson, R., May, R.: Immunisation and herd immunity. Lancet 335(8690), 641–645 (1990)
2. Anderson, R., May, R., et al.: Vaccination and herd immunity to infectious diseases. Nature 318(6044), 323–329 (1985)
3. Barrett, C., Bisset, K., Eubank, S., Feng, X., Marathe, M.: Episimdemics: an efficient algorithm for simulating the spread of infectious disease over large realistic social networks. In: Supercomputing Conference, pp. 1–12. ACM/IEEE (2008)
4. Barrett, C., Bisset, K., Leidig, J., Marathe, A., Marathe, M.: An integrated modeling environment to study the co-evolution of networks, individual behavior, and epidemics. AI Magazine 31(1), 75–87 (2009)
5. Barrett, C., Bisset, K., Leidig, J., Marathe, A., Marathe, M.: Economic and social impact of influenza mitigation strategies by demographic class. Epidemics 3(1), 19–31 (2011)
6. Bisset, K., Chen, J., Feng, X., Kumar, V.A., Marathe, M.: EpiFast: a fast algorithm for large scale realistic epidemic simulations on distributed memory systems. In: Proceedings of the 23rd International Conference on Supercomputing (ICS), pp. 430–439 (2009)
7. Bisset, K., Marathe, M.: A cyber-environment to support pandemic planning and response. In: DOE SciDAC Magazine, pp. 36–47 (2009)
8. Burke, D., Epstein, J., Cummings, D., Parker, J., Cline, K., Singa, R., Chakravarty, S.: Individual-based computational modeling of smallpox epidemic control strategies. Academic Emergency Medicine 13(11), 1142–1149 (2006)
9. Burns, A., Mensbrugghe, D., Timmer, H.: Evaluating the economic consequences of avian influenza (2008)
10. Eubank, S., Guclu, H., Kumar, V.S.A., Marathe, M.V., Srinivasan, A., Toroczkai, Z., Wang, N.: Modelling disease outbreaks in realistic urban social networks. Nature 429, 180–184 (2004)
11. Gilleskie, D.: A dynamic stochastic model of medical care use and work absence. Econometrica, 1–45 (1998)
12. Grabowski, A., Rosinska, M.: The SIS model for assessment of epidemic control in a social network. Acta Physica Polonica Series B 37(5), 1521 (2006)
13. Halloran, E., Ferguson, N.M., Eubank, S., Ira, J., Longini Jr., I.M., Cummings, D.A.T., Lewis, B., Xu, S., Fraser, C., Vullikanti, A., Germann, T.C., Wagener, D., Beckman, R., Kadau, K., Barrett, C., Macken, C.A., Burke, D.S., Cooley, P.: Modeling targeted layered containment of an influenza pandemic in the United States. Proceedings of the National Academy of Sciences 105, 4639–4644 (2008)
14. Hemp, P.: Presenteeism: At work - but out of it. Harvard Business Review (October 2004)
15. Marathe, A., Lewis, B., Barrett, C., Chen, J., Marathe, M., Eubank, S., Ma, Y.: Comparing effectiveness of top-down and bottom-up strategies in containing influenza. PLoS ONE 6(9), e25149 (2011)
16. Meltzer, M., Cox, N., Fukuda, K.: The economic impact of pandemic influenza in the United States: Priorities for intervention. Emerg. Infect. Dis. 5(5), 659–671 (1999)
17. Skåtun, J.: Take some days off, why don't you? Endogenous sick leave and pay. Journal of Health Economics 22(3), 379–402 (2003)

18. Taylor, C., Marathe, A., Beckman, R.: Same influenza vaccination strategies but different outcome across U.S. cities? International Journal of Infectious Diseases 14(9), e792–e795 (2010)
19. Whitley, R., Monto, A.: Seasonal and pandemic influenza preparedness: a global threat. Journal of Infectious Diseases 194(suppl. 2), S65–S69 (2006)

Appendix: Tables for Sensitivity Analysis

Table 7. Pairwise comparison of factor D_{max}. * represents significance at 10%.

grp vs grp	attack rate				medical cost			
	group means		mean diff	HSD-test	group means		mean diff	HSD-test
3 vs 4	-4.03	-4.84	0.81	61.21*	-20.56	-26.59	6.02	94.98*
3 vs 5	-4.03	-5.18	1.16	87.04*	-20.56	-25.31	4.75	74.91*
3 vs 6	-4.03	-5.28	1.25	94.08*	-20.56	-26.95	6.38	100.62*
4 vs 5	-4.84	-5.18	0.34	25.83*	-26.59	-25.31	1.27	20.07*
4 vs 6	-4.84	-5.28	0.43	32.87*	-26.59	-26.95	0.35	5.64*
5 vs 6	-5.18	-5.28	0.09	7.04*	-25.31	-26.95	1.63	25.71*

Table 8. Sensitivity analysis of economic variables. Only interaction terms not significant at 1% level are shown. All factors are significant at 1% level.

Factor	Productivity			Net Benefit		
	DF	SS	F value	DF	SS	F value
B:e	2	3.58E-28	2.23E-27	2	7.25E-28	5.24E-28
D_{max}:B:e	6	4.28E-28	8.89E-28	6	3.43E-28	8.26E-29
C:B:e	2	1.67E-28	1.04E-27	2	1.64E-28	1.19E-28
Di:B:e	2	5.42E-29	3.37E-28	2	4.68E-28	3.38E-28
D_{max}:C:B:e	6	1.31E-27	2.72E-27	6	1.46E-27	3.51E-28
D_{max}:Dis:B:e	6	8.67E-28	1.80E-27	6	7.22E-28	1.74E-27
C:Di:B:e	2	1.19E-27	7.43E-27	2	6.17E-28	4.46E-28
D_{max}:C:Di:B:e	6	4.10E-28	8.52E-28	6	1.56E-27	3.75E-28

Decoding Road Networks into Ancient Routes: The Case of the Aztec Empire in Mexico

Igor Lugo[1] and Carlos Gershenson[2]

[1] Centro Regional de Investigaciones Multidisciplinarias,
Universidad Nacional Autónoma de México,
Av. Universidad s/n, Cto. 2, Col. Chamilpa, CP 62210, Cuernavaca, Morelos, México
`igorlugo@correo.crim.unam.mx`
[2] Instituto de Investigaciones en Matemáticas Aplicadas y en Sistemas,
Universidad Nacional Autónoma de México,
A.P. 20-726, 01000 México, D.F., México
`cgg@unam.mx`

Abstract. Historical evidence in some regions of Latin America has suggested that the system of ancient routes between places could have determined the success or collapse of prehistoric societies. The identification of such routes provided essential information to understand initial conditions in the evolution of the actual road infrastructure. Looking into the increasing technology applied to generate and process geospatial information, we proposed a retrospective spatial analysis for discovering a large-scale network of ancient routes before the conquest of Aztecs by the Spanish around 1520 CE. Such a method consisted in analyzing existing road networks (highways) that connect a system of cities (continuously built-up areas) to deduce routes by using geoprocessing methods, network analysis, and historical evidence. The results of this research support the idea that the retrospective method may be applied to other cases to decipher and to understand initial conditions in the evolution of road infrastructures by combining different types of data and scientific fields.

Keywords: Road networks, ancient routes, Aztec Empire, geoprocessing methods, complex network measures, historical evidence.

1 Introduction

Through time, road networks have been one of the most important physical evidence of social organization and adaptation in heterogeneous landscapes. In particular, historical evidence in some regions of Latin America has suggested that the system of ancient routes between places could have determined the success or collapse of prehistoric societies affecting their evolutionary trajectory. Therefore, the identification of such routes can provide not only essential information to understand early levels of politic, ritual, and economic interactions, but also initial conditions in the evolution of the actual road infrastructure.

K. Glass et al. (Eds.): COMPLEX 2012, LNICST 126, pp. 228–233, 2013.

To identify the network of ancient routes, formal and informal constructions, in a region is not trivial. Historical cartography, aerial photography, and remote sensing techniques have given important contributions to detect and analyze ancient roads [1], but they are limited by the scale and precision of geospatial data. Nowadays, looking into the increasing technology applied to generate and process geospatial information, we consider an alternative method for discovering a large-scale network of ancient routes. It consists in analyzing existing road networks (highways) that connects a system of cities (continuously built-up areas) to deduce such routes by using geoprocessing methods, network analysis, and historical evidence. We called this method a retrospective spatial analysis.

Based on the evidence of the relationship and stability of some cities and roads throughout time [1,2,3], the retrospective method provides us the bases for identifying ancient routes before the conquest of Aztecs by the Spanish around 1520 CE.

2   Data

The geospatial database consisted of vector layers related to highway lines and urban polygons at Mexico in the year of 2005 [4]. Historical evidence about the first layer corresponded to the most representative cartographic information and scientific analyses about ancient routes in the Aztec Empire. This data was divided into three types of scales: a) the city of Tenochtitlan [5,6,7], b) the Basing of Mexico [8], and c) large-scale network of routes [9]. On the other hand, ancient information about the second layer was related to the location of Aztecs settlements around 1520 [10,11].

3   Retrospective Spatial Analysis

Following the work of Burghardt [2] about the relation between road and city networks and the degeneration model proposed by Xie and Levinson [12,13], we defined the retrospective spatial analysis as the process for decoding existing road networks into ancient routes based on geospatial and historical information, where the actual geographic data of cities and roads was modeled by geoprocessing methods, analyzed by complex network measures, and compared to historical evidence (Fig. 1).

3.1   Geoprocessing Methods

Combining the large number of software and programming libraries to process and analyze spatial objects, for example the Geographic Information System (GIS) and third-party libraries of Python respectively, we integrated and modeled different types of spatial information. Urban polygons and road lines were mixed together in order to generate a planar graph, where the information of polygons was attached to superposed line segments by adding the polygon identification code to line segment points. The resulted graph was a modified version

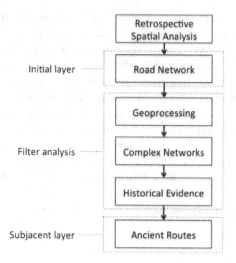

Fig. 1. Method

of the road network with identical topology and additional information in some lines segments. The process to create such a graph was divided into three steps: identifying lines inside polygons, transforming lines segments to individual lines, and defining nodes attributes.

3.2   Complex Network Measures

After modeling a planar spatial network, we used four types of complex network measures to approximate the identification of ancient routes. The first measure was the weighted average nearest-neighbors degree that identified a representative node into a set of nodes with similar polygon identification code. Such a node corresponded to the effective level of cohesiveness and affinity because it is connected with high- or low-degree neighbors [14]. The second was the shortest path defined as the minimum number of edges between a pair of nodes [15]. The third was the circuity measure, which is the ratio of network to Euclidian distance [15,16]. The fourth was the node accessibility that quantified different levels of road infrastructures suggesting efficient ways of how people traveling to their destination in a straight-line subject to other infrastructures and natural barriers.

3.3   Historical Evidence

The first scale of historical evidence concerned with the local organization in the city of Tenochtitlan. In particular, we analyzed two cartographic data: the Nuremberg map of 1524 and its artistic representation. The second scale was a network analysis of the system of roads in the Basing of Mexico. The third scale corresponded to the study of long-distance trade routes in eastern Mexico before the Spanish conquest. In addition, we inspected the cartographic data of the ranking of Aztecs settlements in 1520 to identity their location.

4   Ancient Routes in the Aztec Empire

After applied the retrospective method, results suggested that some highway sections are good approximations to the network of ancient routes in the Aztec Empire (Fig. 2).

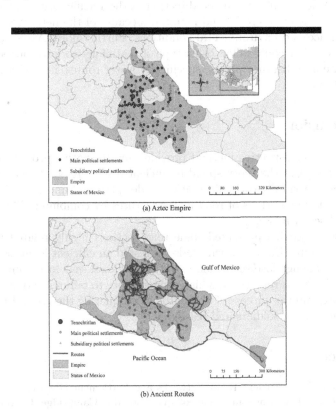

Fig. 2. The Aztec Empire and its Ancient Routes in 1520. Following the work of Solanes and Vela [10] and the Instituto de Geografía-UNAM [11], we generated the area of the empire based on two geographic criteria: altitude and water delimitations; and we located Aztec settlements georeferencing the cartographic data, respectively. Routes in Figure (b) correspond to low (< 70 km) and high (> 1,100 km) weighted shortest paths selected from the total number of highway sections possibly associated with ancient routes.

Figure 2(a) displays hierarchical settlements in the Aztec Empire that is characterized by a non-contiguous territory. Figure 2(b) presents the most likely road sections that correspond to ancient routes based on local and regional historical attributes. Local attributes suggested a large number of short and straight roads that replicated the most efficient way for traveling by foot between hierarchical places. In this case, around the city of Tenochtitlan the network of routes implied a well-connected system among neighboring settlements. On the other hand, the regional attribute, which represents the political control of settlements based on a small number of long-distance routes of commerce, indicated a vulnerable network where few routes provided access to distant places and water resources. Specifically, towards the Golf of Mexico, there were localized routes that spread over southern Veracruz and Tabasco, even further along the coast of Central America [9], but in the direction of the Pacific Ocean, long-distance routes were difficult to create and maintain because of the heterogeneous topographic surface and environmental conditions. Furthermore, Aztec settlements along this coastline could follow other type of spatial network to connect them to Tenochtitlan and distance places, specifically water routes, i.e., river and ocean transportation.

5   Conclusion

The proposed retrospective method provides an alternative process to identify some road networks that correspond to ancient routes. It points out the importance of actual road network information to decipher and understand the initial condition in the evolutions of road infrastructures by combining different types of data and scientific fields.

The case of Mexico presented some road sections highly related to ancient routes in the Aztec Empire. Such roads suggested an efficient local mobility and accessibility among close settlements, but, at regional scale, they implied a low degree of robustness representing high vulnerability to failures and attacks. This result can explain the success of Spaniards in conquering the Aztec Empire with a limited number of soldiers and resources.

References

1. Trombold, C.D.: Ancient Road Networks and Settlement Hierarchies in the New World (New Directions in Archaeology), 1st edn. Cambridge University Press (2011)
2. Burghardt, A.F.: The origins of the road and city network of Roman Pannonia. Journal of Historical Geography 5(1), 1–20 (1979)
3. Strano, E., Nicosia, V., Latora, V., Porta, S., Barthélemy, M.: Elementary processes governing the evolution of road networks. Scientifics Reports 2, 296 (2012), doi:10.1038/srep00296
4. Instituto Nacional de Estadistica y Geografia (INEGI), http://www.inegi.org.mx/geo/contenidos/topografia/InfoEscala.aspx (accessed May 20, 2012)
5. Mundy, B.E.: Mapping the Aztec capital: the 1524 Nuremberg map of Tenochtitlan, its sources and meanings. Imago Mundi 50, 11–33 (1998)

6. Van Tuerenhout, D.R.: The Aztecs: New Perspective. ABC-CLIO, Inc., Santa Barbara (2005)
7. Filsinger, T.: Atlas y Vistas de la Cuenca, Valle, Ciudad y Centro de México a través de los Siglos XIV al XXI. CD Interactivo, Cooperativa Cruz Azul, México (2005)
8. Santley, R.S.: The structure of the Aztec transport network. In: Trombold, C.D. (ed.) Ancient Road Networks and Settlement Hierarchies in the New World, pp. 198–210. Cambridge University Press, UK (2011)
9. Rees, P.W.: Origins of Colonial Transportation in Mexico. Geographical Review 65(3), 323–334 (1975)
10. Solanes, M., del, C., Vela, E.: Cultura Mesoamericana. Arqueología Mexicana, Special Edition: Atlas del México Prehispánico, 64–75 (2000)
11. Instituto de Geografía-UNAM, Nuevo Atlas Nacional de México (2007), http://www.igeograf.unam.mx/web/sigg/publicaciones/atlas/ (accessed May 20, 2012)
12. Xie, F., Levinson, D.: The weakest link: The decline of the surface transportation network. Transportation Research Part E 44, 100–113 (2008)
13. Xie, F., Levinson, D.: Topological evolution of surface transportation networks. Computers, Environment and Urban Systems 33, 211–223 (2009)
14. Barrat, A., Barthélemy, M., Pastor-Satorras, R., Vespignani, A.: The architecture of complex weighted networks. PANAS 101(11), 3747–3752 (2004)
15. Barthélemy, M.: Spatial Networks. Physics Reports 499, 1–101 (2011)
16. Levinson, D., El-Geneidy, A.: The minimum circuity frontier and the journey to work. Regional Science and Urban Economics 39, 732–738 (2009)

A Policy of Strategic Petroleum Market Reserves

Michael D. Mitchell, Walter E. Beyeler, Matthew Antognoli,
Marshall A. Kuypers, and Robert J. Glass

Complex Adaptive Systems of Systems (CASoS) Engineering
Sandia National Laboratories,
Albuquerque, New Mexico, USA
{micmitc,webeyel,mantogn,mkuyper,rjglass}@sandia.gov
http://www.sandia.gov/CasosEngineering/

Abstract. Unexpected price spikes in petroleum can lead to instability in markets and have a negative economic effect on sectors which rely on petroleum consumption. Sudden rises in the price of petroleum do not have to be long-term to cause negative, cascading impacts across the economy. Firms which make futures purchases or hedge against a higher price during a price spike can become insolvent when the price spike deflates. A policy is needed to buffer short-term perturbations in the petroleum market to avoid short-term price spikes. This study looks at the effects of implementing a Strategic Petroleum Market Reserve within a multi-agent Nation-State model which would utilize trading bands to determine when to buy and sell petroleum reserves. Our analysis indicates that the result of implementing this policy is a more stable petroleum market during conditions of resource scarcity.

Keywords: Complex Adaptive Systems, Multi-Agent-Based Modeling, Buffer Stocks, Price Stabilization, Strategic Resource Reserves, Resource Scarcity.

1   Introduction

We conducted a study to determine the effectiveness of a Nation-State having a policy of maintaining a strategic market reserve for a resource whose price can be volatile. The goal of the reserve is to buffer sudden price-spikes by reducing constraints on the availability of a resource in a domestic market. We chose petroleum due to its economic importance to many sectors of the economy and petroleum's history of volatile prices. The study utilized the Nation-State configuration of the Exchange Model developed at Sandia National Laboratories to investigate complex adaptive systems (CAS) [Beyeler et. al. 2011]. This model provides a framework to abstractly represent a system in which interacting specialists (entities) produce and consume resources that flow among entities via continuous markets.

Many countries already have national Strategic Petroleum Reserves (SPR) which are used as a deterrent to organized disruptions in the availability of petroleum in markets. In 1975, the United States (U.S.) Congress passed a bill authorizing the creation of a SPR as an emergency reserve of petroleum for both civilian and military

K. Glass et al. (Eds.): COMPLEX 2012, LNICST 126, pp. 234–243, 2013.

needs in the event of an embargo [Carter 1975]. In the U.S., salt domes are used to store petroleum for the SPR, providing ideal containers because they are large, accessible, self-sealing, and have the virtue of previous excavation by chemical companies. Today the U.S. has roughly 750 million barrels of petroleum stored in salt domes in five different facilities along the Gulf Coast [Blumstein 1996]. The storage facilities are operated by the Department of Energy; disbursement from the SPR has to be authorized by the President of the U.S. The effectiveness of the SPR against embargos and other short-term disruptions to the domestic petroleum supply is not being questioned in this paper. The study presented here addresses the potential efficacy of a new Strategic Petroleum Market Reserve which utilizes trading bands to determine when to buy and sell petroleum with the intention of buffering price volatility resulting from an exogenous event.

The Strategic Petroleum Market Reserve (SPMR) is modeled after the Strategic Petroleum Reserve (SPR) but serves a different function. The SPR was designed and built to deter bad actors trying to disrupt imports to the U.S. Contrariwise, the SPMR is proposed to support a policy of market-based intervention designed to buffer short-term market disruptions by being an active player in the market. Like the SPR, the SPMR would need to acquire petroleum from the market, store the petroleum, and determine when and how much petroleum to release when intervention is required. We use Bollinger-Bands to define the dynamic trading bands used to make systematic buying and selling decisions. The use of Bollinger-Bands gives us a powerful tool to identify the relative highs and lows for the trading price of petroleum and the point at which intervention would be most effective.

In contrast to relying on a governing entity to decide when to buy and sell reserves, the use of a market-driven policy for petroleum reserve maintenance allows trading to be predictable, adaptable, and effective in mitigating short-term price spikes. Predictability results from markets being able to count on a consistent response by the SPMR to price perturbations. Adaptability derives from dynamic trading bands' ability to adjust the SPMR's acquisition and release of petroleum based on short-term and long-term price trends. Effectiveness stems from the SPMR's ability to ease price spikes by alleviating short-term scarcity in the petroleum market. The result of having a Strategic Petroleum Market Reserve is the reduction of short-term price spikes by easing resource contention and providing stability to domestic markets.

2   Model Description

The Exchange Model (ExM) is a hybrid combining system dynamics and agent-based modeling to represent production and consumption sectors, resource flows and market exchanges among interacting specialists (entities) in a system. All entities have a homeostatic process which maintains 'health' via the consumption of resources. The production of resources is a consequence of the consumption of resources. Entities store both the resources needed for consumption and the resources needed for production and control those resource levels through interactions with markets. Markets facilitate the exchange of resources by using a double auction algorithm to match bids and offers. Each market manages the exchanges of a single resource. The price of a proposal (bid or offer) is reflexive and represents the relative scarcity/abundance experienced by the entity creating the proposal.

2.1   Nation-State Configuration

For the purposes of this study, the ExM is configured as a hierarchical system of compound entities (Nation-States) which embed sectors (entities) for the purpose of investigating patterns of system-level exchange of resources necessary for survival. We call this type of the configuration the Nation-State configuration. All resources are exchanged in domestic markets and some resources are also exchanged in an international market. Resources are exchanged via proposals to buy or sell in markets. Figure 1 illustrates a single Nation-State configured with two entities (each producing and consuming two resources), two domestic markets and one international market.

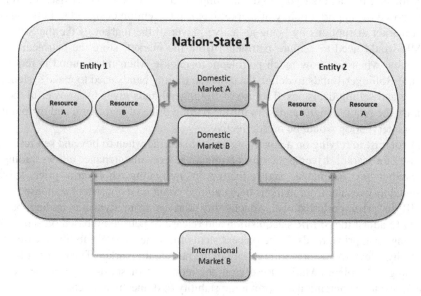

Fig. 1. Diagram of a single Nation-State depicting relationships among producing/consuming entities and markets

2.2   Trading Bands

In this model, trading bands provide a metric to determine when an observed price diverges significantly from normal price trends. Upper and lower trading bands are used to determine the significance of a price divergence. We define the upper and lower trading bands as Bollinger Bands (BB) configured with a Simple Moving Average (SMA) rule. Bollinger Bands were chosen for their ability to capture abrupt deviations from normal price movements and their prevalent use as a technical training tool [Leung, et. al. 2003]. Bollinger Bands are created from three curves drawn in relation to a series of observed prices. The middle curve is a Simple Moving Average (SMA) which serves as a base for the upper and lower bands: band spacing is wide when prices are volatile, narrowing as price becomes less volatile. The typical configuration of the bands is the use of an SMA with 20 observations and two standard deviations from the SMA to create the upper and lower bands. This configuration captures 95% of price movements.

The SMA function is defined as follows:

$$SMA_N(t) = \frac{\sum_{i=t-N+1}^{t} P(i)}{N} \tag{1}$$

where $SMA_N(t)$ is defined as the mean of the prices in the past N observations at time t, and P(i) is the price for a given observation i.

The Bollinger Band function to determine upper and lower bands is described as follows:

$$BB_N^k(t) = SMA_N(t) \pm k \times \sqrt{\frac{\sum_{i=t-N+1}^{t} [P(i) - SMA_N(i)]^2}{N}} \tag{2}$$

where k controls the width of the Bollinger Bands due to price fluctuations around the mean in terms of the standard deviation at time t for N prior observations.

Figure 2 illustrates three Bollinger Bands (BB) calculated using historical crude oil spot price data.

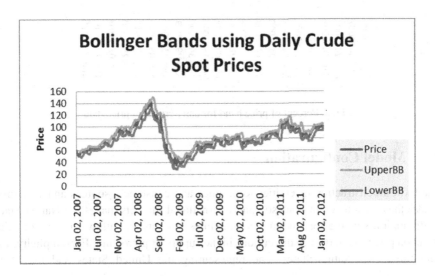

Fig. 2. Bollinger Band using daily crude oil spot prices from 01/ 2007 – 01/ 2012

Determining when to buy and sell using Bollinger Bands requires the calculation of the observed price in relation to the price bands. The following equation calculates the observed price $P(t)$ in relation to the upper $BB_N^{upper}(t)$ and lower $BB_N^{lower}(t)$ trading bands. The result of the calculation is used by the SPMR to determine whether to buy, sell, or do nothing.

$$percentB = \frac{P(t) - BB_N^{lower}(t)}{BB_N^{upper}(t) - BB_N^{lower}(t)} \tag{3}$$

Figure 3 illustrates the percentB calculation on the daily crude oil spot prices for a three year period.

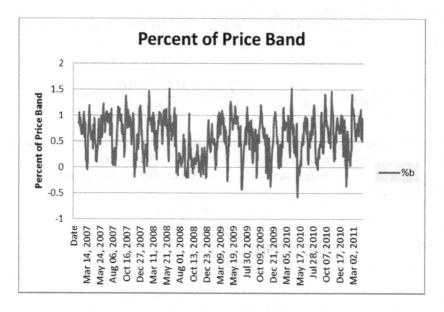

Fig. 3. Percent of price band for daily crude oil price data

3   Model Configuration

The ExM's construction of entities to produce and consume resources and exchange (trade) those resources in domestic and international markets makes the Nation-State configuration suitable for studying commodities markets and allows us to study the effect on price of short-term disruptions in the supply of petroleum. For simplicity, we configured two Nation-States: one representing the United States and the other representing the rest of the world. Table 1 lists the resources, production, and consumption rates used to configure the U.S. Nation-State; Table 2 lists the resources, production, and consumption rates used to configure the World Nation-State. The only stoichiometry imbalanced in the model, meaning that the nation-state cannot wholly obtain that resource, is Oil. The U.S. Nation-State its domestic oil consumption requirements through domestic markets and trade with international markets. All resources have domestic markets: Oil, Food, Goods, Energy, and Labor. In addition to domestic markets, there are international markets for three resources: Oil, Food, and Goods.

Table 1. U.S. Nation-State Configuration

United States	Consumed Resources					Produced Resources				
Sectors	Oil	Food	Goods	Energy	Labor	Oil	Food	Goods	Energy	Labor
Household	0	2	2	2	0	0	0	0	0	3.6
Manufacturing	0	0	0	2	1	0	0	1.8	0	0
Farming	0	0	0	1	1	0	1.8	0	0	0
Oil Production	0	0	0	0	1	1.8	0	0	0	0
Energy from Oil	4	0	0	0	1	0	0	0	4.5	0

Table 2. World Nation-State Configuration

World	Consumed Resources					Produced Resources				
Sectors	Oil	Food	Goods	Energy	Labor	Oil	Food	Goods	Energy	Labor
Household	0	8	8	8	0	0	0	0	0	14.4
Manufacturing	0	0	0	8	4	0	0	7.2	0	0
Farming	0	0	0	4	4	0	7.2	0	0	0
Oil Production	0	0	0	0	4	16.2	0	0	0	0
Energy from Oil	16	0	0	0	4	0	0	0	18	0

We ran each simulation for a total time of 7.E+4. A disruption was scripted at time 5.E+4 to reduce the international availability of petroleum causing a short-term resource constraint and resulting in a sharp increase in the spot trading price for petroleum. We configured ten different environments of petroleum scarcity. For each environment, we ran two simulations: one with a SPMR and one without an SPMR to compare the impacts to price. Table 3 describes the environments configured for this study.

Table 3. Levels of disruption to international oil production analyzed in this study

Environment	International Oil Production Disruption
1	5%
2	10%
3	15%
4	20%
5	25%
6	30%
7	35%
8	40%
9	45%
10	50%

3.1   Strategic Petroleum Market Reserve

An SPMR was added to the model to mediate short-term disruptions in the supply of petroleum to the U.S. domestic market. The SPMR was endowed with an initial level of petroleum and was configured with the trading band rules needed to govern when it buys and sells petroleum from the U.S. domestic petroleum market. Table 1 lists the parameters used to configure the SPMR.

Table 4. Strategic Petroleum Market Reserve Parameters

Parameter	Value	Description
monitorPeriod	10	A time constant used to determine when to sample prices from the market and make a decision to buy, sell, or do nothing.
tAveragePrice	2000	A time constant for how long, in terms of time, is the simple moving average window
reserveLevel	1000	Initial amount of resource endowed to the reserve.
proposalAmount	50	The amount of resources to propose when buying or selling from the SPMR.

4   Analysis

We are modeling the SPMR with an initial reserve capacity which can increase or decrease depending on whether or not the SPMR is buying or selling petroleum. There is not limit on the total amount of petroleum which can be bought or sold only the quantity proposed. All of our simulations configure the SPMR to have 1000 units of petroleum in the reserve. This initial amount is somewhat arbitrary and represents a large buffer for U.S. consumption requirements.

Output results from the Nation-State ExM are stated in terms of the percent deviation in price during the period of resource supply disruption. The magnitude of the resource disruption is stated as a fraction of the SPMR petroleum reserve capacity. For each environment configured, a simulation was run with and without a U.S. SPMR. The SPMR was designed to automatically buy and sell petroleum from the market when the trading bands indicated intervention was warranted.

Figure 4 illustrates the percent change in price for the various environments of international petroleum supply disruption (see Table 3). The y-axis lists the percent change in price from the pre-disruption average spot price to the maximum price during the price spike. The x-axis lists the magnitude of the resource disruption stated as a fraction of the SPMR resource level. For example, the first data points are for a simulation where the total international resource loss was 50% of the SPMR capacity and the resulting price was driven up 2.5%.

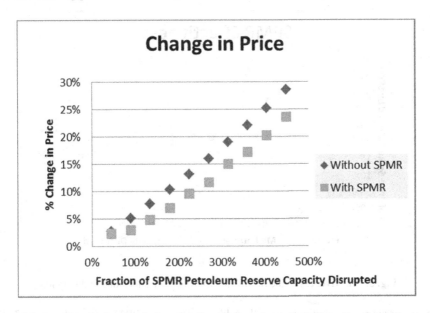

Fig. 4. Comparison of the change in price due to disruption of petroleum for simulations with and without a SPMR

Modeling results show a linear relationship between the price spike and the level of disruption to petroleum supply and indicate that having a SPMR reduces a price spike but did not eliminate it. At the smallest modeled disruption, the impact of the SPMR on the price spike is minimal; as the magnitude of disruption increases, the SPMR buffers the spike more effectively. The implementation of buying and selling rules in accordance with trading bands explains this behavior. Small changes in the average spot trading price are not that far outside of normal price fluctuations and thus require little or no intervention from the SPMR. As disruptions become larger, more dramatic changes in price signal a larger response from the SPMR. In our study, the response of the SPMR is limited by its resource level, which is why we stated the disruption as a fraction of the SPMR reserve capacity.

Figure 5 depicts the effectiveness of the SPMR at buffering price spike. The y-axis shows how much of the price spike was buffered by the SPMR. The x-axis lists the magnitude of the resource disruption stated as a fraction of the SPMR resource level.

The results shown in Figure 5 demonstrate that the SPMR is most effective when mitigating a disruption in supply which is relative in size to the SPMR capacity. The disruption scenarios we implemented affected the international production of petroleum, so at first glance these results were puzzling. Because the U.S. market purchases only a fraction of the international production of petroleum (see Table 1), we anticipated that the SPRM would be able to mitigate disruptions much larger than

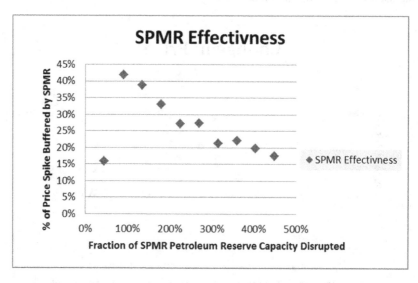

Fig. 5. Effectiveness of the SPMR in buffering price shock due to disruption

the resources it has available to sell. Additionally, the SPRM sells only to the U.S. domestic market and the impact of a disruption to international petroleum production would only fractionally affect the U.S. On further investigation we found that the open access of U.S. petroleum producers and consumers to both domestic and international markets causes prices to equalize in both markets. The effect of the SPMR selling petroleum on the U.S. domestic market is a reduction in the spot trading price for the domestic market. U.S. producers selling petroleum do not have an incentive to sell in the domestic market if they can get higher prices in the international market. Therefore both the U.S. domestic market and the international market will always have the same spot trading price for petroleum.

5   Conclusion

The results of this study indicate that implementation of a national policy for maintenance of a Strategic Petroleum Market Reserve would effectively buffer short-term disruptions to petroleum resource availability and provide some level of price stability that could not be achieved without it. Utilizing trading bands to determine when to buy and sell from such a reserve allows the petroleum market to function without intervention unless spot prices deviate from normal trends. By configuring the trading bands to use a Simple Moving Average to determine normal price trends and setting the upper and lower bands two standard deviations from the SMA, 95% of the price movement will fall within those bands. The SPMR would only interact with the market when prices rise or fall outside of the trading bands. When domestic and international markets are frictionless, the amount of resource needed to respond to a disruption in supply is not limited to the relative disruption to the domestic consumers, but is relative to the absolute amount of resources disrupted.

References

1. Carter, L.J.: Energy: A Strategic Oil Reserve as a Hedge against Embargoes. Science 189(4200), 364–366 (1975), doi:10.1126/science.189.4200.364.
2. Blumstein, C., Kamor, P.: Another look at the Strategic Reserve: Should its oil holdings be privatized? Journal of Policy Analysis and Management 15(2), 271–275 (1996)
3. Leung, J.M.-J., Chong, T.T.-L.: An empirical comparison of moving average envelopes and Bollinger Bands. Applied Economics Letters 10(6), 339–341 (2003)
4. Beyeler, W.E., Glass, R.J., Finley, P.D., Brown, T.J., Norton, M.D., Bauer, M., Mitchell, M., Hobbs, J.A.: Modeling Systems of Interacting Specialists. In: 8th International Conference on Complex Systems (June 2011)

Binary Consensus via Exponential Smoothing

Marco A. Montes de Oca[1], Eliseo Ferrante[2,3], Alexander Scheidler[4],
and Louis F. Rossi[1]

[1] Department of Mathematical Sciences, University of Delaware, USA
{mmontes,rossi}@math.udel.edu
[2] IRIDIA, CoDE, Université Libre de Bruxelles, Belgium
[3] Socioecology and Social Evolution Lab, Katholieke Universiteit Leuven, Belgium
eferrant@ulb.ac.be
[4] Fraunhofer IWES, Germany
alexander.scheidler@iwes.fraunhofer.de

Abstract. In this paper, we reinterpret the most basic exponential smoothing equation, $S^{t+1} = (1 - \alpha)S^t + \alpha X^t$, as a model of social influence. This equation is typically used to estimate the value of a series at time $t + 1$, denoted by S^{t+1}, as a convex combination of the current estimate S^t and the actual observation of the time series X^t. In our work, we interpret the variable S^t as an agent's tendency to adopt the observed behavior or opinion of another agent, which is represented by a binary variable X^t. We study the dynamics of the resulting system when the agents' recently adopted behaviors or opinions do not change for a period of time of stochastic duration, called latency. Latency allows us to model real-life situations such as product adoption, or action execution. When different latencies are associated with the two different behaviors or opinions, a bias is produced. This bias makes all the agents in a population adopt one specific behavior or opinion. We discuss the relevance of this phenomenon in the swarm intelligence field.

Keywords: Consensus, Collective Decision-Making, Self-Organization, Swarm Intelligence.

1 Introduction

The old adage "When in Rome, do as the Romans do" summarizes the intuition that it is sometimes wise to imitate the behavior of seemingly more knowledgeable individuals, especially when we are in a new environment. Recent research has provided evidence that imitating is indeed the best thing to do even in situations previously thought to require individuals to rely more on themselves [15]. When agents[1] are under pressure to choose an adaptive action, that is, an action that provides them with some benefit, imitation can be seen as a filtering process that allows agents to collectively discard actions that provide the lowest

---

[1] We use the term *agent* to refer to an entity, be it an animal or an artifact, such as a robot or a piece of software, capable of autonomous perception and action.

K. Glass et al. (Eds.): COMPLEX 2012, LNICST 126, pp. 244–255, 2013.
© Institute for Computer Sciences, Social Informatics and Telecommunications Engineering 2013

rewards [15]. This phenomenon may explain why in nature doing what others do is a strategy frequently used by different groups of animals. For example, ants choose the same paths other ants choose by following pheromone trails [8], sheep move in the same direction other sheep move [14], and we humans tend to cross the road whenever we see other people do so [5].

Given that imitation is a strategy used by animal groups as different as insect colonies and human crowds, we wonder to what extent individual imitation induces a good collective decision-making mechanism for groups of artificial agents. In particular, we would like to know whether imitation is an individual strategy that allows a large group of agents to make good collective decisions by consensus. As a step toward answering this question, in this paper we introduce and study the dynamics of an agent-based model in which individual agents tend to perform the action most commonly performed by other agents. In our model, presented in detail in Section 2, each agent's behavior is governed by a rule similar to the basic exponential smoothing equation used for data filtering and time series forecasting [6,10]:

$$S^{t+1} = (1 - \alpha)S^t + \alpha X^t \tag{1}$$

where S^t is the estimate of a time series at time t, X^t is the actual value of the time series at time t, and α determines the strength of the error correction.

In this paper, we interpret the variable S^t as the tendency of an agent to imitate an observed behavior, perform an observed action, or adopt an opinion externalized by another agent.[2] The variable X^t encodes the behavior, action, or opinion of another agent. The parameter $\alpha \in [0, 1]$ controls the strength of the imitation tendency. At the two extremes, if α is equal to zero, an agent is insensitive to any social influence; if α is equal to one, an agent directly copies the behavior, action, or opinion of another agent. Since the behavior of all the agents is governed by the same rule, the collective dynamics of the group is determined by the dynamics of a system of coupled exponential smoothing equations.

In Section 3, we study the dynamics of the system in the cases where agents may or may not influence other agents at all times. An example of a situation where an agent may influence other agents at all times is when someone decides to buy a product that is visible to others (e.g., someone buys a tablet computer that is frequently used). The very fact that the bought item is visible, gives information to an observer agent (e.g., that the tablet is useful) that may increase its tendency to buy the same product. There are also situations where agents may not influence others at all times. For example, in a robotics application, a robot that decides to go from one point to another is only visible to other robots while it is in their close vicinity. Thus, robot actions or choices may be observed by other robots only in a confined region of the robots' operating environment. The time that passes between two product adoptions or two executions of an action, is modeled as *latency*, which is a period of time of stochastic duration during which agents do not update their tendency or state. In our examples, a

---

[2] In the text, we use the terms behavior, action, or opinion depending on the context.

person who is evaluating a product or a robot executing an action are modeled as latent agents.

Our results are organized into two parts (visible *vs.* not visible latent agents). We show that in both cases a population of agents adopts the same behavior, that is, they reach a consensus. However, the actual consensus state when latent agents are visible is the opposite than when latent agents are not. These results are discussed in Section 4 in which we also describe the similarities and differences that exist between our work and previously published works. Final conclusions are given in Section 5.

2   Exponential Smoothing as a Social Influence Model

The model proposed in this paper is based on the notion that the actions and opinions of others influence our own actions and opinions. When an agent is exposed to the behavior, actions or the opinion of another agent, the observing/listening agent may be more likely to perform the observed behavior or action, or adopt the opinion of the other agent. In this paper, we focus on the case where there are only two observable behaviors/opinions. This model captures real-life scenarios where two options are available to the members of a population, but only one can be chosen. For example, a person has to choose between an Android or an iOS phone, a French voter must choose one of the two candidates in the second round of the elections, and in laboratory conditions, ants have to choose one of two paths between their nest and a food source.

Our model consists of a set of agents, each of which is in one of two possible states, which represent the agent's current behavior or opinion. We use the binary variable $X_i \in \{0, 1\}$ to represent an agent i's state. This variable is in turn governed by an internal real-valued variable S_i and a threshold θ. The variable S_i can be thought of as the tendency of agent i to be in one of the two possible states (hereafter, we refer to S_i simply as agent i's tendency). The threshold θ is constant and common to all agents, while S_i is variable and private to each agent.

At each time step t of the system's evolution, an agent i might be able to observe the state of another random agent $j \neq i$. When an agent observes the state of another agent, the observing agent updates its tendency as follows:

$$S_i^{t+1} = (1 - \alpha)S_i^t + \alpha X_j^t, \tag{2}$$

where $\alpha \in [0, 1]$ determines how much importance is given to the agent's latest observation (X_j^t) as opposed to the agent's accumulated experience (S_i^t). After updating its tendency, an agent updates its state as follows:

$$X_i^{t+1} = \begin{cases} 1, & \text{if } S_i^{t+1} > \theta \\ 0, & \text{if } S_i^{t+1} < 1 - \theta \\ X_i^t, & \text{if } 1 - \theta \leq S_i^{t+1} \leq \theta, \end{cases} \tag{3}$$

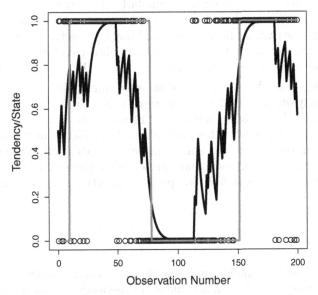

Fig. 1. Single agent behavior. Starting with an initialization of $S^0 = 0.5$ and $X^0 = 0$, an agent observes a stream of state values plotted as dots with values 0 or 1. The black line shows the evolution of the agent's tendency and the gray line shows the evolution of the agent's state. In this simulation, $\alpha = 0.2$ and $\theta = 0.8$.

where $\theta \geq \frac{1}{2}$ due to the symmetry of the actual threshold value that triggers the adoption of one or another state. Thus, an agent's state is a function of its tendency and the threshold θ.

In Fig. 1, we show an example of the behavior of a single agent controlled by the rules defined in Equations 2 and 3. In this example, the agent is exposed to a controlled stream of state observations that repeatedly switches between 1 and 0. We can observe how an agent's tendency follows the stream of observations and the threshold rule makes the agent's state behave like a comparator with hysteresis. Combined, these two rules make an agent adopt a state that agrees with the most commonly observed state at any point in time. This conclusion can be reached if one expands Eq. 2 iteratively as

$$S_i^n = (1 - \alpha)^n S_i^0 + \alpha \sum_{k=0}^{n-1} (1 - \alpha)^{n-k-1} X_{\mathbf{I}(k)}^k, \tag{4}$$

where S_i^n is the value of agent i's tendency after n updates, S_i^0 is agent i's initial tendency, and $X_{\mathbf{I}(k)}^k$ is the state of the agent observed at time step k. A multivariate random variable $\mathbf{I}$ is used to represent the fact that observed agents are chosen at random, and thus $\mathbf{I}(k)$ is the index of the agent observed at time step k. When $0 < \alpha < 1$, an agent's tendency is a weighted moving average of the agent's observations with exponentially decreasing weights. Any possible bias introduced by an initial tendency $S_i^0 \neq \frac{1}{2}$ vanishes when an agent performs a sufficiently large number of observations. Similarly, the weight of old observations approaches zero as the number of observations increases. This means that only the most recent observations have a significant influence on the observing agent's state. Thus, if the majority of the most recently observed states, that is, if the most commonly observed behaviors, are encoded by $X_{\mathbf{I}(k)}^{k} = 1$, the observing agent will have a state $X_{i}^{t+1} = 1$. The same reasoning applies if the majority of the most recently observed states are encoded by $X_{\mathbf{I}(k)}^{k} = 0$.

The collective behavior of a population of agents controlled by the rules given in Equations 2 and 3 is much richer and difficult to predict than the behavior of a single agent. In this paper, we explore this system's dynamics through Monte Carlo simulations. The simulation study, presented in the next Section, is focused on the effects that agent visibility, as explained in Section 1, has on the system's collective dynamics.

3   Simulations

We perform two sets of simulations. In both sets, when an agent changes state from 0 to 1, or *vice versa*, it does not update its tendency or state for a period of time of stochastic duration, called *latency* period [12,13]. During this time, we say that an agent is *latent*. In the first set of simulations, latent agents are always visible to other agents, that is, they can influence other agents while being latent. This scenario models situations similar the iOS vs. Android example mentioned earlier. For example, after adopting one of the two competing brands, a buyer will not change immediately to the other brand, but rather evaluate the newly-bought product for some time (which we model with the latency period). During this evaluation time, other persons will observe the buyer and may be more inclined to adopt the same product as a result of our natural tendency to do what others do. In the second set of simulations, latent agents are not visible to other agents, and therefore, cannot influence them. This scenario models situations like the robotics example mentioned in Section 1. In a robotics application, states may be interpreted as robot actions. The latency period models the duration of an action execution during which the robot may not be able to interact with other robots. Thus, latent robots cannot be observed by other robots.

An extra element in our simulations is known as *differential latency* [13]. This scheme associates a different average duration of the latency period to each of the two states of an agent. This is done in order to model situations where two products have different sets of features, causing a buyer to spend more time evaluating one of the competing products, or actions which have similar results but that take different times to perform. With differential latency we expect to alter the frequency with which one state is observed by the population of agents, inducing a consensus on one specific state. The simulation of differential latency proceeds as follows: While an agent is latent, it does not change state or tendency. The difference between the two sets of simulations that are reported in this section lies in the actions taken by the agents when their latency period ends. In the observable latent agents case (see Section 3.1), an agent chooses, from the whole population, a random agent to observe after its latency period ends. Then, the agent updates its tendency and becomes latent again for a period whose duration depends on its (possibly new) state. In the case of nonobservable latent agents (see Section 3.2), an agent whose latency period ends observes the state of a randomly chosen non-latent agent. If there is none available, that is, if all agents are latent, it waits until one becomes observable. This waiting time also allows other agents that switch from latent to non-latent state to observe this agent. As before, a state observation is used for updating an agent's tendency and state. The process of observing, updating tendency, updating state, and becoming latent is repeated until the time steps counter reaches a certain limit.

In our simulations, the duration of latency periods are normally distributed. The mean duration and standard deviation of the latency period associated with state 1, denoted by μ_1 and σ_1 respectively, are equal to 100 and 10 time steps. We keep the standard deviation of the latency period associated with the state 0 constant ($\sigma_0 = \sigma_1$) and test three different values for the mean duration of the latency period $\mu_0 = \{\mu_1, 1.2\mu_1, 2\mu_1\}$. A ratio $\mu_0/\mu_1 = 1$ is used to observe the system's dynamics when both states are associated with equal latency periods. The ratios $\mu_0/\mu_1 = 1.2$ and $\mu_0/\mu_1 = 2$ are used to observe the system's dynamics with high and low degrees of latency overlap respectively. We also explore the effects of different combinations of values of α and θ on the system's dynamics. The maximum time step count is set to 10,000. A summary of our results is presented next.

3.1   Observable Latent Agents

Let us first present the results obtained when latent agents are always visible. In Fig. 2, we present a typical example of the evolution of the average tendency across a population of 100 agents under the three tested latency period conditions.

When the latency periods associated with the two states have equal average duration, the final average tendency is approximately 0.5. This is the result of approximately half of our simulations converging to a consensus on state 0 and the other half on state 1. Thus, we conclude that independently of the values of α, when $\mu_0/\mu_1 = 1$ the population of agents reaches a consensus on one of the two states with equal probability.

When the latency periods have different average duration, agents whose state is associated with the longest latency period are more likely to spread their state to the rest of the population because their state remains "frozen", and therefore visible, for longer periods of time. This phenomenon induces a positive feedback process whereby the state associated with the longest latency period is observed more often, and therefore copied more rapidly than the other state until every agent adopts the same state. In our simulations, state 0 is associated with the longest average latency period. In Fig. 2 parts (b) and (c), we can see that state 0 is the state that is more often adopted by the population. The effect of the parameter α depends on the overlap of the duration distribution of latency periods. With large overlaps (in our simulations, this is modeled with the case

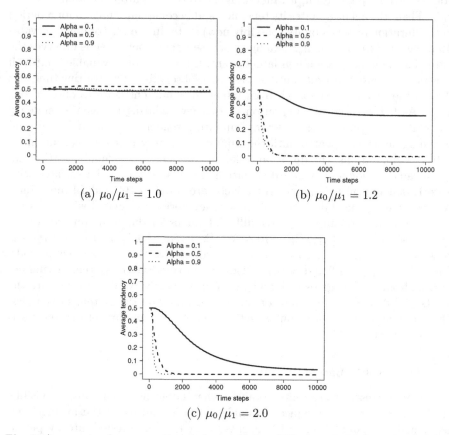

Fig. 2. Average tendency over the whole population (100 agents). In these simulations, $\mu_1 = 100$, $\sigma_1 = \sigma_0 = 10$ time steps. The acceptance threshold θ is equal to 0.6 in all cases. Averages obtained through 500 independent runs of a Monte Carlo Simulation.

$\mu_0/\mu_1 = 1.2$), larger values of α have larger effects on the system's dynamics. In our simulations, with $\mu_0/\mu_1 = 1.2$ and $\alpha = 0.1$, in 69% of the cases the population reaches consensus on state 0, while with $\mu_0/\mu_1 = 1.2$ and $\alpha = 0.9$, the population reaches consensus on state 0 in 100% of the cases. With low overlaps (in our simulations, when $\mu_0/\mu_1 = 2$), the parameter α affects more the speed of convergence to consensus than the actual state on which the consensus is reached.

3.2 Nonobservable Latent Agents

We now present the results obtained when latent agents are not visible to other agents. In Fig. 3, we show the evolution of the average tendency across a population of 100 agents under the three tested latency period distributions.

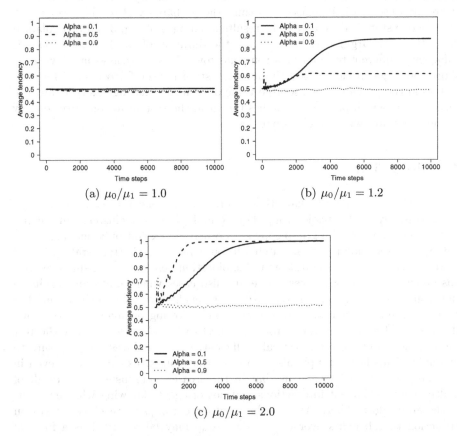

Fig. 3. Average tendency over the whole population (100 agents). In these simulations, $\mu_1 = 100$, $\sigma_1 = \sigma_0 = 10$ time steps. The acceptance threshold θ is equal to 0.6 in all cases. Averages obtained through 500 independent runs of a Monte Carlo Simulation.

In Fig. 3 part (a), it is apparent that when the average duration of the latency periods associated with each of the two states is equal ($\mu_0/\mu_1 = 1$), the probability of the population to reach consensus on any of the two states is 0.5. Thus, irrespective of whether latent agents are visible all the time or only when they are in a nonlatent state, when the ratio $\mu_0/\mu_1 = 1$, a population of agents governed by Eqs. 2 and 3 initialized with $S_i^0 = 0.5$ will reach a consensus on any of the two states with equal probability.

In Fig. 3 parts (b) and (c), we see the results when $\mu_0/\mu_1 = 1.2$ and $\mu_0/\mu_1 = 2$, respectively. When $\mu_0/\mu_1 = 1.2$, the smaller the value of the parameter α, the higher the probability of the population to reach consensus on state 1, which is associated with the shorter average latency period. This phenomenon occurs because large values of α make agents favor copying over using their past experience. Therefore, when the average duration of the latency periods is similar, the agents switch from one state to the other more rapidly with large values of

α than with small ones. The result is that the frequency with which different states are observed is increased and thus, the probability of reaching consensus on only one state is decreased. The results in part (c) of Fig. 3 can be explained using the same argument. In this case, the duration of the latency periods is sufficiently different to induce a strong bias toward the state associated with the shorter average latency period (in our case, state 1) even for $\alpha = 0.5$. However, just as in the previous case, a large value of α means that agents copy any observed state, which produces switches that lead the system to consensus on any of the two states with equal probability.

4   Discussion

In some situations, doing what others do may be beneficial to some agents but not necessarily to the whole group. For example, if food is clustered in patches and all the members of a group copy each other and exploit only one discovered patch, very soon food in that patch will be depleted. A better strategy would be to switch between exploitation and exploration behaviors to ensure everyone has enough food [7]. However, there are also cases where imitation results in the group reaching a state of consensus that benefits everyone in the group. For example, the aforementioned trail-laying and following behavior of ants allows them to find the shortest route from their nest to a food source [8]. The shortest route is not only more energetically efficient, it also reduces the exposure to predators. This last example also shows us that consensus can emerge even in very large groups, which means that in these cases, consensus is the result of multiple agent-to-agent interactions and not of agents knowing what everyone in the group does. Therefore, the ability of large groups to reach consensus on an action that benefits everyone in the group may be regarded as a form of collective intelligence (also called swarm intelligence [2]).

In this paper, we explored some circumstances under which imitation makes a population of agents reach a consensus. In particular, we studied the effects of latency periods of different duration on the system's dynamics. We saw how imitation and latency induce a positive feedback process that eventually leads the population to a consensus state. Imitation and positive feedback have been already used successfully in some software-based swarm intelligence systems for optimization. For example, in ant colony optimization (ACO) [4] algorithms, very simple agents simulate ants following pheromone trails laid by other ants. In ACO, these artificial ants move on a graph that represents adjacency relations between solution components to a combinatorial optimization problem. At each decision point, ants are attracted toward nodes with higher levels of "pheromone", which is a numerical variable associated with each pair of solution components i and j, denoted by τ_{ij}. The result is an "emergent" optimization process. At each iteration of an ACO algorithm, these pheromone variables are updated using a rule that resembles the following equation:

$$\tau_{ij}^{t+1} = (1 - \rho)\tau_{ij}^t + \Delta\tau\,, \tag{5}$$

where $0 \leq \rho \leq 1$ is a parameter that simulates "pheromone evaporation", and $\Delta\tau$ is a quantity that reinforces a pheromone value. The similarity between Equations 2 and 5 is apparent. However, in ACO algorithms, the quantity $\Delta\tau$ is different across problems and ACO variants. Such heterogeneity makes a unified approach to the study of the imitation dynamics that occur in ACO algorithms practically impossible. Nevertheless, our model enables the study of the abstract process of imitation and thus, any insights gained from its study may shed light onto the operation of ACO algorithms for specific problems.

Imitation is also an important idea behind the particle swarm optimization (PSO) [11] algorithm, which is another successful swarm intelligence algorithm for optimization. In a PSO algorithm, agents move in an n-dimensional space. Their positions represent candidate solutions to a continuous optimization problem. The stochastic rules used to update the particles' position make particles move toward the position of (i.e., imitate) their most successful neighbors. It can be shown that the expected value of the i-th particle's position obeys the equation:

$$E(\mathbf{x}_i^{t+1}) = w\mathbf{x}_i^t + \frac{\mathbf{L} + \mathbf{G}}{2} - w\mathbf{x}_i^{t-1}, \tag{6}$$

where $0 \leq w \leq 1$ is a parameter called inertia weight in the PSO jargon, $\frac{\mathbf{L}+\mathbf{G}}{2}$ is the midpoint between the position of the particle's best local neighbor, denoted by $\mathbf{L}$, and the position of the best particle of the swarm, denoted by $\mathbf{G}$. Despite the fact that the PSO algorithm is clearly a second order system, we can write $w = (1 - \alpha)$ for some $0 < \alpha < 1$, which makes Equation 6 resemble Equation 2. Thus, our model seems to capture certain commonalities present in several swarm intelligence systems with the advantage that our model is not focused on any specific application. Future work should be aimed at determining the extent to which the higher level of abstraction of Equation 2 is useful for the design and analysis of swarm intelligence systems.

As a model aimed at understanding the dynamics of collective decision and action, our model is not unique, and in fact, it shares many features with previous models. For example, in Granovetter's models [9] agents also use a threshold to decide whether to do what others do, and in the Bass model [1] agents are subject to social influence to decide whether to buy a product or not. Our model, however, is more similar to models known as *opinion formation models* [3]. In the large body of literature dealing with these kinds of models, we can find two that are closely related to the model proposed in this paper. One of these models was proposed by Scheidler *et al.* [16]. Their model consists of a population of agents each of which has a memory of fixed size, which stores the values of the last k observed states. If all of these states are all equal to, say state 1, then the agent adopts state 1. The authors called this state update mechanism the k-*unanimity* rule. Our model and the k-unanimity model are similar in that the k-unanimity rule can be seen as a special case of our tendency mechanism. In our case, we sum the values of the observed states with exponentially decreasing weights tunable via the parameter α. The k-unanimity rule, on the other hand, can be interpreted as a sum of the values of the observed states with equal

254 M.A. Montes de Oca

weights. In both cases, a threshold determines whether an agent changes state or not. The other model that is related to ours, is the so-called majority rule model [13]. In the majority rule model, teams of three agents are repeatedly formed, each time with different agents. When three agents form a team, they exchange their states and the state of the team's majority (that is, the state of at least two of these agents) is adopted by all the agents in the team. The majority rule may be seen as a distributed implementation of the k-unanimity rule with $k = 2$. Another common element between our model and the majority rule model is the concept of differential latency, which was first introduced in [13].

The model presented here may be seen as a more general model than the aforementioned models because in many cases it is possible to find a value for α that reduces Equation 2 to a simple average of the value of the last k observed states. Therefore, it is possible to reduce our model to the k-unanimity model. The majority model is also subsumed within our model because, as we discussed earlier, the majority rule model is a distributed implementation of the k-unanimity model. Thus, even though the dynamics at the individual level are different, one can find a configuration of the model presented here that mimics the dynamics of the majority rule at the collective level.

5 Conclusions

When an animal is part of a group, its behavior is influenced by the behavior of other animals that are also members of the same group. The fact that this phenomenon occurs across very diverse animal groups (including human groups), seems to indicate that there are intrinsic benefits to imitation, or at least action exploiting socially acquired information. In this paper, we introduce a simple social influence model that can be seen as a system of coupled exponential smoothing equations. The model is a binary decision model in which the tendency of an agent to adopt one of the two available behaviors/opinions is reinforced by the observation of another agent exhibiting a particular behavior or holding a certain opinion. We explored the dynamics of the system when agents can always be observed or when agents are observable only at certain times. Both of these circumstances model real-life situations. We observe that consensus is reached in both cases; however, the consensus state in one case is the opposite of the consensus state in the other case.

Future work includes a more thorough study of the system's dynamics and parameters both in simulation and analytically. An application area that benefits from our work is swarm intelligence. We discussed how our model captures the essential dynamics of two families of swarm intelligence algorithms for optimization, ACO and PSO, as well as collective decision-making processes in swarms of robots.

Acknowledgments. Eliseo Ferrante acknowledges support from the Research Foundation Flanders (FWO) of the Flemish Community of Belgium through the H2Swarm project. Louis Rossi would like to acknowledge the support of NSF grant CCF 0916035.

References

1. Bass, F.M.: A new product growth for model consumer durables. Management Science 15(5), 215–227 (1969)
2. Bonabeau, E., Dorigo, M., Theraulaz, G.: Swarm Intelligence: From Natural to Artificial Systems. Santa Fe Institute Studies on the Sciences of Complexity, Oxford University Press, New York (1999)
3. Castellano, C., Fortunato, S., Loreto, V.: Statistical physics of social dynamics. Reviews of Modern Physics 81(2), 591–646 (2009)
4. Dorigo, M., Stützle, T.: Ant Colony Optimization. Bradford Books, MIT Press, Cambridge, MA (2004)
5. Faria, J.J., Krause, S., Krause, J.: Collective behavior in road crossing pedestrians: the role of social information. Behavioral Ecology 21(6), 1236–1242 (2010)
6. Gardner Jr., E.S.: Exponential smoothing: The state of the art–Part II. International Journal of Forecasting 22(4), 637–666 (2006)
7. Giraldeau, L.A., Valone, T.J., Templeton, J.J.: Potential disadvantages of using socially acquired information. Philosophical Transactions of the Royal Society of London. Series B: Biological Sciences 357(1427), 1559–1566 (2002)
8. Goss, S., Aron, S., Deneubourg, J.L., Pasteels, J.M.: Self-organized shortcuts in the argentine ant. Naturwissenschaften 76(12), 579–581 (1989)
9. Granovetter, M.: Threshold models of collective behavior. The American Journal of Sociology 83(6), 1420–1443 (1978)
10. Hyndman, R., Koehler, A., Ord, K., Snyder, R.: Forecasting with Exponential Smoothing. The State Space Approach. Springer, Berlin (2008)
11. Kennedy, J., Eberhart, R.: Particle swarm optimization. In: Proceedings of IEEE International Conference on Neural Networks, pp. 1942–1948. IEEE Press, Piscataway (1995)
12. Lambiotte, R., Saramäki, J., Blondel, V.D.: Dynamics of latent voters. Physical Review E 79(4), 046107, 6 pages (2009)
13. Montes de Oca, M.A., Ferrante, E., Scheidler, A., Pinciroli, C., Birattari, M., Dorigo, M.: Majority-rule opinion dynamics with differential latency: A mechanism for self-organized collective decision-making. Swarm Intelligence 5(3-4), 305–327 (2011)
14. Pillot, M.H., Gautrais, J., Gouello, J., Michelena, P., Sibbald, A., Bon, R.: Moving together: Incidental leaders and naïve followers. Behavioural Processes 83(3), 235–241 (2010)
15. Rendell, L., Boyd, R., Cownden, D., Enquist, M., Eriksson, K., Feldman, M.W., Fogarty, L., Ghirlanda, S., Lillicrap, T., Laland, K.N.: Why copy others? Insights from the social learning strategies tournament. Science 328(5975), 208–213 (2010)
16. Scheidler, A., Brutschy, A., Ferrante, E., Dorigo, M.: The k-unanimity rule for self-organized decision making in swarms of robots. Tech. rep., TR-IRIDIA/2011-23, IRIDIA, CoDE, Université Libre de Bruxelles, Brussels, Belgium (October 2011)

Identifying Critical Infrastructure Interdependencies for Healthcare Operations during Extreme Events

Jukrin Moon and Taesik Lee[*]

Department of Industrial and Systems Engineering, Korea Advanced Institute of Science and Technology, 335 Gwahangno, Yuseong-gu, Daejeon, 305-701, Republic of Korea
taesik.lee@kaist.edu

Abstract. Critical infrastructures provide vital functions for sustaining our society, and failure in a critical infrastructure leads to massive economic losses and even human casualties. As such, protecting critical infrastructure from disasters is the highest priority task for all countries. One of the key challenges is to understand and manage interdependencies between critical infrastructures. Failure in one infrastructure can cause unanticipated disruptions in others causing a cascade, and the degree and extent of damage could far exceed the initial prediction. In this paper, a critical infrastructure is viewed as a function that satisfies relevant need from a society. In fulfilling its function, a critical infrastructure may rely on resources and services that other infrastructures provide. With this view, we propose a conceptual definition for interdependency between critical infrastructures: interdependency via demand and capability. Using this definition, an interdependency matrix for critical infrastructures can be constructed, with which potential cascading scenarios can be identified. For an illustration purpose, a pilot interdependency matrix at an abstract level is presented, and a few cascading scenarios are identified and compared to those reported in prior literatures on real cases.

Keywords: infrastructure interdependency modeling, healthcare infrastructure, rare disaster management, critical infrastructure protection, cascading failure.

1 Introduction

Every year, many disasters, natural and man-made, bring catastrophic losses worldwide. In 2011, there were 332 reported natural disasters which caused more than 30,770 deaths and 244.7 million victims at a cost of US$ 366.1 billion [1]. In order to reduce the scale of economic losses and human casualties, it is very important to protect critical infrastructures of a society, and much research is dedicted to the topic of critical infrastructure protection (CIP) [2]. Critical infrastructures are vital systems and assets of a society that must be protected [3]. This paper tackles the issues of modeling interdependencies in large, complex infrastructure networks.

---

[*] Corresponding author.

K. Glass et al. (Eds.): COMPLEX 2012, LNICST 126, pp. 256–266, 2013.

1.1 Definition of Interdependencies and Cascading Failures

For most part, research on CIP tends to center around the modeling of individual infrastructure. To date, few studies have attempted to consider interdependencies between critical infrastructures. Rinaldi et al. [4] defines interdependency as a bidirectional relationship between infrastructures. Due to the interdependencies between critical infrastructures, failures in one infrastructure can cause unanticipated disruptions in others and accordingly, the degree and extent of damage may far exceed the initial expectations from the initiating event [5]. This can be referred to as "cascading failure". As infrastructures of a modern society have become increasingly interdependent, it is becoming more common to face these unanticipated cascading failures, so-called rare disasters.

1.2 Network-Based Interdependency Modeling

Although research on identifying rare disasters is still in its early stage, a number of U.S. national laboratories, including Los Alamos National Laboratory and Argonne National Laboratory, and research programs at universities have taken on this role. They have tried to understand the behavior of infrastructures by developing their own modeling and analysis tools [6]. Literature on this field fall into several categories: Agent-based modeling [7-8], System Dynamics [9-10], Input-output Model [11-14], and Petri-net [15-16].

As far as modeling is concerned, network-based modeling approaches form the basis for most researches. In network-based modeling, interlinked infrastructures are modeled as a network of nodes and edges, where nodes represent individual infrastructures and edges represent their relationship. National Infrastructure Simulation and Analysis Center (NISAC), a program within the U.S. department of homeland security (DHS), has applied network theory in an attempt to clarify interdependencies between eighteen critical infrastructure sectors. There are several types of NISAC network-based models: network flow models, system dynamics models, agent-based models and combinations of these models [17].

Critical Infrastructure Modeling System (CIMS) proposed by Dudenhoeffer [7] uses an agent-based modeling approach (ABM), where each physical infrastructure is modeled as an agent. The physical entities are displayed as nodes in infrastructure networks. Min et al. [9] developed a system dynamics model employing a functional modeling methodology, IDEF0, in order to show material flow relationships among infrastructures. Unlike other network-based approach, they model functions of infrastructures, instead of existing physical entities.

In this paper, a critical infrastructure is viewed as a function that satisfies relevant need from a society. In fulfilling its function, a critical infrastructure may rely on resources and services that other infrastructures provide. With this view, we propose a conceptual definition for interdependency between critical infrastructures: interdependency via demand and capability. Using this definition, an interdependency matrix for critical infrastructures can be constructed, with which potential cascading scenarios can be identified. Our framework is similar to IDEF0 model in that it focuses on each infrastructure's function. The main difference between our framework and IDEF0 [9] is the definition of function and the classification of dependency edges. This would be explained further in section 2.

1.3   Significance of Identifying Cascading Scenarios for Healthcare Operations

Since healthcare is directly related to human survival, it is important to guarantee the continuity of healthcare services, especially in extreme events. Healthcare operations are inevitably affected by disruptions in other infrastructures, and we certainly need to identify cascading scenarios showing how failures cascade through various infrastructures' functionalities and finally affect the provision of healthcare. So far, very little has been done in this direction. Arboleda et al. [18-19] used a network to represent interdependencies. However, they only dealt with physical healthcare facilities and lacked practical usage in other entities. Using the conceptual definitions and ensuing framework proposed in this paper, we represent infrastructure interdependencies related to healthcare domain. The identified cascading scenarios for healthcare operations of our framework will be presented in section 3.

2   Overview of Proposed Framework

This framework is motivated by the need for infrastructure interdependency networks. Although the above mentioned literatures deal with this topic, they primarily focused on existing physical entities [7, 18-19]. A few of them looked into this topic with top-down approach, functional modeling in terms of nodes [9], but did not define the edges according to the characteristics of infrastructure interdependencies. Thus, we propose a framework with both functional modeling and redefined dependency edges.

In section 2.1, we will define nodes and classify dependency edges between infrastructures into two types – Capability (C) and Demand (D). In section 2.2, an interdependency matrix for critical infrastructures will be constructed. Based on the completed matrix, the benefits of the framework will be discussed in section 3.

2.1   Infrastructure as a Function and Classification of Dependencies

In IDEF0 diagram, a function or activity has four types of edges connected to it. There are three incoming edges (input, control, mechanism) and one outgoing edge (output). A function in the IDEF0 diagram serves to transform or change the inputs in some way in order to create the outputs. Based on this IDEF0 representation, Min et al. [11] used the diagram to describe the material flows among infrastructures. We also model an infrastructure as a function, with a slightly different perspective. A critical infrastructure is viewed as a function that satisfies relevant need from a society. Thus, an infrastructure is defined as a function in the context of 1) its purpose, i.e., the need or demand from a society, 2) resources or services it uses, and 3) outputs it produces (Fig. 1). With this notion, an infrastructure may affect other infrastructures by 1) influencing their needs or 2) failing to provide necessary resources/services, thereby affecting their capability. In short, an infrastructure as a function serves to meet its demand with its capability. The failure of a function means that the function could not meet its demand with its capability. Accordingly, a function could be affected by other functions in two possible ways: the capability or the demand. We call each of these dependences as C-dependency and D-dependency (Table 1). Fig. 2(a) depicts these relationships.

For a simpler representation, we use graph representation of an infrastructure network. A node in a graph represents a function that an infrastructure serves, and a node can have two types of directed edges to indicate the type of dependency. Interdependencies are made up of multiple dependencies [4], so the two types of dependency are basic elements of our interdependency modeling framework.

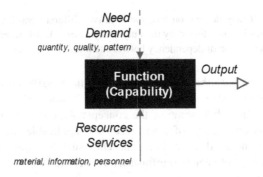

Fig. 1. In this paper, a *function* is defined by the role it serves in the middle of many interacting functions. The role of a function (*the exact quantity, quality and pattern of the demand*) is determined by the need of other functions. The ability of a function (*the capability*) may rely on *resources and services* that other functions provide.

Table 1. The classification of dependencies between *function A* and *function B*

Classification	Definition
C-dependency	'C-dependency exists from function A to function B' means that function A affects the *capability* of function B.
D-dependency	'D-dependency exists from function A to function B' means that function A affects the *demand* of function B.

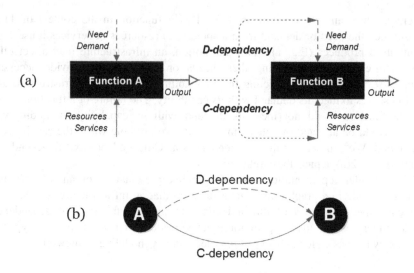

Fig. 2. (a) *Function B* may depend on *function A* in two different ways, depending on which component of *function B* was affected by the output of *function A*. (b) An edge is unidirectional and represents a specific type of dependency between two function nodes.

It is worthwhile to note that the definition and the classification of dependencies found in the previous literatures, shown in Table 2, imply that infrastructures are physical entities. Thus, while some of the concepts share common aspects with our dependency definition, many of them are not applicable to our case where infrastructures are modeled as a function. One such example is geographic or geospatial dependency between two infrastructure facilities [4].

Table 2. Previous classifications of dependencies between two nodes (*physical entities*)

Literature	Classification of dependencies
Rinaldi et al. [4]	Geographic, Physical, Cyber, Logical
Brown [17]	Geographic, Physical, Logical
Buhne et al. [20]	Requires-dependency, Exclusive-dependency, Hints-dependency, Hinders dependency
Dudenhoeffer et al. [7]	Physical, Geospatial, Societal, Policy/Procedural, Informational
Zhang et al. [21]	Physical, Functional, Budgetary, Market, Information, Environmental

2.2   Construction of an Interdependency Matrix

Based on the nodes and edges we defined in section 2.1, we will construct an interdependency matrix which can be transformed into a network.

Determining Infrastructure Nodes – Construct an Empty Matrix. Among eighteen critical sectors (CIKR) defined by the U.S. department of Homeland Security [22], we chose several relevant sectors and regrouped them into five infrastructures as shown in Fig. 3. We will occasionally call them using the abbreviation of their name: H, E, T, I[1], W.

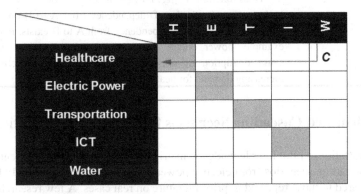

Fig. 3. The rows and columns of the interdependency matrix shows the infrastructure functions (*H, E, T, I, W*) that we chose to include in the network. The alphabet *C* written at the first row means that *there exists a C-dependency from function W to function H*. In other words, Water infrastructure affects the capability of Healthcare.

Determining Dependency Edges – Filling in Blanks in the Matrix. Determining which dependency exists from the column node to the row node, deserves our full attention and thoroughness. We need to ask ourselves whether any one of two dependencies exists between two nodes. Table 3 illustrates how to determine the existence of dependency types between two functions through a series of questions.

Table 3. The existence of C-dependency and D-dependency

Dependency Type	How to determine the existence of each dependency type between two functions?
C-dependency	We can determine whether or not 'C-dependency exists from function A to function B' by asking the following question: "Does function A *provide the resources or services* required for the capability of function B?" If the answer is *no*, then C-dependency from A to B does not exist. If it is *yes*, C-dependency from A to B exists. For example, Transportation infrastructure affects healthcare infrastructure via C-dependency because emergency medical service function requires proper functioning of transportation infrastructure.

---

[1] ICT : Information and Communications Technology.

Table 3. (*continued*)

D-dependency	We can determine whether or not 'D-dependency exists from function A to function B' by asking the following question:

"Does function A *affect the demand* of function B?"

If the answer is *no*, then D-dependency from A to B does not exist. If it is *yes*, D-dependency from A to B exists. For example, power outage in electricity infrastructure may cause disruption in self-care at patients' home thereby increasing demand for healthcare infrastructure.

3   Identified Cascading Scenarios for Healthcare Operations

As an illustration, a pilot interdependency matrix at an abstract level is presented (Fig. 4), and cascading scenarios from electric power to healthcare are identified (Table 4) and compared to those reported in prior literature on real cases. A few researches have

	H	E	T	I	W
Healthcare		CD	CD	CD	CD
Electric Power	C		C	CD	C
Transportation	CD	CD		CD	D
ICT	C	CD	C		D
Water	C	C	C	CD	

Fig. 4. The completed interdependency matrix contains the information of C and D dependencies

Table 4. The list of identified cascading scenarios (from E to H) *based on C and D-dependency*

Order	How does failure in electric power cascade to healthcare?		
1	E-H		
2	E-T-H	E-I-H	E-W-H
3	E-T-I-H	E-T-W-H	E-I-T-H
	E-I-W-H	E-W-T-H	E-W-I-H
4	E-T-I-W-H	E-T-W-I-H	E-I-T-W-H
	E-I-W-T-H	E-W-T-I-H	E-W-I-T-H

tried to identify those cascading scenarios by analyzing the data of previous power outages. Prezant et al. examined the effects of August 2003 blackout in the United States and Canada [23] on New York City's healthcare system [24]. Chang et al. [25] analyzed the data for the power outage consequences in the 1998 Ice Storm in Canada and identified impacted systems.

[E-H] – Direct Cascade from Electric Power. Imagine how easily the *capability* of healthcare (e.g., operating room or intensive care unit) can be affected. Chang et al. [25] provided more examples: medical staff taking the stairs due to malfunctioning elevators, patients tying up beds (refusing to return to blacked-out homes), hospitals calling for volunteers with no medical experience. On the other hand, Prezant et al. [24] gives an example that the *demand* of healthcare is affected by power outage: increased patients (especially respiratory patients due to electrically powered respiratory device failures) and exceptionally high call volume for 911.

[E-T-H] – Cascade through Transportation. The *capability* of transportation was affected in the power outage of 1998 Ice Storm [25]: traffic lights went off, metro stations closed and gas stations unable to pump fuel. The 2003 blackout [24] also caused transportation failure and healthcare support (*capability*) was not delivered to an increasing volume of patients (*demand*).

[E-I-H] – Cascade through ICT. Communication was the most time consuming problem in the power outage of 1998 Ice Storm [25]. Emergency lines became flooded, affecting the *capability* of ICT. Without ICT's help, without computers (*capability*), it was impossible to access vital information needed in healthcare.

The real cases from two previous power outages matched with simple cascading scenarios (e.g., [E-H], [E-T-H], [E-I-H] and [E-W-H]), but cascading scenarios going through more than two infrastructures could not be identified. This is because the prior researches [24, 25] focused on the impact and extent of cascading failures instead of the cascading process. Even though our framework is not based on the data of previous disasters, it can generate, by thought-experiment, the same scenarios as other researches. On top of that, we could come up with even more complicated scenarios like [E-I-T-W-H] based on the interdependency matrix.

[E-I-T-W-H] – Cascade through More Than Two Infrastructures. From other researches [24, 25], we got real examples of simple cascading scenarios like [E-I-H], [E-T-H], [E-W-H]. Based on those examples, we could deduce that much more complicated scenarios (e.g. [E-I-T-W-H]) could have been occurring simultaneously. It can be explained by using the examples in [25]. Firstly, ICT was affected by power outage. This resulted in a cut of electronic communications, and unknown road conditions. As transportation infrastructure's capability was affected, bottled water supplies to communities could not be delivered. Finally, because the capability of water infrastructure was affected, many shelters and other healthcare facilities were negatively affected.

The exhaustiveness of the scenarios in the prior researches [24, 25] cannot be guaranteed because investigating the cascading failures already happened in earlier events allows limited imagination. On the other hand, the exhaustiveness of identified cascading scenarios in this paper (Table 4) can be guaranteed on the basis of C- and D- dependencies. Especially, checking the existence of D-dependency is highly advantageous for an exhaustive enumeration of possible cascading scenarios. Table 5 shows the cascading scenarios identified from only C-dependency. It shows that many scenarios with high order (e.g., [E-W-T-H], [E-W-I-H], [E-T-W-I-H], [E-I-W-T-H], [E-W-T-I-H], [E-W-I-T-H]) were omitted from Table 4. Thus, we can say with confidence that our framework is highly beneficial in that it saves our efforts to analyze massive data.

Table 5. The list of identified cascading scenarios (from E to H) based on *only C-dependency*

Order	How does failure in electric power cascade to healthcare?		
1	E-H		
2	E-T-H	E-I-H	E-W-H
3	E-T-I-H	E-T-W-H	E-I-T-H
	E-I-W-H		
4	E-T-I-W-H	E-I-T-W-H	

4   Conclusion

Disasters have a big influence and will continue to impact our infrastructures. It is in everyone's interest to understand how we can manage them effectively and efficiently. One of the key challenges is to understand and manage interdependencies between critical infrastructures. In this paper, a critical infrastructure is viewed as a function that satisfies relevant need from a society. An infrastructure may affect other infrastructures by 1) influencing their needs or 2) failing to provide necessary resources/services, thereby affecting their capability. With this view, we propose a conceptual definition for interdependency between critical infrastructures: interdependency via demand and capability. Using this definition, an interdependency matrix for critical infrastructures can be constructed, with which potential cascading scenarios can be identified.

The strengths of the framework come from the simplicity of construction and the exhaustiveness of redefined nodes and edges. It gives an opportunity to identify cascading scenarios thoroughly without handling with massive data. As an illustration, a pilot interdependency matrix at an abstract level was presented, and the identified cascading scenarios were compared to those reported in prior literatures on real cases.

Further studies are needed to explore the dynamics of cascading failure. We are planning to generate a system dynamics model to show cause-and-effect relationships between capabilities and demands of infrastructures. Also, in order to incorporate more detailed level of knowledge of each individual infrastructure, we are going to decompose infrastructures further and conduct a Delphi survey to confirm the existence and the types of dependencies between functions. Consensus-driven edges would make a thorough investigation on critical infrastructures in the Republic of Korea. It is expected to draw meaningful insights on disaster management.

Acknowledgments. This research was supported by the Public welfare & Safety research program through the National Research Foundation of Korea (NRF) funded by the Ministry of Education, Science and Technology (2011-0029883).

References

1. Guha-sapir, D., Vos, F., Below, R., Ponserre, S.: Annual Disaster Statistical Review 2011: The Numbers and Trends. CRED, Brussels (2012)
2. Brunner, E.M., Suter, M.: International CIIP Handbook 2008/2009: An Inventory of 25 National and 7 International Critical Information Infrastructure Protection Policies, Center for Security Studies, ETH Zurich (2008)
3. United States Congress, U.S.A. Patriot Act (2001)
4. Rinaldi, S.M., Peerenboom, J.P., Kelly, T.K.: Identifying, understanding, and analyzing critical infrastructure interdependency. IEEE Control Systems Magazine 21, 11–25 (2001)
5. Holt, M., Campbell, R.J., Nikitin, M.B.: Fukushima Nuclear Disaster. In: Congressional Research Service 7-570 (2012)
6. Pederson, P., Dudenhoeffer, D., Hartley, S., Permann, M.: Critical Infrastructure Interdependency Modeling: A Survey of U.S. and International Research, Idaho National Laboratory (2006)
7. Dudenhoeffer, D., Permann, M., Manic, M.: CIMS: A framework for infrastructure interdependency modeling and analysis. In: Proceedings of 2006 Winter Simulation Conference, pp. 478–485. IEEE, Piscataway (2006)
8. Lee, E.E., Mendonca, D., Mitchell, J.E., Wallace, W.A.: Restoration of services in interdependent infrastructure systems: A network flows approach. IEEE Transactions on Systems, Man, and Cybernetics – Part C: Applications and Reviews 37(6) (2007)
9. Min, H.-S.J., Beyeler, W., Brown, T., Son, Y.J., Jones, A.T.: Toward modeling and simulation of critical national infrastructure interdependencies. IIE Transactions 39(1), 57–71 (2007)
10. Homer, J.B., Hirsch, G.B.: System Dynamics Modeling for Public Health: Background and Opportunities. American Journal of Public Health 96(3) (2006)
11. Marti, J.R., Hollman, J.A., Ventura, C.E., Jatskevich, J.: Dynamic recovery of critical infrastructures: real-time temporal coordination. International Journal of Critical Infrastructures 4(1/2), 17–31 (2008)
12. Haimes, Y.Y., Horowitz, B.M., Lambert, J.H., Santos, J.R., Crowther, K.G.: Inoperability input-output model for interdependent infrastructure sectors. I. Theory and Methodology Journal of Infrastructure Systems 11(2), 67–79 (2005)
13. Rahman, H.M.: Modeling and Simulation of Interdependencies between the communication and information technology infrastructure and other critical infrastructures. A thesis submitted to the University of British Columbia (2009)
14. Rahman, H.M., Armstrong, M., Marti, J.R., Srivastava, K.D.: Infrastructure Interdependencies simulation through matrix partitioning technique. International Journal of Critical Infrastructures 7(2) (2011)
15. Ge, Y., Xing, X., Cheng, Q.: Simulation and analysis of infrastructure interdependencies using a Petri net simulator in a geographical information system. International Journal of Applied Earth Observation and Geoinformation 12, 419–430 (2010)

16. Sultana, S., Chen, Z.: Modeling flood induced interdependencies among hydroelectricity generating infrastructures. Journal of Environmental Management 90, 3272–3282 (2009)
17. Brown, T.: Dependency Indicators. Technical article (ID: SS31) in Wiley handbook of science and technology for homeland security
18. Arboleda, C.A., Abraham, D.M., Richard, J.P., Lubitz, R.: Impact of interdependencies between infrastructure systems in the operation of health care facilities during disaster events. In: Joint International Conference on Computing and Decision Making in Civil and Building Engineering, pp. 3020–3029 (2006)
19. Arboleda, C.A., Abraham, D.M., Lubitz, R.: Simulation As a Tool to Assess the Vulnerability of the Operation of a Health Care Facility. J. Performance of Constructed Facilities 21(4), 302–312 (2007)
20. Buhne, S.G., Halmans, G., Pohl, K.: Modelling Dependencies between Variations Points in Use Case Diagrams. In: Proceeding of 9th International Workshop on Requirements Engineering, pp. 43–54 (2003)
21. Zhang, P., Peeta, S.: Dynamic Game Theoretic Model of Multi-Layer Infrastructure Networks. Networks and Spatial Economics 5(2), 147–178 (2005)
22. U.S. Department of Homeland Security: National Infrastructure Protection Plan: Partnering to enhance protection and resiliency (2009)
23. U.S.-Canada Power System Outage Task Force: Final Report on the August 14, 2003 Blackout in the United States and Canada: Causes and Recommendations (2004)
24. Prezant, D.J., Clair, J., Belyaev, S., Alleyne, D., Banauch, G., Davitt, M., Vandervoorts, K., Kelly, K.J., Currie, B., Kalkut, G.: MPH: Effects of the August 2003 blackout on the New York City healthcare delivery system: A lesson for disaster preparedness. Critical Care Medicine 33(1), 96–101 (2005)
25. Chang, S.E., McDaniels, T.L., Mikawoz, J., Peterson, K.: Infrastructure failure interdependencies in extreme events: power outage consequences in the 1998 Ice Storm. Natural Hazards 41(2), 337–358 (2007)

Information Dynamic Spectrum Predicts Critical Transitions

Kang-Yu Ni and Tsai-Ching Lu

HRL Laboratories, Malibu, CA 90265, USA
{kni,tlu}@hrl.com

Abstract. This paper addresses the need of predicting system instability toward critical transitions occurred in complex systems. A novel information dynamic spectrum framework and a method for automated prediction of system trajectories are proposed. Our framework goes beyond unidirectional diffusion dynamics to investigate heterogeneously networked dynamical systems with transient directional influence dynamics. Our method automatically analyzes the input time series of system instability to predict the instability trajectories toward critical transitions.

Keywords: Transfer entropy, prediction, information dynamics.

1 Introduction

In complex social-technological systems, self-organized emerging interactions provide the benefits of exchanging information and resources effectively, yet at the same time increase the risk and pace of spreading attacks or failures. A small perturbation on a complex system operating in a high-risk unstable region can induce a critical transition that leads to catastrophic failures. In this work, we propose to avoid catastrophic failures by detecting early warnings of critical transitions and predicting the likelihood of system trajectories.

A review on early warning signals for critical transitions, originated in the ecological domain, can be found in [8]. Signals such as increased temporal correlation, skewness, and spatial correlation in population dynamics are used to quantify the phenomena of critical slowing down as early warning indicators of critical transitions. The reviewed methods in [8] have only been applied to homogeneous lattices, not heterogeneously networked dynamical systems. The very recent review in [9] highlights the critical role of heterogeneous network structures in anticipating critical transitions.

There are a few recent works of early warning signals for heterogeneously networked dynamical systems [6,7]. A spectral early warning signals (EWS) theory is developed to detect the approaching of critical transitions and estimate the system structure and network connectivity near critical transitions using the covariance spectrum. Although Spectral EWS quantifies how much entities of a system are moving together, the symmetric nature of covariance spectrum does not permit the analysis of directional influences among entities.

K. Glass et al. (Eds.): COMPLEX 2012, LNICST 126, pp. 267–280, 2013.

There have been attempts in identifying directional influence in complex systems using transfer entropy [10]. In [11], symbolic transfer entropy (STE) is proposed to analyze brain electrical activity data for the detection of the asymmetric dependences, and to identify the hemisphere containing the epileptic focus without observing actual seizure activity. In [4], the transfer entropy matrix is used on financial market data to analyze asymmetrical influence from mature markets to emerging markets. Although global transfer entropy quantified in [4,11] showed promising results in analyzing financial and neurophysiological data, they fail short to quantify local transfer entropy [5] changing in structure and time. Moreover, these papers do not predict system instability.

To detect critical transitions and predict instability trajectories, we propose an information dynamic spectrum framework to quantify directional influences in heterogeneously networked dynamical systems. Our framework is based on a novel Associative Transfer Entropy (ATE) measure which decomposes the pairwise directional influence of transfer entropy to associative states of asymmetric, directional information flows. We transform multivariate time series of complex systems into the spectrum of Transfer Entropy Matrix (TEM) and Associative Transfer Entropy Matrix (ATEM) to capture information dynamics of the system. We develop the novel spectral radius measure of TEM and ATEM to detect early warning signals of source-driven instability, and induce directional influence structure to identify the source and reveal dynamics of directional influences. The nature of convex growth of spectral radius of TEM and ATEM prior to critical transitions enables us to generate the probabilistic light cones of system trajectories using natural logarithmic curve modeling. We demonstrate our methods on (1) early detection and prediction of instability of non-Foster circuit, (2) changes of directional influences in Latin America stock indices during 2008 financial crisis, and (3) asymmetric derivers in Wikipedia editing behaviors.

2   Information Dynamics Spectrum

Given the time series of a system of m elements: $X(t) = [x_1(t), x_2(t), , x_m(t)]^T$, the Transfer Entropy (TE) from source x_j to destination x_i is defined as [10],

$$T_{x_j \to x_i} = \sum_{(x_{i,t+\tau}, x_{i,t}^{(k)}, x_{j,t}^{(\ell)}) \in D} p(x_{i,t+\tau}, x_{i,t}^{(k)}, x_{j,t}^{(\ell)}) \log \frac{p(x_{i,t+\tau} | x_{i,t}^{(k)}, x_{j,t}^{(\ell)})}{p(x_{i,t+\tau} | x_{i,t}^{(k)})}. \tag{1}$$

where D is the set of all possible values of $(x_{i,t+\tau}, x_{i,t}^{(k)}, x_{j,t}^{(\ell)})$. TE quantifies the amount of information transferred from x_j to x_i and is asymmetric. More details can be found in [10].

2.1   Associative Transfer Entropy

The idea of the proposed associative transfer entropy is to decompose TE by constraining associated states of processes. It is often important to distinguish the types of the information flow, in addition to the amount of information flow. We propose a new measure, Associative Transfer Entropy (ATE) at state D_k, for $i \neq j$ defined by:

$$T^{D_k}_{x_j \to x_i} = \sum_{(x_{i,t+\tau}, x^{(k)}_{i,t}, x^{(\ell)}_{j,t}) \in D^{D_k}} p(x_{i,t+\tau}, x^{(k)}_{i,t}, x^{(\ell)}_{j,t}) \log \frac{p(x_{i,t+\tau}|x^{(k)}_{i,t}, x^{(\ell)}_{j,t})}{p(x_{i,t+\tau}|x^{(k)}_{i,t})}, \quad (2)$$

where D_k is a subset of D that represents a certain associated state between x_i and x_j. The purpose of ATE is to capture information transfer between two variables for a particular state association. For the simplest example, we can decompose TE into two influence classes: one positive ATE+ and the other negative ATE−. Therefore ATE in this case is able to distinguish two situations where the amount of information transfer is the same but have opposite effects: one source drives the destination the same direction and the other source drives the destination the opposite direction.

For a simple illustration, we simulate two time series $x(t)$ and $y(t)$ with a binary difference between the current value and the next value, e.g. 1 represents an increment and 0 represents a decrement. The probabilities of increment/decrement are conditioned by the previous increment/decrement. Fig. 1 illustrates a few simulated data with the following setup. Let $\dot{x}(t) = x(t) - x(t-1)$ and $\dot{y}(t) = y(t) - y(t-1)$. We fix conditional probabilities $Pr(\dot{x}(t+1) = a|\dot{x}(t) = b) = 0.5$, $Pr(\dot{y}(t+1) = a|\dot{y}(t) = b) = 0.5$, and $Pr(\dot{x}(t+1) = a|\dot{y}(t) = b) = 0.5$, where a and b are all combinations of 0 and 1. The only bias comes from $Pr(\dot{y}(t+1) = a|\dot{x}(t) = b) = p$. When $p < 0.5$, an increment in x is more likely to cause an decrement in y, and an decrement in x is more likely to cause an increment in y. Fig. 1 (a)-(f) show a few such simulated time series of length 1000 from arbitrary initializations with $p = 0, 0.2, 0.4, 0.6, 1$, respectively.

Fig. 2 plots the TE and ATEs as functions of p, for $0 \leq p \leq 1$, averaged over 100 trials for each p. In this binary case, we decompose TE in positive association ATE+ and negative association ATE−, summing over the sets $D_+ = \{(\dot{y}(t+1), \dot{x}(t), \dot{y}(t)) = (0,0,0), (0,0,1), (1,1,0), \text{or}(1,1,1)\}$ and $D_- = \{(\dot{y}(t+1), \dot{x}(t), \dot{y}(t)) = (0,1,0), (0,1,1), (1,0,0), \text{or}(1,0,1)\}$, respectively. ATE+ sums over positive association: $\dot{y}(t+1)$ and $\dot{x}(t)$ are both 0 or both 1. ATE− sums over negative association: one of $\dot{y}(t+1)$ and $\dot{x}(t)$ is 0 and the other is 1. Therefore, as p increases, ATE+ increases and ATE− decreases. One the other hand, TE does not distinguish the types of influence between two time series. One can further carry out the expectation values for ATE+ and ATE− in the case of (a): ATE+ $= p \log_2(p/0.5)$ and ATE− $= (1 - p) \log_2((1 - p/0.5)$. Near $p = 1$, ATE+ in (c) is much smaller than ATE+ in (b), even though the positive influence from $\dot{x}(t) \to \dot{y}(t + 1)$ is high. The explanation of this is that TE/ATE± negate the amount of influence by itself: $\dot{y}(t) \to \dot{y}(t+1)$. The self-influence in (c) is 0.9, much larger than that of (b), which is 0.6. Alternatively, TE/ATE± measures the net influence, which is the external influence subtracted by the internal dynamics.

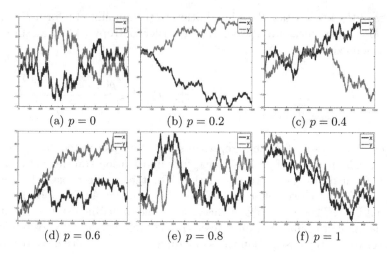

Fig. 1. Examples of simulated data generated from the following probabilities. The conditional probabilities of $Pr(\dot{x}(t+1)|\dot{x}(t))$, $Pr(\dot{y}(t+1)|\dot{y}(t))$, and $Pr(\dot{x}(t+1)|\dot{y}(t))$ are fixed and unbiased. The only bias is from $x'(t) \rightarrow \dot{y}(t+1)$: $Pr(\dot{y}(t+1) = 0|\dot{x}(t) = 0) = p$ and $Pr(\dot{y}(t+1) = 1|\dot{x}(t) = 1) = p$. The influence from x to y is negative when $p < 0.5$, as seen in (a-c). Similarly, the influence from x to y is positive when $p > 0.5$, as seen in (d-f).

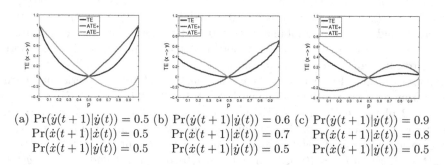

(a) $Pr(\dot{y}(t+1)|\dot{y}(t)) = 0.5$ (b) $Pr(\dot{y}(t+1)|\dot{y}(t)) = 0.6$ (c) $Pr(\dot{y}(t+1)|\dot{y}(t)) = 0.9$
$\quad\;\; Pr(\dot{x}(t+1)|\dot{x}(t)) = 0.5 \quad\;\;\; Pr(\dot{x}(t+1)|\dot{x}(t)) = 0.7 \quad\;\;\; Pr(\dot{x}(t+1)|\dot{x}(t)) = 0.8$
$\quad\;\; Pr(\dot{x}(t+1)|\dot{y}(t)) = 0.5 \quad\;\;\; Pr(\dot{x}(t+1)|\dot{y}(t)) = 0.5 \quad\;\;\; Pr(\dot{x}(t+1)|\dot{y}(t)) = 0.5$

Fig. 2. ATE is able to distinguish the types of influence in information transfer. The TE, ATE+ and ATE− curves are functions of $p = Pr(\dot{y}(t + 1) = 0|\dot{x}(t) = 0) = Pr(\dot{y}(t + 1) = 1|\dot{x}(t) = 1)$.

2.2   Local TE/ATE

We next consider dynamic data, by which we mean the amount of information flow from one node to another is not constant. For dynamic data, we calculate the TE/ATE of time series in a local time window, so that TE/ATE becomes a function of time. For illustration, we consider a two-node network, which produces two time series $x(t)$ and $y(t)$, $t = 1, ..., 1000$. Let $\dot{x}(t) = x(t) - x(t-1)$ and $\dot{y}(t) = y(t) - y(t-1)$. Consider the binary case $n = 2$ for $\dot{x}(t)$ and $\dot{y}(t)$, where $\dot{x}(t) = 1$ represents increment and $\dot{x}(t) = 0$ represents decrement. We then simulate the data according to $Pr(\dot{y}(t+1) = 0|\dot{x}(t) = 0) = Pr(\dot{y}(t+1) = 1|\dot{x}(t) = 1) = p_1$ and $Pr(\dot{y}(t+1) = 0|\dot{x}(t) = 1) = Pr(\dot{y}(t+1) = 1|\dot{x}(t) = 0) = p_2$. In the simulated data, the probabilities of associative influence change at $t = 300$ and 600. For $1 \le t < 300$, $p_1 = [0.9, 1]$ and $p_2 = [0, 0.2]$; for $300 \le t < 600$, $p_1 = [0, 0.2]$ and $p_2 = [0.9, 1]$; and for $600 \le t \le 1000$, $p_1 = [0.9, 1]$ and $p_2 = [0, 0.2]$. Therefore, initially x has a strong positive influence, then a strong negative influence during the middle period, finally a strong positive influence again at the end. For simplicity, we fix $Pr(\dot{x}(t+1) = 0|\dot{y}(t) = 0) = Pr(\dot{x}(t+1) = 1|\dot{y}(t) = 1) = p_1$ and $Pr(\dot{x}(t+1) = 0|\dot{y}(t) = 1) = Pr(\dot{x}(t+1) = 1|\dot{y}(t) = 0) = p_2$. For larger number of states, for example $n = 4$, one way to define the increment and decrement is: 0 and 1 represent decrement by 2 and 1, respectively and 2 and 3 represent increment by 1 and 2, respectively. Similarly, for $n = 6$ levels, 0, 1, 2, 3, 4, 5 represent -3, -2, -1, 1, 2, 3, respectively. The top row of Fig. 3 shows the simulated data, and the bottom row shows TE/ATE$\pm$ as functions of time, which are calculated with sliding window size = 100. We can see that ATE+ and ATE− are able to capture the types of influence besides the amount of influence that change over time. Note that for $n > 2$, one can decompose ATE into more than two states.

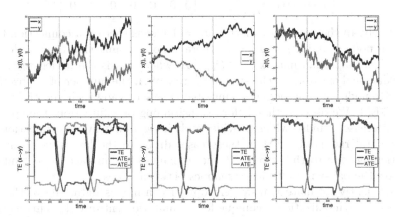

Fig. 3. TE/ATE as functions of time in the dynamic case. $Pr(\dot{x}(t+1)|\dot{x}(t),\dot{y}(t))$ is fixed and unbiased for all cases and $Pr(\dot{y}(t+1)|\dot{y}(t),\dot{x}(t))$ is switched at time $t = 300$ and 600, indicated by the blue vertical lines. The number of increment and decrement levels are $n = 2, 4, 6$ for (a), (b), and (c), respectively.

2.3   Spectral Radius of TEM/ATEM

The ATE Matrix (ATEM) T^{D_k} of a system of m elements is an $m \times m$ matrix with ij^{th} entry $(T^{D_k})_{ij} = T^{D_k}_{x_i \to x_j}$. Similarly, the ij^{th} entry of the $m \times m$ TE Matrix (TEM) is $D_{x_i \to x_j}$. TEM has been used in [4] to reveal the asymmetric influences from mature markets to emerging markets. For dynamic data, we calculate pairwise local TE/ATEs to form TEM/ATEM. Since TE/ATE is directional, TEM/ATEM is non-symmetric. To identify the amount of information transfer in the system, we calculate the spectral radius of the TEM/ATEM, the largest absolute eigenvalue of the matrix.

Since the TEM matrix is nonsymmetric, its eigenvalues are complex-valued. We use the spectral radius of the TEM matrix to measure the total amount of information flow of the entire network. Fig. 4 shows spectral radius of TEM as a function of time for each simulated data. The data is simulated according to the following pitchfork bifurcation equation:

$$x_t = \tanh(ct - 10)x - x^3 + \alpha \triangle x + \sigma d\omega, \tag{3}$$

where $\triangle$ represents graph Laplacian. Fig. 4 shows a few canonical structures of graph: (a) chain, (b) directed chain, and directed binary trees, (c) downward and (d) upward. The adjacency matrix A and Laplacian matrix L of (d) are:

$$A = \begin{bmatrix} 0 & 1 & 1 & 0 & 0 & 0 & 0 \\ 0 & 0 & 0 & 1 & 1 & 0 & 0 \\ 0 & 0 & 0 & 0 & 0 & 1 & 1 \\ 0 & 0 & 0 & 0 & 0 & 0 & 0 \\ 0 & 0 & 0 & 0 & 0 & 0 & 0 \\ 0 & 0 & 0 & 0 & 0 & 0 & 0 \\ 0 & 0 & 0 & 0 & 0 & 0 & 0 \end{bmatrix} \text{ and } L = \begin{bmatrix} 2 & -1 & -1 & 0 & 0 & 0 & 0 \\ 0 & 2 & 0 & -1 & -1 & 0 & 0 \\ 0 & 0 & 2 & 0 & 0 & -1 & -1 \\ 0 & 0 & 0 & 0 & 0 & 0 & 0 \\ 0 & 0 & 0 & 0 & 0 & 0 & 0 \\ 0 & 0 & 0 & 0 & 0 & 0 & 0 \\ 0 & 0 & 0 & 0 & 0 & 0 & 0 \end{bmatrix}, \tag{4}$$

respectively. In the adjacency matrix, $a_{ij} = 1$ means there is a connection from node j to node i, and $a_{ij} = 0$ means there is not a connection from node j to node i. The Laplacian matrix $L = D_{\text{in}} - A$, where D_{in} is the in-degree matrix whose diagonal entry d_{ii} is the sum of ith row of A, the number of connections going into node i. Fig. 4 plots the simulated time series and their spectral radius of the TEM. We observe that before transitioning or bifurcation, the spectral radius decreases rapidly, which provides early indication of system transitioning. The spectral radius of TEM in these cases drops to the lowest point during transitioning, due to the strong internal dynamics within each node, and TE measures the pair-wise influence dynamics between nodes.

To efficiently estimate transfer entropy of continuous data, the method in [11] uses the symbolization technique for permutation entropy [1]. This method to estimate transfer entropy is robust and computationally fast. We adapt the symbolization technique to calculate ATE. Specifically, the continuous-valued time sequence $\{x(t)\}_{t=1}^{N}$ is symbolized by first ordering the values of $\{x(t), x(t + 1), x(t + m)\}$ with $1, 2, , m$ and denoting $\hat{x}(t) =$ associated permutation of order m the symbol of $x(t)$ at t. Then we estimate the ATE of $\{x(t)\}_{t=1}^{N}$ by calculating ATE of $\{\hat{x}(t)\}_{t=1}^{N}$.

3 Numerical Results of Information Dynamic Spectrum

3.1 ATE Early Indication of Instability of non-Foster Circuit Data

Fig. 5 shows an ATE analysis of a non-Foster network [12]. The circuit is initially operated in the stable region, where there is no oscillation. Then a small perturbation is added. It is unknown whether the circuit will become oscillatory

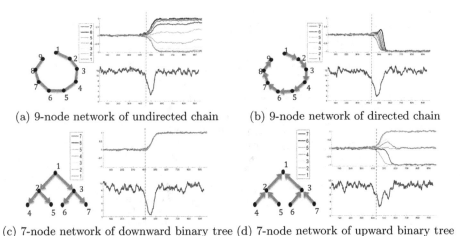

(a) 9-node network of undirected chain   (b) 9-node network of directed chain

(c) 7-node network of downward binary tree (d) 7-node network of upward binary tree

Fig. 4. Simulated pitchfork bifurcations with difference canonical graph structures. The spectral radius of TEM decreases before bifurcation, which provides an early indication for phase transition.

(unstable) or stay stable. We perform an ATE analysis to detect if the circuit will become synchronized (unstable). The top plot of Fig. 5 is the circuit in voltage over time. The bottom plot shows the TE/ATE± curves over time. The curves are obtained from absolute sum of spectrum of TEM/ATEM±, respectively. We found that the spectral radius of TEM/ATEM± also show similar outcome, but since TEM is the sum of ATEM+ and ATEM−, the ATE+ curve will never cross the TE curve. Thus, the absolute sum of the spectrum is more informative. As shown in the plot, the ATE+ curve crosses over the TE curve near 800, indicating the increasing in synchronization. In addition, the ATE+ curve reaches peak right before full synchronization, while the ATE− curve flattens because there is no negative association.

3.2 TEMs Infer Directional Influences in Latin America Stock Indices

We use TEMs of different periods to analyze the dynamics of the stocks indices around a critical event and discover the directional structures in different periods. Fig. 6 shows an analysis of a 9-node network in which each node represents a Latin America stock market index. The indices are detailed in Table 1. The top row of Fig. 6 shows the TEMs before, during, and after the 2008 October Crash from left to right. The red color corresponds to a large value, while blue corresponds to a small value. During crash, the total amount of information transfer decreases. After the crash, the TEM came back to TEM before the crash. Visualization of the network structure shows that Panama is strongly influenced by Columbia and Brazil before and after the crash, yet during the crash Panama is primarily driven by Mexico and Venezuela.

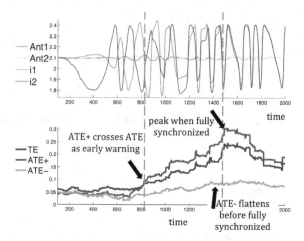

Fig. 5. The system takes multivariate time series of observed system behavior as inputs, and outputs the warnings and trends toward critical transitions with sources and paths of propagations

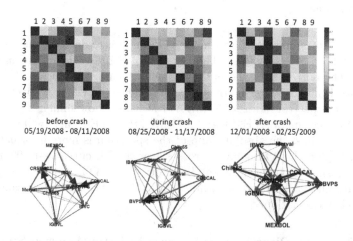

Fig. 6. TEMs infer directional influences in Latin America stock indices before, during, and after the 2008 October Crash

3.3   ATE Reveals Wikipedia Motifs That Drive the Changes

The dynamics of editor behaviors of Wikipedia's content is explored in [3] using temporal motifs, which are temporal bipartite graphs with multiple node and edge types for users and revisions. The first two rows or Fig. 7 shows 12 most frequent motifs. For example, in the second row, the first motif from the left, shows a minor edit of a Wikipedia page by an anonymous author, followed by a revert from a registered user, then followed by a minor edit from an anonymous author. The bottom row from left to right are global ATEM+, ATEM−, and TEM 2001 to 2011. We see that by using ATEM+ and ATEM−, we can identify

Table 1. Latin American stock market indices

1	2	3	4	5	6	7	8	9
BVPSBVPS	Chile65	COLCAP	CRSMBCT	IBOV	IBVC	IGBVL	Merval	MEXBOL
Panama	Chile	Columbia	Costa Rica	Brazil	Venzuela	Peru	Argentina	Mexico

important motifs that drive the changes in other motifs. In particular, ATEM+ shows that motifs 1, 8 and 12 have the most positive influence on other motifs, observing that "major add", "revert", and consecutive minor add by registered users encourage Wikipedia's content growth. ATEM− shows that motif 9 has the most negative influence on other motifs. TEM shows the asymmetric influence among the motifs but the contrast is not as strong as ATEM+ and ATEM−. Fig. 8 shows the TE/ATE± curves, spectral radii of local TEM/ATEM±, in blue/red/green, respectively. There is a significant increase in the ATE+ curve near 55, as an early indication of rapid growth in Wikipedia's contents.

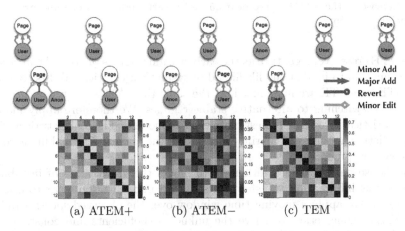

    (a) ATEM+   (b) ATEM−   (c) TEM

Fig. 7. Most frequent motifs of Wikipedia from left to right and first row to second row. (a) ATEM+ shows motifs 1, 8 and 12 have the most positive influence on other motifs. (b) ATEM− shows motif 9 has the most negative influence on other motifs. (c) TEM shows asymmetric influence among the motifs but the contrast is not as strong as ATEM+ and ATEM−.

4   Probabilistic Cones for Trajectories Prediction

The heart of the prediction method is the model-based probabilistic light cone prediction of instability trajectories using the spectral radius (TE/ATE±) of associative transfer entropy matrix, TEM/ATEM±. A MCMC method to predict citation growth based on preferential attach models is described in [2]. However, critical transitions is not discussed. Our method differs in model-based probabilistic light cone generation using natural logarithmic curve modeling. Given

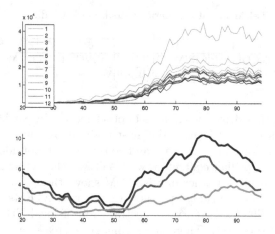

Fig. 8. Information dynamics of Wikipedia motifs. Top: Wikipedia motif occurrence in time. Bottom: TE/ATE± curves in blue/red/green, respectively. There is a significant increase in the ATE+ curve near 55, as an early indication of rapid growth in Wikipedia's contents.

ATE+/TE cross-over signature as the early warning signals of critical transition, we set our goal to predict the likelihood of system trajectories. We first observe that ATE+ and TE are maximized, at the same time ATE− flattened, prior to the system switching to alternative stable regimes. We therefore apply model-based statistical forecasting to estimate the likelihood of TE/ATE± trajectories. Our prediction method will continuously output the update of probabilistic cones as the system progresses.

We choose natural logarithm curves to model the growth rate of instability trajectory as information transfer grows toward maximization prior to critical transitions. We apply a moving time window over the observed spectral radius (TE/ATE±) time series to derive the unknown coefficients and constants for natural logarithm curves. For a given prediction time point, we generate the probabilistic light cone based on 95% confidence intervals of predicted instability trajectories with fitted natural logarithm curves.

4.1   Model-Based Forecasting

We propose a model-based statistical forecasting to estimate the TE/ATE trajectories. An advantage of analyzing the TE/ATE curves is that TE/ATE removes the spikes of the raw data. We have observed that the TE curve can be well approximated by the natural logarithm with an unknown coefficient a and a constant c:

$$g(t) = a \ln(t) + c. \tag{5}$$

Our observations of the TE/ATE curve in different types of data show that the curve approaches its maximum gradually, instead of a spiking and sudden approach. The latter, convex increase, is very common, as in Schaffer's auto-correlation, variances, and other signals. We believe that TE/ATE serves as a better a prediction function because the increase is more concave.

To estimate the coefficient a, instead of fitting the TE/ATE curves deterministically with the logarithmic function in (5), we estimate the rate of change with various discrete time steps. This generates a prediction cone. Taking the derivative of (5), we have $\hat{g}'(t) = \frac{a}{t}$. We obtain the following discretized version for the unknown a :

$$\frac{\triangle g}{\triangle t} = a\frac{1}{t}. \tag{6}$$

For a fixed timestep $\triangle t$ in a window $[T_{\text{start}}, T_{\text{end}}]$, we use the least squares to solve the unknown a. First, we write the following matrix equation:

$$
\begin{bmatrix}
\frac{1}{t_1+\triangle t/2} \\
\frac{1}{t_2+\triangle t/2} \\
\vdots \\
\frac{1}{t_k+\triangle t/2}
\end{bmatrix}
a =
\begin{bmatrix}
\frac{g(t_1+\triangle t)-g(t_1)}{\triangle t} \\
\frac{g(t_2+\triangle t)-g(t_2)}{\triangle t} \\
\vdots \\
\frac{g(t_k+\triangle t)-g(t_k)}{\triangle t}
\end{bmatrix}, \tag{7}
$$

where $t_1,, t_k+\triangle t \in [T_{\text{start}}, T_{\text{end}}]$. The approximation of $a_{\triangle}t$ for the fixed timestep $\triangle t$ is then obtained by

$$
a_{\triangle t} = \text{argmin}_a
\left\|
\begin{bmatrix}
\frac{1}{t_1+\triangle t/2} \\
\frac{1}{t_2+\triangle t/2} \\
\vdots \\
\frac{1}{t_k+\triangle t/2}
\end{bmatrix}
a -
\begin{bmatrix}
\frac{g(t_1-\triangle t)-g(t_1)}{\triangle t} \\
\frac{g(t_2-\triangle t)-g(t_2)}{\triangle t} \\
\vdots \\
\frac{g(t_k-\triangle t)-g(t_k)}{\triangle t}
\end{bmatrix}
\right\|_2, \tag{8}
$$

The constant corresponding to this timestep is then $c_{\triangle t} = g(T_{\text{end}}) - a_{\triangle t} \ln(T_{\text{end}})$. Therefore, the predicted value for the future time $T_{\text{end}} + t_d$ is

$$F_{\triangle t}(t_d) = a_{\triangle t} \ln(T_{\text{end}} + t_d) + c_{\triangle t}. \tag{9}$$

4.2 Probabilistic Light Cone and Error Estimation

Now that we have established a method to estimate the constants c and a, we can generate a probabilistic light cone at each time as we vary the δt to obtain multiple estimations of c and a. Therefore, a probabilistic light cone at a given time consists of a collections of natural logarithm curves starting from that point.

To estimate the error of this prediction for the immediate next timestep, we calculate the following:

$$\text{error}_{\triangle t}(i) = g(t_i + \triangle t) - [a_{\triangle t} \ln(t_i + \triangle t) + c(i)], \tag{10}$$

where $c(i) = g(t_{i-1}) - a_{\triangle t} \ln(t_{i-1})$.

Let E be the collection of all error$_{\triangle t}$ over all timesteps $\triangle t$ and let σ = standard deviation of E. Let G be the collection of all predicted value $G_{\triangle}t(t_d)$ over all

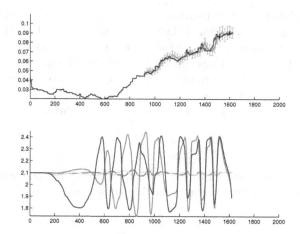

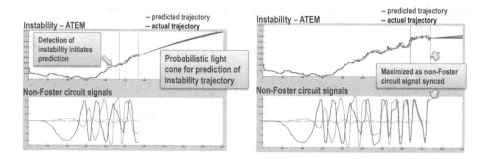

Fig. 9. Dark blue: actual trajectory. Green: predicted trajectory. Light blue: 95% confidence interval of predicted trajectory. Magenta: actual trajectory points outside the 95% confidence interval.

Fig. 10. Red: prediction light cone. Blue: actual TE trajectory. Green: predicted value.

timesteps Δt and $\mu = \text{mean}(G)$. Therefore, the 95% confidence interval for the future time $T_{end} + t_d$ is

$$CI = [\mu - 1.96 \frac{\sigma}{\sqrt{N}}, \mu + 1.96 \frac{\sigma}{\sqrt{N}}], \tag{11}$$

where N is the size of E.

Fig. 9 shows the 95% confidence interval in light blue. Two out of 36 points of the actual trajectory were outside the 95% confidence interval, right before the non-Foster circuits are fully synched around time $t = 1500$. Fig. 10 shows two snapshots of the light cone produced according to the method described above and the predicted trajectory.

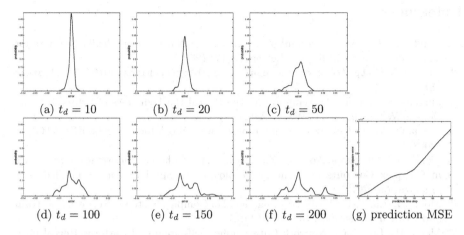

Fig. 11. Error distribution for each prediction leap time step. As expected, a smaller prediction leap time step t_d gives more accurate prediction, and as t_d increases, the error distribution flattens out. (g) MSE of prediction values increases as prediction timestep increases.

To see how far ahead in tim t_d one can predict the value in relation with error, the distribution of errors for each time t_d is plotted in Fig. 11. As the time leap t_d increases, the distribution of errors starts to flatten out, because the error increases. This is shown in Fig. 11 (g), where the curve is the mean squared error of the prediction values versus time leap t_d.

5   Conclusions

We proposed a novel information dynamic spectrum framework for automated detection of critical transitions and identification of directional influences. We have shown that the framework is able to: (1) provide an effective measure fro quantifying associative, asymmetric directional influence rather than symmetric influence, (2) provide an effective formulation that capture system-wise directional influence, rather than pair-wise influence and (3) provide an effective measure for detecting instability in systems with directional influence dynamics. Within this framework, we further proposed a method that analyzes the time series of complex systems to predict instability trajectory toward critical transitions. This enables the advancement of predicting dynamics of complex systems. The advantages include: (1) provide system trajectory prediction using instability signals (TE/ATE±) which capture information that cannot be discerned by looking at (or predicting) individual signal alone and (2) provide model-based prediction with $x\%$ (e.g. 95%) confidence interval prediction of system trajectories beyond conventional deterministic predictions or complicated probability state space predictions.

References

1. Bandt, C., Pompe, B.: Permutation Entropy: A Natural Complexity Measure for Time Series. Physical Review Letters 88(17) (2002)
2. Clauset, A.: http://www.cs.unm.edu/ aaron/blog/archives/2012/03/milestone.htm
3. Jurgens, D., Lu, T.-C.: Temporal Motifs Reveal the Dynamics of Editor Interactions in Wikipedia. In: ICWSM (2012)
4. Kwon, O., Yang, J.-S.: Information Flow between Stock Indices, 2008 EPL 82 68003 (2008)
5. Lizier, J.T., Prokopenko, M., Zoomaya, A.Y.: Coherent Information Structure in Complex Computation. Theory in Biosciences Special Issue on Guided Self-organization (2012)
6. Moon, H., Lu, T.-C.: Early Warning Signal of Complex Systems: Network Spectrum and Critical Transitions. In: WIN (2010)
7. Moon, H., Lu, T.-C.: Network Catastrophe: Self-Organized Patterns Reveal both the Instability and the Structure of Complex Networks (2012) (preprint)
8. Scheffer, M., Bascompte, J., Brock, W.A., Brovkin, V., Carpenter, S.R., Dakos, V., Held, H., van Nes, E.H., Rietkerk, M., Sugihara, G.: Early-warning signals for critical transitions. Nature 461 (2009)
9. Scheffer, M., Carpenter, S.R., Lenton, T.M., Bascompte, J., Brock, W., Dakos, V., van de Koppel, J., van de Leemput, I.A., Levin, S.A., van Nes, E.H., Pascual, M., Vandermeer, J.: Anticipating Critical Transitions. Science 338, 344 (2012)
10. Schreiber, T.: Measuring Information Transfer. Phys. Rev. Lett. 85, 461 (2000)
11. Staniek, M., Lehnertz, K.: Symbolic Transfer Entropy. Physical Review Letters 100, 15801 (2008)
12. White, C.R., Colburn, J.S., Nagele, R.G.: A Non-Foster VHF Monopole Antenna. IEEE Antennas and Wireless Propagation Letters 11 (2012)

Self-image and the Emergence of Brand Loyalty in Networked Markets

Andrzej Nowak[1,*], Paul Ormerod[2], and Wojciech Borkowski[3]

[1] Advanced School for Humanities and Social Sciences, University of Warsaw and Florida Atlantic University
[2] University of Durham, UK and Volterra Partners
[3] Institute for Social Studies, University of Warsaw
andrzejn232@gmail.com

Abstract. Brand loyalty consists of a consumer's commitment to repurchase or otherwise continue using a given brand and is demonstrated by repeated buying of a product or service, or other positive behaviours such as word of mouth advocacy. Standard models of the emergence of brand loyalty consider the behaviour of autonomous individuals who are essentially reacting to the objective attributes of the brand. Here, we show that brand loyalty can be regarded as a social construct, which emerges when the fundamental psychological principle of self-image is combined with agents reacting to each others' decisions in social network markets. Brand loyalty can emerge even when agents find it hard to distinguish between brands in terms of their objective attributes. We illustrate the principles in the context of the well-known model of binary choice with externalities. We endogenise the behaviour of agents using the principle of self-image, and illustrate the consequences in situations where consumers face not a one-off choice of adopting or not adopting, but a chain of mutually dependent decisions about complex products over a period of time.

Keywords: brand loyalty, self-image; evolving thresholds; cascades, binary choice with externalities.

1 Introduction

Brand loyalty consists of a consumer's commitment to repurchase or otherwise continue using a given brand and is demonstrated by repeated buying of a product or service, or other positive behaviours such as word of mouth advocacy. It is key concept in marketing.

In this paper, we consider the emergence of brand loyalty in situations where consumers face not a one-off choice of adopting or not adopting, but a chain of mutually dependent decisions about complex products over a period of time. So, for example, in electronic consumer durables, over a relatively short period of time, an agent may face the choice to buy or replace his or her cell phone, personal computer, MP3 player and so on.

---

[*] Corresponding author.

K. Glass et al. (Eds.): COMPLEX 2012, LNICST 126, pp. 281–290, 2013.

We analyse the emergence of brand loyalty when consumers use the behavioural choice heuristic of copying, combined with the fundamental psychological concept of self-image. Here, we show that brand loyalty can emerge regardless of the objective attributes of the alternatives choices available to consumers.

In section 2 we set the question in context and motivate the model. Section 3 describes the model and section 4 presents results and a brief discussion.

2   Background and Model Motivation

In the marketing literature, there are two very influential models of the process by which brand loyalty emerges. These were developed by Aaker [1] and by Dyson, Farr and Hollis [2]. (For convenience, we refer to these as A and DFH below). Both these models are based upon a pyramid structure, through which the loyalty of a consumer may evolve, with the most committed, the most loyal, to a brand being at the top of the pyramid.

Although there are differences between the two approaches, they have several important principles in common:

- Marketing efforts are required in order to get the consumer into the bottom layer of the pyramid
- Progression through the middle layers of the pyramid depends upon the objective attributes of the brand
- Emotional commitment to the brand only occurs at the higher levels of the pyramid

So, for example, in the DFH model, the bottom layer of the pyramid is designated as 'presence', which indicates that the consumer is aware of the brand. This awareness is achieved by, for example, advertising or by making the brand available at a wide range of outlets. In the A model, this layer of consumers are referred to as 'switchers' and marketing activity is required to raise brand awareness, a necessary condition of a consumer moving further up the pyramid.

In this latter model, the middle of the pyramid is occupied by 'satisfied buyers with switching costs'. These consumers are satisfied with the brand, and realise that switching incurs costs, either in terms of time or money, or in reduced quality. They perceive the quality of the brand to be better than the market average. At level 3 of the DFH model, 'performance', the attributes of the brand need only be as good as those of the market average, although to progress to the next level, 'advantage', consumers need to be convinced that its qualities are superior to the average. In both the models, the attributes of the brand are important.

Finally, at the highest levels, an emotional bond develops between the consumer and the brand. In the DFH model, consumers can in fact progress to the highest level if they have a strong rational belief in the superiority of the brand over its competitors, but emotional belief can bring about the same behavioural effect. In the A model, emotional benefits are associated both with level 4 ('brand likers') and the highest level, 5 ('committed buyer').

Many modern products are often complex and difficult to evaluate, even when large amounts of information are available in the form of expert reviews in specialised journals, consumer reviews on the internet and so forth. This is especially the case with markets created by new technologies, developing a stream of products which hitherto did not exist.   In such circumstances, it may be very difficult for the consumer to differentiate between products in terms of their attributes

Consumers in such circumstances often pay attention to the decisions of others and use these as the basis of their own decisions, rather than attempting to evaluate the objective attributes of the product. They may, for example, do so either because they have limited information about the problem itself or limited ability to process even the information that is available. Simon, in an article which is the foundation for all modern developments in behavioural economics [3], argues that such circumstances may obtain in many actual circumstances.

Hauser [4] suggests that it is ecologically rational in many circumstances for consumers to use simple heuristics as a basis for decision making. The heuristic principle of making a choice by copying the decisions of others is well established empirically in a range of areas such as popular culture, financial markets and crowd behaviour. Evolutionary anthropologists and psychologists (for example Dunbar and Shultz in 2007[5]) have argued that that the anomalously large brain (neo-cortex) size in humans evolved primarily for the purposes of copying, or social learning as it is referred to in this context.

Schelling [6] offers a classic exposition of a copying heuristic, which he refers to as being one of 'binary choice with externalities'. 'Binary choice' means a situation in which a consumer faces one of two alternatives, in this instance being the choice to buy or not to buy a particular brand. 'Externalities' mean that the decision of any given consumer may have consequences for the decisions of others. If a consumer decides to buy the brand, for example, then other consumers may also decide to buy it.

It is a fundamental insight of psychology that the self represents the primary structure responsible for decision making and action   ([7]; [8]; [9]; [10]). The regulatory function of the self is governed by several principles, of which self-enhancement and self-verification are regarded as the most important (for example, Martin and Tesser [11]). Self-verification (self-consistency) describes the tendency of people to act in a way that is consistent with their self-image.

The classical notion of cognitive dissonance [12] is interpreted as a tendency to avoid beliefs or actions that are contradicting opinions held with respect to the self [13]. Confirming the self-image or the self-view represents an important motive of human behaviour as put forward in self-verification theory [14]. People have a strong tendency to act in a way consistent with their self-views. According to self-perception theory [15] individuals under conditions of uncertainty build their self-view by observing their own behaviour.

Here, we combine the concept of binary choice with externalities and the psychological concept of self-image. We show that in very general circumstances, brand loyalty emerges even when consumers, by deliberate assumption, do not attempt to evaluate the attributes, real or perceived, of the brand, but base their decisions solely on the principle of copying/social learning.

Marketing activity remains important, indeed in some ways its importance is enhanced compared to its role in the classic models of brand loyalty referred to above.

But the heuristic of copying/social learning combined with the evolving self-image of the individual is sufficient to explain the emergence of brand loyalty over a sequence of mutually dependent decisions.

3   Model Specification

The basic model which we use endogenises the behavior of agents in the context of the well-known model of binary choice with externalities (op. cit. [16]). N agents are connected on a network. The agents can be in one of two states of the world (0 and 1 for purposes of description), and initially all agents are in state 0. To start, a number of agents are chosen at random as 'seeds' to switch to state 1. Each agent is allocated a 'threshold' drawn at random from a uniform distribution on [0, 1]. The agents are therefore heterogeneous in behavior. But the thresholds, once drawn, are fixed. An agent switches from state 0 to state 1 according to the state of the world of the agents to which it is connected. If the proportion of these which are in state 1 exceeds the threshold of the agent, the agent also switches to state.

The literature on this, and related models, makes the assumption that the behaviour of agents is time invariant. In many practical contexts, agents do not face a purely one-off decision of, for example, whether or not to buy a new consumer product, but instead encounter a sequence of such decisions over time.

The behaviour of agents, the nodes of the networks, in such situations is not time invariant, but evolves over time in ways which are based on their previous decisions. The principles of our analysis of endogenous, dynamic node behaviour are built on well-established, fundamental principles of psychology.

The concept of the self-image implies that the attitudes of agents towards adoption to evolve in ways which depend upon the previous decisions of the agent on adoption. At one level, this could be thought of as amounting to simply saying that agents have a memory of what they done previously. Whilst this is obviously true, the statement does not give any scientific basis for determining how agent behavior is affected by memory. Several rules could readily be written down on a purely *a priori* basis. However, the fundamental psychological concept of self-image provides a clear, empirically grounded for how agent behavior evolves with respect to previous decisions. If an agent adopts an innovation, he or she is more likely to adopt the innovation the next time he or she is confronted with this choice. The self-image of the agent towards adoption has been altered, in a specific way. Nowak [17] show that the agent's concept of self leads to very different distributions of cascade sizes than those observed when the standard assumption is made that agent behavior is fixed.

We initialize the model as described above, and obtain a solution in the standard way. A proportion, π, switch to state 1 of the world. This solution corresponds to the situation where a new technology is introduced, and agents decide whether or not to adopt a particular brand of the technology.

During the course of a solution when an agent is called upon to make a decision, if it does not switch to state 1 i.e. purchase the brand, it sees itself as less willing to adopt products of that company in future, and it therefore increases its threshold. Similarly, if an agent switches it sees itself as more willing to switch or adopt and its threshold is reduced. The rule for adjusting thresholds ensures that they remain bounded in [0, 1]. If an agent is not called upon to decide, which is often the case in situation where the cascade size, π, is small, its threshold remains unchanged.

More formally, the adjustment to the threshold of an agent is made as follows. Denote the existing threshold of an agent by α. If the agent switches, in the next solution its threshold is set at a value drawn at random from a uniform distribution on [$(\alpha - \beta\alpha)$, α], where β is an input parameter to the model and $0 \leq \beta \leq 1$. The results are robust with respect to the assumption of a uniform distribution, which is the most convenient to implement. With the assumption of a normal distribution, for example, we have to impose over-rides to prevent the threshold from falling below zero or rising above 1. If $\beta = 1$, for example, then the new value is drawn from the range [0, α]. The higher the chosen value of β, the more sensitive are agents to their choice not to switch. If the agent does not switch, in the next solution its threshold is increased to a value drawn from a uniform distribution on [α, $(\alpha + \beta(1 - \alpha))$], so again if $\beta = 1$, the draw is made from the range [α, 1]. In the standard models with fixed thresholds, $\beta = 0$ of course.

We then re-initialize the model, with all agents again in state 0. This corresponds to the situation in which the company introduces the next variant of the technology, in the way, for example, that the cell phone and smart phone markets have evolved. There are, however, two fundamental differences to the standard model of binary choice with externalities:

- the thresholds are no longer distributed at random, but take the values which emerge during the first solution of the model
- the seeds are not drawn completely at random, but with a bias towards those who adopted the brand in the first solution

In terms of the latter point, an agent is drawn at random, and is checked for whether or not it adopted (i.e. was in state 1 of the world) in the initial solution. If it did, it becomes a seed. If not, another agent is drawn at random and the same checking takes place. This task is carried out $2N$ times, where N is the number of agents. If the required number of seeds has not been found, then the remainder is selected at random. This process is realistic in that agents which purchase the first version of the brand are more likely to purchase the second, given that the initial values of their thresholds are reduced. However, it does not bias the solution too strongly, which could possibly be the case if seeds were only chosen from those agents which adopted during the first solution of the model.

We repeat the process ten times. In the results described below, we set $N = 1000$, β which was a random number from a uniform distribution from 0 to 1, and in the first solution the thresholds are drawn at random from a uniform distribution. We consider three random networks in which the probability of any pair of agents being connected is, respectively, 0 (not connected agents), 0.01, and 0.10. The qualitative nature of the results is robust with respect to these assumptions [17].

We examine results in which the number of seeds is 20 and 100. These can be thought of as corresponding to situations in which the initial marketing effort varies, acquiring only 2 per cent of potential consumers in the former case, and 10 per cent in the latter.

The whole process described above is repeated 1,000 times.

4   Results and Brief Discussion

The thresholds of agents, indicating their willingness to be persuaded to purchase a particular brand, initially follow a random uniform distribution in [0, 1]. These are endogenous to the model, based upon the principle of the self-image of the agent. We examine how these evolve during the process of ten successive variants of a brand being introduced.

A very general feature of the results is that a group of agents emerges with thresholds which are very close to zero. In other words, they become willing to adopt the next variant of the brand almost regardless of the behavior of the agents to which they are connected on the network. Only when literally none of their social network purchases the current variant of the brand will they not adopt it. This is the case even in densely connected networks when the probability of any given pair of agents being connected is 0.1. Equally, during this process of evolving thresholds, a group of consumers emerges which is very resistant to subsequent purchases of the particular brand.

This phenomenon takes place even though, by assumption, no attention is paid to the attributes of the brand. As noted in the Introduction, the main approaches in the marketing literature to the emergence of brand loyalty do require consumers to relate to the objective qualities of the brand. We are not saying that in practice consumers pay no attention to these at all. However, brand loyalty emerges even in situations when brands are difficult to compare in terms of their attributes. As we noted above, many new technology products such as smart phones exhibit these characteristics.

The results suggest that brand loyalty is essentially a social construct, in which social network market effects [18] combined with the psychological principle of self-image are the key factors). Marketing activity remains important, indeed in some ways its importance is enhanced compared to its role in the standard models of brand loyalty. But the heuristic of copying/social learning combined with the evolving self-image of the individual is sufficient to explain the emergence of brand loyalty over a sequence of mutually dependent decisions.

Figures 1(a) to (c) plot the results across 1,000 solutions, where each solution consist of 10 decisions to adopt the product of the same brand, with networks of different degrees of connectivity with 20 seeds in each solution, and Figures 2(a) to (c) show the results with 100 seeds. In each case, a group of 'brand loyalists' emerges. But the results suggest that the initial marketing effort does have an important effect on the overall outcome of the process. This effect gets stronger with higher network connectivity.

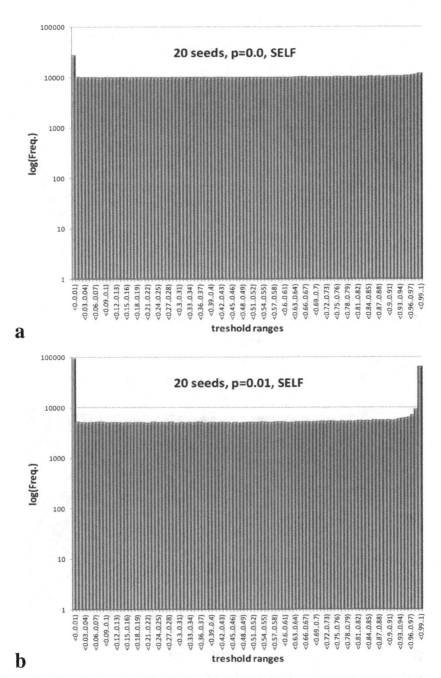

Fig. 1. The final distribution of thresholds after 10 adoptions for networks of 1000 agents and 20 seeds. Figure 1a corresponds to no connections, 1b to the network average density of 10 connections per individual, and 1c to the average network density of 100 connections per individual. The frequency is displayed on a logarithmic scale. Simulations involved 1000 agents and 1000 repetitions, so each graph shows distributions of 1 000 000 thresholds.

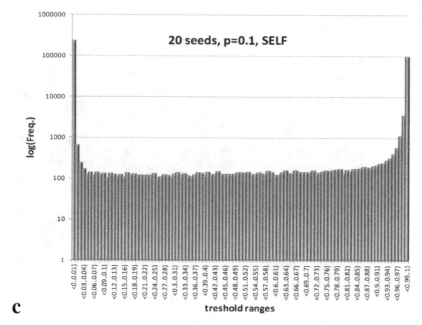

c

Fig.1. *continued*

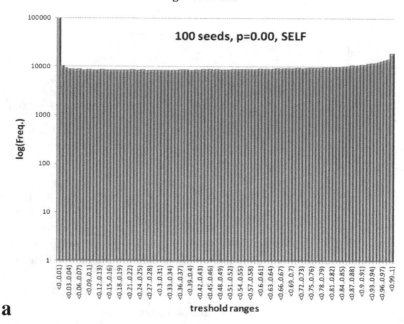

a

Fig. 2. The final distribution of thresholds after 10 adoptions for networks of 1000 agents and 100 seeds. Figure 2a corresponds to no connections, 2b to the network average density of 10 connections per individual, and 2c to the average network density of 100 connections per individual. The frequency is displayed on a logarithmic scale. Simulations involved 1000 agents and 1000 repetitions so each graph shows distributions of 1 000 000 thresholds.

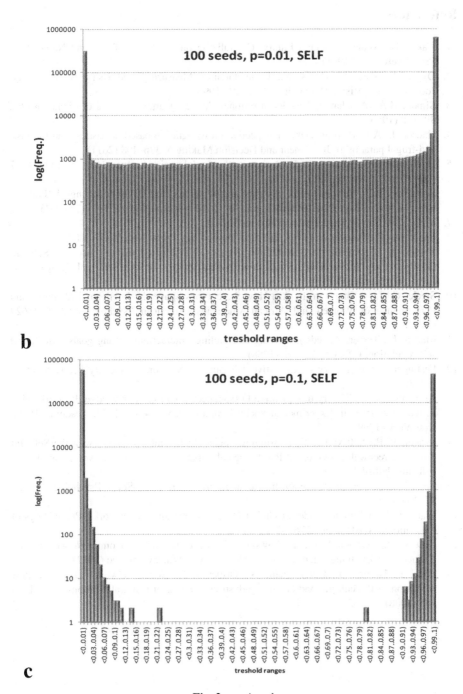

b

c

Fig. 2. *continued*

References

1. Aaker, D.: Managing Brand Equity: Capitalizing on the Value of a Brand Name. Free Press, New York (1991)
2. Dyson, P., Farr, A., Hollis, N.: Understanding, Measuring, and Using Brand Equity. Journal of Advertising Research 36(6), 9–21 (1996)
3. Simon, H.A.: A behavioral model of rational choice. Quarterly Journal of Economics 69, 99–118 (1955)
4. Hauser, J.: A marketing science perspective on recognition-based heuristics (and the fast-and-frugal paradigm). Judgement and Decision Making 6, 396–408 (2011)
5. Dunbar, R.I.M., Shultz, S.: Evolution in the social brain. Science 317, 45, 1344–1347 (2007)
6. Schelling, T.C.: Hockey Helmets, Concealed Weapons, and Daylight Saving: A Study of Binary Choices with Externalities. Journal of Conflict Resolution 17, 381–428 (1973)
7. Meyers, R.: Psychology. Pearson, Boston (2010)
8. Sedikides, C., Spencer, S. (eds.): The Self. Psychology Press, New York (2007)
9. Tesser, A., Campbell, J.: Self-definition and self-evaluation maintenance. In: Suls, J., Greenwald, A. (eds.) Social Psychological Perspectives on the Self, vol. 2, pp. 1–31. Erlbaum, Hillsdale (1983)
10. Steele, C.M.: A Threat in the Air: How Stereotypes Shape the Intellectual Identities and Performance of Women and African-Americans. American Psychologist 52, 613–629 (1997)
11. Martin, L., Tesser, A. (eds.): Striving and feeling: interactions among goals, affect, and self-regulation. Erlbaum, Mawah (1988)
12. Festinger, L.: A Theory of Cognitive Dissonance. Stanford University Press, Stanford (1957)
13. Aronson, E.: The Theory of Cognitive Dissonance: A Current Perspective. In: Berkowitz, L. (ed.) Advances in Experimental Social Psychology, vol. 4, pp. 1–34. Academic Press, New York (1969)
14. Swann, W.B.: Self-verification: Bringing Social Reality into Harmony with the Self. In: Suls, J., Greenwald, A.G. (eds.) Psychological Perspectives on the Self, vol. 2, pp. 33–66. Erlbaum, Hillsdale (1983)
15. Bem, D.: Self-perception theory. In: Advances in Experimental Social Psychology, pp. 2–57 (1972)
16. Watts, D.J.: A Simple Model of Global Cascades on Random Networks. Proceedings of the National Academy 99, 5766–5771 (2002)
17. Nowak, A., Ormerod, P., Borkowski, W.: Cascades on Economic Networks with Endogenous, Evolving Agent Behavior, University of Warsaw, mimeo (2012)
18. Potts, J., Hartley, J., Cunningham, S., Ormerod, P.: Social network markets: a new definition of the cultural and creative industries. Journal of Cultural Economics 32, 167–185 (2008)

Modeling Emergence in Neuroprotective Regulatory Networks

Antonio P. Sanfilippo[1], Jereme N. Haack[1], Jason E. McDermott[1],
Susan L. Stevens[2], and Mary P. Stenzel-Poore[2]

[1] Pacific Northwest National Laboratory, Richland, WA, USA
[2] Oregon Health & Sciences University, Portland, OR, USA
{antonio,jereme,jason.mcdermott}@pnnl.gov,
{stevensu,poorem}@ohsu.ed

Abstract. The use of predictive modeling in the analysis of gene expression data can greatly accelerate the pace of scientific discovery in biomedical research by enabling *in silico* experimentation to test disease triggers and potential drug therapies. Techniques such as agent-based modeling and multi-agent simulations are of particular interest as they support the discovery of emergent pathways, as opposed to other dynamic modeling approaches such as dynamic Bayesian nets and system dynamics. Thus far, emergence-modeling techniques have been primarily applied at the multi-cellular level, or have focused on signaling and metabolic networks. We present an approach where emergence modeling is extended to regulatory networks and demonstrate its application to the discovery of neuroprotective pathways. An initial evaluation of the approach indicates that emergence modeling provides novel insights for the analysis of regulatory networks which can advance the discovery of acute treatments for stroke and other diseases.

Keywords: regulatory networks, emergence, complex systems, agent-based modeling, neuroprotection, stroke.

1   Introduction

Stroke is the third leading cause of death and the major cause of disability in the United States. Each year, approximately 795,000 people suffer a stroke, and more than 140,000 die; those who survive are subject to recurrent attacks and long-term disability [1]. Injury due to ischemic stroke occurs as a result of a sequence of events that involve complex interactions across fundamental cell injury mechanisms. The potential for neuroprotective stroke therapy through agents that interfere in the ischemic cascade of cell injury is therefore enormous.

Preclinical evaluations of neuroprotectants have fostered high expectations of clinical efficacy and a large number of neuroprotective agents have been designed to interrupt the ischemic cascade. However, neuroprotective clinical trials run over the past 30 years have failed to provide a strategy to improve outcome after acute ischemic stroke. A 2006 review of 1026 experimental treatments in acute stroke

K. Glass et al. (Eds.): COMPLEX 2012, LNICST 126, pp. 291–302, 2013.
© Institute for Computer Sciences, Social Informatics and Telecommunications Engineering 2013

revealed that no particular drug mechanism taken forward to clinical trial has shown superior efficacy in animal models of focal ischemia [2]. These conclusions are corroborated by more recent studies [3-5]. New approaches to stroke neuroprotection are needed to break this impasse. Dynamic modeling approaches such as agent-based modeling have great potential in providing new ways of understanding neuronal ischemic injury as they support an active systems-biology analytical framework through the discovery of emergent pathways in complex biological systems. The goal of this paper is to explore and evaluate the use of agent-based modeling for the discovery of emergent neuroprotective pathways in regulatory networks.

2   Background

A major weakness in current neuroprotective approaches to stroke therapy is due to a poor understanding of the complexity of interconnections across molecular pathways induced in brain cells by ischemia. For example, [2] observe that current approaches to stroke therapy tend to frame drug activity exclusively in terms of the dominant schema of stroke damage (e.g. excitotoxicity, free radical damage). The failure of these approaches may "reflect the multifaceted nature of the sequelae of ischemic stroke" [2, p.474]. A dynamic network analysis of ischemic stroke that provides an active systems biology framework for understanding neuronal ischemic injury would therefore be better suited to address the complexity of ischemic stroke. Agent-based modeling techniques are of great interest in this regards as they support the discovery of emergent pathways in complex biological systems. The focus on modeling emergence is of particular interest as it supports the discovery of new network pathways that emerge iteratively from self-organizational properties of gene and gene clusters, as opposed to other dynamic modeling approaches such as dynamic Bayesian nets and system dynamics where network structure remains unchanged through the simulation process.

In systems biology research, agent-based and multi-agent simulations have been primarily applied at the multi-cellular level to study topics such as tumor growth [6, 7] and immune responses [8, 9]. At the molecular level, agent-based modeling efforts have focused on signaling networks [10, 11, 12] and metabolic networks [13]. So far, most modeling work on regulatory networks has relied on algorithms other than agent-based modeling [14, 15].

3   Data Selection and Network Creation

During the last decade, substantial effort has been devoted to understanding the systems biology of neuroprotection in stroke by researching the effect of preconditioning on the genomic response to cerebral ischemia [16, 17]. This work has yielded rich gene expression data that provides evidence about the genomic dynamics of neuroprotection in diverse contexts and can be used to train dynamic pathway models of neuroprotection in stroke. We use the gene expression data generated by these studies as our point of departure. These consist of microarray results from blood of mice in a transcriptional study of a mouse model of preconditioning-induced neuroprotection against stroke injury [18]. The dataset comprises five treatments: ischemic preconditioning; lipopolysaccharide (LPS) injection; CpG injection, and two control treatments (saline injection and sham surgery). Microarray data were taken at 3, 24 and 72 hours post treatment, and 3 and 24 hours post-stroke. We focus on the two drug treatments documented in these data: LPS and CpG injection.

We selected 7352 significant gene probes from the dataset described in [18] (see previous paragraph) using the normalized probe intensities obtained with the robust multi-array average algorithm [19] to evaluate significantly changing probes, and filtering for p-value < 0.05 and fold changes greater than 2.0 compared to a baseline group. Next, we identified 25 functional gene modules encompassing the selected 7352 gene probes using hierarchical clustering, as shown in Figure 1. We then applied a modified version of the *Inferelator* algorithm [20] to learn ordinary differential equations (ODEs) between clusters. This algorithm uses an approach called L1 error regression (also known as "lasso") to choose a parsimonious set of regulatory influences that can model the expression of each cluster [21, 22]. The relation between the expression of a target (y) and the expression levels of regulators with non-null influences on y (X) is expressed by the equation in (1), where τ is the time step used in model construction, and β is the weight for relationship X on y, as determined by L_1 shrinkage using least angle regression [22]. This process as a whole yields the network model in Figure 1, where the expression of a target cluster can be predicted given the expression levels of the input regulatory clusters linked to the target cluster. Details of this work are provided in [23].

$$\tau\frac{dy}{dt}=-y+\Sigma\beta_jX_j \tag{1}$$

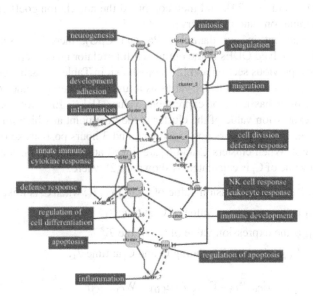

Fig. 1. Regulatory influence network model of neuroprotection in ischemia

We used simulated annealing to optimize the ODEs between clusters. Simulated annealing randomly perturbs variables in the model, then compares the performance of the perturbed model with the original using a fitness function. Perturbations that improve performance are retained in the model and perturbations that decrease performance can be retained based on a probability that is decreased over the simulation, resulting in more and more conservative changes to the model. We used a fitness function to evaluate the performance of each test matrix by the correlation of its simulated expression values with the observed expression values. The optimized ODEs yield a high correlation coefficient between the observed and simulated cluster connections (about 0.8). For further details of this work see [24].

4   Modeling Emergent Neuroprotective Regulatory Networks

Our hypothesis is that neuroprotective biological networks emerge from self-organizational regulatory properties of genes, akin to how complex systems arise in nature from swarming behaviors – e.g. food source selection and cooperative transport in ant colonies, hive construction in termites, formation of slime mold colonies [25]. More specifically, genes or functional gene clusters may be systemically driven to form new pathway connections during an ischemic event driven by a systemic push to maintain their expression values at levels that characterize a healthy organism. The neuroprotective pathways emerging from these gene networking activities may be enabled by specific treatments such as LPS or CpG injection. To test this hypothesis, we developed agent-based models (ABM) for two drug treatments in the data described in [18], LPS and CpG, using the functional clusters and ODEs in [23, 24], and then computed the correlation coefficient between the models' simulations and the observed data.

For each treatment in our dataset (LPS and CpG), there is a set of weights $\{W_1,...,W_n\}$ from solved ODEs in the optimized Inferrelator model across gene clusters $\{C_1,...,C_{25}\}$ (see previous section and [24]), as shown in Table 1. Each weight specifies how the expression level of a cluster varies as a function of being connected to another cluster. Each cluster has a set of expressions values $\{E_1,...,E_n\}$, for each treatment; each E_j is the expression value of the cluster at the time point at which microarray data were taken – e.g. 3 and 72 hours post-treatment, and 3 hours post-stroke (hour 75) – as shown in Table 2. When clusters C_i and C_j are linked at time T_k, indicated as $C_i \rightarrow^{Tk} C_j$, the expression value of C_i is calculated as shown in (2) where:

- $E_{Ci \rightarrow}{}^{Tk}{}_{Cj}$ is the expression value of C_i at time T_k, when C_i is linked to C_j
- E_{CiTk} is the expression value of C_i at time T_k
- E_{CjTk} is the expression value of C_j at time T_k
- $W_{Ci \rightarrow}{}^{Tk}{}_{Cj}$ is the weight relating C_i and C_j at time T_k.

$$E_{Ci \rightarrow}{}^{Tk}{}_{Cj} = E_{CiTk} + (E_{CjTk} * W_{Ci \rightarrow}{}^{Tk}{}_{Cj}) \qquad (2)$$

For example, the expression value of **cluster_1** when linking to **cluster_3** at hour 3 (*H3*) is calculated as shown in (3). The model uses the equation in (2) to calculate the expression value of each cluster in the simulation process.

$$E_{C1 \rightarrow}{}^{H3}{}_{C3} = -1.486 + (1.686 * -0.738) = -2.730 \tag{3}$$

Table 1. Weights relating gene cluster (LPS treatment)

	cluster_1	*cluster_2*	*cluster_3*	...
cluster_1	0.124	0	−0.738	...
cluster_2	−0.032	0	−0.283	...
cluster_3	−0.060	0	−0.197	...
...	...	...	...	...

Table 2. Cluster expression levels by time point (LPS treatment)

	Hour 3	*Hour 72*	*Hour 75*	...
cluster_1	−1.486	0.034	−0.738	...
cluster_3	1.686	0.500	1.573	...
...	...	...	...	...

In the agent-based model we have developed, the post-stroke gene-cluster network (hour 75, 3 hour post-stroke) provides the initial network state for the simulation. The expression values of the clusters in pre-stroke network (hour 72) provide the target "healthy" expression values that the clusters in the post-stroke network try to achieve. The objective of the simulation is to observe how the system behaves post-stroke in terms of forming regulatory gene networks that offer neuroprotection with reference to two treatments, LPS and CpG injection.

At every simulation tick, each cluster whose expression value is different from the target "healthy" expression value, attempts to improve its expression value by selecting a cluster to link up to at random. If the new expression value for the cluster, calculated as shown in (1), is closer to the target expression value, then it is used to replace the old expression value for the cluster; otherwise, it is rejected. A steady state in the simulation is reached when either all clusters achieve their target "healthy" expression values, or when clusters can no longer improve their expression values by establishing new network connections.

We used NetLogo [25] to implement the agent-based model that simulates the neuroprotective dynamics of the post-stroke regulatory network. Figure 2 provides a graphical description of the simulation environment with reference to the simulated LPS neuroprotective network dynamics. The left quadrant contains the clusters in the initial network state. The center quadrant represents the steady state of the cluster network that emerges through simulation. Shades of color from red to green indicate how close a cluster's expression value is to the target expression value. Green indicates that the post-stroke and target clusters' expression are the same. Red indicates that the post-stroke and target clusters' expression are as far apart as they can be. At each simulation tick, clusters can establish or remove links with other clusters. The number of clusters connections added (blue) and removed (red) is represented in the "Connection" graph window (top right). As new links are added and old ones removed, the clusters can come closer to or further from the target expression value, as indicated by the "good-bad" black line in the "Gene Status" graph. As they do so, the clusters' expression levels change, as indicated by the "exp-change" blue line in the "Gene Status" graph window (bottom right). When there is no more change (i.e. the "exp-change" blue line in the "Gene Status" graph goes down to 0), the simulation reaches a steady state. In the simulation shown in Figure 2, the system reaches a steady state after establishing 201 connections, in approximately one thousand simulation ticks. Given the nature of the randomness of the way cluster interact establishing or severing network connections, rerunning a simulation is unlikely to yield the same network. This variation provides a rich set of alternative neuroprotective scenarios that are possible from the same network premises.

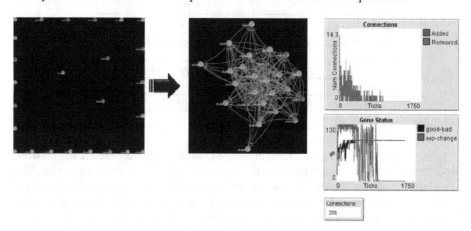

Fig. 2. LPS neuroprotective regulatory network simulation in NetLogo

5   Results and Evaluation

We run two simulations, one for each of the two treatments targeted for exploration: LPS and CpG injection. To assess the validity of the simulations generated, we correlated the predicted expression values of the 25 clusters from the simulation results with the observed expression values of the 25 clusters in the dataset at 24 hours post-stroke. We also established 25 different baseline simulation results by randomizing weight values in the input matrix of cluster to cluster weights (see Table 1). Each set of baseline simulation results included some 10,000 ticks/iterations. The rationale in using a baseline with random weight values is to verify that "real" correlation results obtained with the non-random weights are significantly better than results obtained by chance.

Figure 3 shows the correlation coefficients of real and randomized simulation results with observed expression values for the LPS treatment. The mean correlation coefficient for the real simulation (e.g. averages across the 25 clusters) is 0.76 while the mean correlation coefficient for the 25 sets of randomized simulation results is

0.22, with a p-value below 2.2e-16. These results provide a clear indication that our model generates simulations for the LPS treatment that have a significant correlation with observed data and are not due to chance.

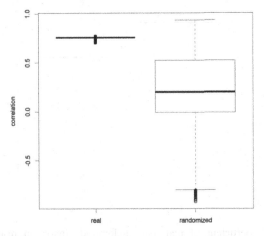

Fig. 3. Correlation coefficient of real and randomized simulation results with observed expression values for LPS treatment (p < 2.2e-16)

We also assessed cluster by cluster performance by measuring the distance between predicted (*EP*) and observed (*EO*) cluster expression values as indicated in (4). As shown in Figure 4, while some clusters perform better than others, all predicted and observed cluster expression values aside from cluster 1 are less than 10% apart. Further improvements can be achieved by addressing cluster by cluster performance – e.g. improving the accuracy of the ODEs that provide the weights and/or the ODE optimization algorithm (see section 3).

$$\text{dist}\left(EP_i, EO_j\right) = \left|EP_i / \left(EP_i + EO_j\right)\right|, \qquad where\ i, j, k, l \geq 1 \qquad (4)$$

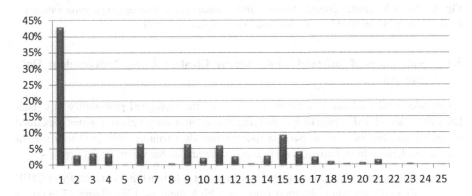

Fig. 4. Cluster by cluster distance between real-simulated and observed expression values for LPS. Distance values have been normalized. 0% indicates perfect fit.

The evaluation of the CpG-treatment simulations corroborates the LPS results, as shown in Figure 5 and 6.

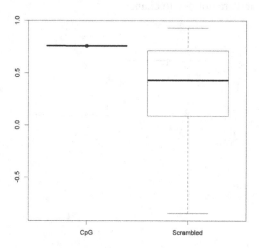

Fig. 5. Correlation coefficient of real and randomized simulation results with observed expression values for CpG treatment (p < 2.2e-16)

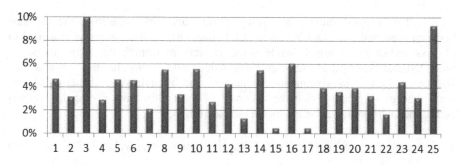

Fig. 6. Cluster by cluster distance between real-simulated and observed expression values for CpG. Distance values have been normalized. 0% indicates perfect fit..

5.1   Validation of Inferred Links across Clusters Using Independent Data Sources

As an additional evaluation measure, we assessed the biological plausibility of cluster links in the simulated networks by verifying that relationships between genes contained in connected clusters were attested in independent data sources. We used four sets of references derived from known gene-interaction databases for verification purposes:

1. Regulatory interactions (regulator to target relationships) derived from CHIP experiments were obtained from the ChEA database [26] giving 2506 edges between 1587 genes included in our model

2. Protein-protein interactions (PPI) were obtained from the Human Proteome Research Database (HPRD) [27], and identifiers were mapped to mouse using gene symbols giving 2974 edges between 1393 genes included in our model. Though some interactions identified in human may not be preserved in mouse, overall they are likely to be consistent across organisms [28]

3. Known gene regulatory interactions (Reg) were obtained from the Molecular Signatures Database [29] giving 1895 edges between 570 genes in our model

4. Functional interactions derived from computational integration of multiple data source (Functional) were obtained from high-confidence (score > 0.5) interactions made by MouseNet [30] giving 2307 edges between 1023 genes in our model.

For each cluster in our model, we determined the number of known interactions (i.e. those attested in the four databases described above) between a gene/gene product in the cluster and genes in each of the other clusters. We used the network model derived through the *Inferelator* algorithm discussed in section 3 (ODE Model) as baseline. The results of this evaluation are shown in Table 3.

We run 30 different simulations with the agent-based model (ABM) described in section 4 using different random seeds so as to maximize the difference across simulation results, and counted the number of times an edge was found between two clusters in all the simulated networks. The model was considered to have a valid edge between two clusters if it occurred in 25 or more simulations. To determine a p-value for the interactions, we counted interactions gathered by randomizing the known edges for each external interaction dataset 1000 times – for undirected edges (HPRD and MouseNet), relationships were counted for both directions. For each interaction dataset, those cluster-to-cluster relationships with a p-value of less than 0.05 were considered to be true positive (TP) matches if there was a corresponding edge in our inferred model, and false positive (FP) matches otherwise. If the p-value was greater than 0.05 and there was no matching inferred edge, we would count a true negative (TN) match; we would count a false negative (FN) match otherwise. Accuracy was calculated as shown in (5).

$$accuracy = \frac{TP + TN}{TP + TN + FP + FN} \tag{5}$$

As shown in Table 3, the results of this evaluation show that the ABM regulatory network models for LPS and CpG are well supported by all independent interaction data sources, and yield a better match than the network model derived through the *Inferalator* algorithm (ODE model). The overlap between inferred and independently documented gene-to-gene links ranges from 67.9% 80.6%. However, if we combine results across each interaction dataset by counting a match as a true positive or true negative if it was validated by any interaction dataset, the combined accuracy is around 90% for both the inferred LPS and CpG models. The number of edges in each of these models at this threshold (considering 25 or more occurrences as an edge) was less than half that in the ODE-based model. This is because using the ABM results we were able to filter the edges based on confidence, as determined by frequency. This indicates that the ABM is producing accurate interactions, and that the more frequently an edge is seen across simulations, the more accurate it is.

Table 3. Validation of cluster-to-cluster interactions from simulated regulatory networks through comparison with independent data sources. The network model derived through the *Inferalator* algorithm (ODE Model) serves as baseline.

Datasets	ODE Model (baseline)	Agent-Based Model for LPS	Agent-Based Model for CpG
1. CHIP	60.9%	80.6%	79.5%
2. PPI	66.8%	73.1%	74.2%
3. Reg	64.3%	69.0%	67.9%
4. Functional	64.3%	75.6%	76.7%
1-4	82.0%	90.0%	89.8%
Number of network edges	133	50	58

6   Conclusions

The methodology we have described in this paper provides the first step towards using the notion of emergence to model regulatory gene networks. The ensuing approach can be used to uncover novel therapies for a variety of diseases by simulating how protective pathways may endogenously form from a preconditioning stimulus. The specific embodiment presented shows how neuroprotective pathways emerge from preconditioning treatment with LPS and CpG by letting gene clusters search and establish links that maximize proximity to the expression level of a healthy organism. Due to the self-organizational dynamics of the emergent approach to modeling, there are potentially many ways in which neuroprotective regulatory networks may emerge. Therefore, different simulation runs with the same cluster-to-cluster weights, and clusters target and initial expression levels is likely to output different networks. For example, alternative steady states resulting from the same simulation setup shown in Figure 2 may have less/more and different connections, as shown in Figure 7. This variety is of great potential interest in understanding how differently the organism may respond to the same treatment in order to overcome disease.

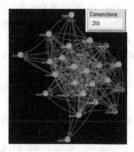 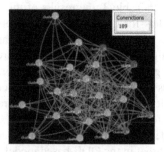

Fig. 7. Alternative simulations of LPS neuroprotective regulatory network

Acknowledgements. The research described in this paper was supported under NIH/NINDS grant R01NS057484-05.

References

1. The Internet Stroke Center, http://www.strokecenter.org/patients/~about-stroke/stroke-statistics (accessed on June 7, 2012)
2. O'Collins, V.E., Macleod, M.R., Donnan, G.A., Horky, L.L., van der Worp, B.H., Howells, D.W.: 1,026 experimental treatments in acute stroke. Ann. Neurol. 59(3), 467–477 (2006)
3. Savitz, S.I.: A critical appraisal of the NXY-059 neuroprotection studies for acute stroke: a need for more rigorous testing of neuroprotective agents in animal models of stroke. Exp. Neurol. 205(1), 20–25 (2007)
4. Fisher, M., Feuerstein, G., Howells, D.W., Hurn, P.D., Kent, T.A., Savitz, S.I., Lo, E.H., STAIR Group: Update of the stroke therapy academic industry roundtable preclinical recommendations. Stroke 40(6), 2244–2250 (2009)
5. Sahota, P., Savitz, S.I.: Investigational therapies for ischemic stroke: neuroprotection and neurorecovery. Neurotherapeutics 8(3), 434–451 (2011)
6. Zhang, L., Athale, C., Deisboeck, T.: Development of a three-dimensional multiscale agent-based tumor model: simulating gene-protein interaction profiles, cell phenotypes and multicellular patterns in brain cancer. J. Theor. Biol. 244(1), 96–107 (2007)
7. Engelberg, J., Ropella, G., Hunt, C.: Essential operating principles for tumor spheroid growth. BMC Syst. Biol. 2(1), 110 (2008)
8. Lollini, P., Motta, S., Pappalardo, F.: Discovery of cancer vaccination protocols with a genetic algorithm driving an agent based simulator. BMC Bioinf. 7(1), 352 (2006)
9. Li, N., Verdolini, K., Clermont, G., Mi, Q., Rubinstein, E., Hebda, P., Vodovotz, Y.: A patient-specific in silico model of inflammation and healing tested in acute vocal fold injury. PLoS ONE 3(7), e2789 (2008)
10. Gonzalez, P., Cardenas, M., Camacho, D., Franyuti, A., Rosas, O., Lagunez-Otero, J.: Cellulat: an agent-based intracellular signalling model. Biosystems 68(2-3), 171–185 (2003)
11. Pogson, M., Smallwood, R., Qwarnstrom, E., Holcombe, M.: Formal agent-based modelling of intracellular chemical interactions. Biosystems 85(1), 37–45 (2006)
12. Pogson, M., Holcombe, M., Smallwood, R., Qwarnstrom, E.: Introducing Spatial Information into Predictive NF-κB Modelling-An Agent-Based Approach. PLoS ONE 3(6), e2367 (2008)
13. Klann, M., Lapin, A., Reuss, M.: Agent-based simulation of reactions in the crowded and structured intracellular environment: Influence of mobility and location of the reactants. BMC Syst. Biol. 5(1), 71 (2011)
14. Schlitt, T., Brazma, A.: Current approaches to gene regulatory network modelling. BMC Bioinf. 8(suppl. 6), S9 (2007)
15. Karlebach, G., Shamir, R.: Modelling and analysis of gene regulatory networks. Nat. Rev. Mol. Cell Biol. 9(10), 770–780 (2008)
16. Stenzel-Poore, M.P., Stevens, S.L., Xiong, Z., Lessov, N.S., Harrington, C.A., Mori, M., Meller, R., Rosenzweig, H.L., Tobar, E., Shaw, T.E., Chu, X., Simon, R.P.: Effect of ischaemic preconditioning on genomic response to cerebral ischaemia: similarity to neuroprotective strategies in hibernation and hypoxia-tolerant states. Lancet 362(9389), 1028–1037 (2003)

17. Stevens, S.L., Ciesielski, T.M., Marsh, B.J., Yang, T., Homen, D.S., Boule, J.L., Lessov, N.S., Simon, R.P., Stenzel-Poore, M.P.: Toll-like receptor 9: a new target of ischemic preconditioning in the brain. J. Cereb. Blood Flow Metab. 28(5), 1040–1047 (2008)
18. Marsh, B., et al.: Systemic lipopolysaccharide protects the brain from ischemic injury by reprogramming the response of the brain to stroke: a critical role for IRF3. J. Neurosci. 29, 9839–9849 (2009)
19. Irizarry, R.A., Hobbs, B., Collin, F., Beazer-Barclay, Y.D., Antonellis, K.J., Scherf, U., Speed, T.P.: Exploration, Normalization, and Summaries of High Density Oligonucleotide Array Probe Level Data. Biostatistics 4(2), 249–264 (2003)
20. Bonneau, R., et al.: The Inferelator: an algorithm for learning parsimonious regulatory networks from systems-biology data sets de novo. Genome Biol. 7, R36 (2006)
21. Efron, B., Johnstone, I., Hastie, T., Tibshirani, R.: Least angle regression. Annals of Statistics 32, 407–499 (2003)
22. Tibshirani, R.: Regression shrinkage and selection via the lasso. J. Royal Statist. Soc. B 58, 267–288 (1996)
23. McDermott, J.E., Archuleta, M., Stevens, S.L., Stenzel-Poore, M.P., Sanfilippo, A.: Defining the players in higher-order networks: predictive modeling for reverse engineering functional influence networks. In: Pac. Symp. Biocomput., pp. 314–325 (2011a)
24. Mcdermott, J., Jarman, K., Taylor, R., Lancaster, M., Stevens, S., Vartanian, K., Stenzel-Poore, M., Sanfilippo, A.: Modeling Cumulative Change of Dynamic Regulatory Processes in Stroke. PLoS Computational Biology (forthcoming)
25. Camazine, S., Deneubourg, J., Franks, N., Sneyd, J., Theraulaz, G., Bonabeau, E.: Self-Organization in Biological Systems. Princeton University Press (2011)
26. Wilensky, U.: NetLogo. Center for Connected Learning and Computer-Based Modeling, Northwestern University, Evanston, IL (1999), http://ccl.northwestern.edu/netlogo/
27. Lachmann, A., Xu, H., Krishnan, J., Berger, S.I., Mazloom, A.R., et al.: ChEA: Transcription Factor Regulation Inferred from Integrating Genome-Wide ChIP-X Experiments. Bioinformatics (2010)
28. Peri, S., Navarro, J.D., Kristiansen, T.Z., Amanchy, R., Surendranath, V., et al.: Human protein reference database as a discovery resource for proteomics. Nucleic Acids Res. 32, D497–D501 (2004)
29. Yu, H., Luscombe, N.M., Lu, H.X., Zhu, X., Xia, Y., et al.: Annotation Transfer Between Genomes: Protein-Protein Interologs and Protein-DNA Regulogs. Genome Res. 14, 1107–1118 (2004)
30. Liberzon, A., Subramanian, A., Pinchback, R., Thorvaldsdottir, H., Tamayo, P., et al.: Molecular signatures database (MSigDB) 3.0. Bioinformatics 27, 1739–1740 (2011)
31. Kim, W.K., Krumpelman, C., Marcotte, E.M.: Inferring mouse gene functions from genomic-scale data using a combined functional network/classification strategy. Genome Biology 9(suppl. 1), S5 (2008)

Detecting Demand-Supply Situations of Hotel Opportunities: An Empirical Analysis of Japanese Room Opportunities Data[*]

Aki-Hiro Sato[**]

Department of Applied Mathematics and Physics, Graduate School of Informatics,
Kyoto University, Yoshida-Honmachi, Sakyo-ku, 606-8501 Kyoto Japan
and Chair for Systems Design, D-MTEC, ETH Zurich, CH-8092 Zurich Switzerland
`sato.akihiro.5m@kyoto-u.ac.jp`

Abstract. This study analyzes the availability of room opportunity types collected from a Japanese hotel booking site. The status of opportunity type is empirically analyzed from a comprehensive point of view. We characterize demand-supply situations of room prices at each region with both room availability and average room rate. The average room rate decreases in terms of the room availability in many districts. However, it is found that the average room rate increases with respect to the room availability in some districts. This is an evidence that the theory of demand and supply is not always satisfied in Japanese hotel industry.

Keywords: Japanese Hotel Booking Data, Demand-Supply Situation, Comprehensive Analysis.

1 Introduction

Recent technological developments enable us to purchase various kinds of items and services via E-commerce systems. The emergence of Internet applications has had an unprecedented impact on our life style regarding the purchase of goods and services. From the availability of items and services at such E-commerce platforms, one can estimate utilities of agents in socio-economic systems.

Demand and supply drive the price of goods or services. If there are a plenty of people who want to buy goods or services, then the price normally goes up. Namely, both demand-supply curves determine the equilibrium price. This results in an economic equilibrium for price and quantity in microeconomics. However, some assumptions are necessary for the validate of the standard model of demand and supply: that supply and demand are independent and that supply is constrained by a fixed resource. This is known as Sraffa's critique [1]. Also if goods have several kinds of properties, then the theory of demand and supply

---

[*] This manuscript is prepared based on the presentation at Complex 2012 at Santa-Fe, NM, USA on 6 December, 2012.

[**] Corresponding author.

K. Glass et al. (Eds.): COMPLEX 2012, LNICST 126, pp. 303–315, 2013.

may not always be true. It is probable to observe that they are purchased with low price in the low supply.

Here we study the demand for rooms related to temporal migration. I examine that stay capacities of hotels included in areas may provide insights on relationship between the social wealth and the migration process. In this article, I also investigate regional dependence of social wealth based on the data on Japanese hotel industry with geographical information. By using data on room capacities as proxy variables of the regional dependence of demand-supply situations, I propose a method to characterize a spatial density of Japanese economy.

In Japan, there are over 54,000 accommodations [2], which are rich in various types: from the largest hotel having about 3,600 rooms to the highest class Japanese inn with a few rooms. Their types and capacities also depend on a district. This availability of the hotel bookings may indicate the future demand of tourists and supply of hotel managers. If it contains availability of all the rooms of hotels, we may detect demand-supply situations in each region from comprehensive data.

According to the study of tourism management [3], there are push and pull factors, so that tourism motivation is determined by the situation of the travelers (push) and the situation of the destination (pull). The idea behind this two-dimensional approach is that people travel because they are pushed by their own internal forces and pulled by the external forces of the destination attributes [4]. The pull factors originate from the destination properties (supply). More recently, Tkaczynski et al. applied the stake-holder theory, a management theory proposed by Freeman (1984) [5], to a destination in tourism [6]. The existence of hotel accommodations implies that pull factors are present in the district where they are located. In the context of economics, this means that the demand-supply situation is generated by both consumers and suppliers. Namely, they can be dependent on the area and the season [7].

Moreover, a problem for estimating demand from censored booking data has been recognized for many years in the hotel industry. Liu et al. developed parametric regression models that consider not only the demand distribution, but also the conditions under which the data were collected [8]. Sato investigated regional patterns of Japanese travel behavior by using the EM algorithm for finite mixtures of Poisson distributions [9].

Deep and wide knowledge on econophysics developed in the last decade [10,11,12,13,14,15] will also help us to understand price mechanism. As well as financial markets, we can assume that balance between demand and supply in the hotel industry reflects both the social and economic situations. In the present article, I propose a method to detect available opportunity types of demand-supply situations from data on room opportunities collected from internet hotel booking site. The hotel industry is highly inhomogeneous. The quality, stuff, and services are very unique, but there are no rooms with the completely same properties. The anormality of the relationships between demand-supply and the market price seems to be recognized. This is the motivation of this study. As we show as the main result, the averaged market prices of some districts do not follow the normal relationship between excess supply and the market price.

This paper is organized as follows. In Sec. 2, the data description will be shown. In Sec. 3, the overview of Japanese hotel industry will be briefly explained. In Sec. 4 the empirical analysis will be conducted. Sec. 5 is devoted to concluding remarks.

2   Data Description

The data are sampled from the Jalan net web site (*http://www.jalan.net*) every day. This data set contains plans for two adults to stay at the hotel per night. Each plan also contains the sample date when the data were sampled, the target date when the stay at the hotel is consumed, the regional sequential number, the hotel identification number, the hotel name, the postal address, the URL of the hotel web page, the geographical position, the plan name, and the room rate. The data period is from 1 January 2010 to 15 May 2011. The data is missing from 19 to 29 March 2011 due to the Great East Earthquake.

In the dataset, there exist about 100,000 opportunity types in about 14,000 hotels every day. Table 1 gives an example of the contents included in the data set. Since the data contain regional information, it is possible for us to analyze regional dependencies of hotel rates.

We define two kinds of dates (a target date and a sample date). Let us denote D as the difference between the target and sample dates. It is inferred that as D decreases the remaining number of opportunity types decreases. Furthermore the regional dependence of the remaining number of opportunity types on D may be related to the supply-demand situation in each district.

Figure 1 (top) shows an example of the regional distribution and representative rates on 15 April 2010 ($D = 6$). The data to draw the map is sampled on 9 April 2010. Green dots represent hotel plans cost 50,000 JPY per night. Black dots represent hotel plans cost 1,000 JPY per night. Red dots represent hotel plans cost over 50,000 JPY per night. As one can see, there is a strong regional concentration. Table 2 shows the total number of rooms of accommodations located in each prefecture. Tokyo (JP-13) has the largest number of rooms (116,542 rooms). Hokkaido (JP-01) has the second largest (72,327 rooms). Osaka (JP-27) has the third largest (57,082 rooms).

Table 1. Structure of our dataset

Variable Name	Meaning	Example Content
Sample date	Date of collection	from 2009-12-24
Target date	Date of Stay	Up to 6 days before
Hotelid	Hotel identification number	300000
Hotelname	Hotel name	Hotel ABC
Postcode	Postal code	066-8520
Hoteladdress	Address	Honmachi, Chitose City
Hoteldetailurl	URL	http://www.jalan...
X	Latitude	509943536
Y	Longtitude	154132695
Planname	Opportunity name	Wonderful travels!
Meal	Meal availability	no meal
Sample rate	the latest best rate	3500 (JPY)
Rate	Rate per night	7000 (JPY)

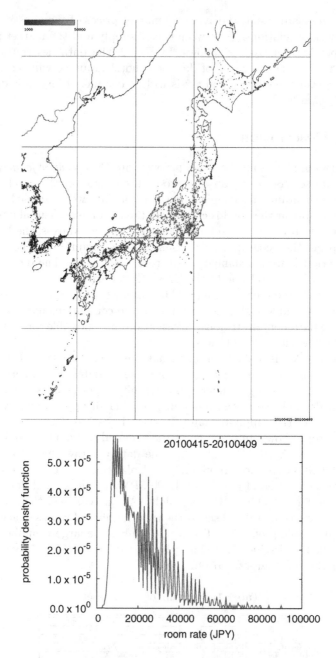

Fig. 1. (top) Example of the regional distribution of opportunity types over Japan on 15th April 2010 ($D = 6$). Color dots correspond to the room rates per night. The green dots represent room opportunities with 50,000 JPY. Black dots represent room opportunities with 1,000 JPY. The red dots represent room opportunities with prices more than 50,000 JPY. (bottom) The probability density function.

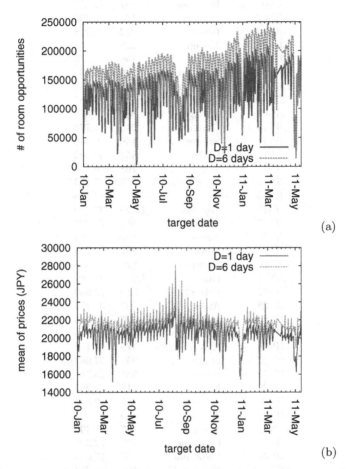

(a)

(b)

Fig. 2. (a) Temporal development of the number of hotel rooms where two adults would be able to stay one night during the period from 1 January 2010 to 15 May 2011. D represents the duration between target date and sample date (in days). $D = 1$ means 1 day before room use, and $D = 6$ 6 days before use. (b) Time series of the average rates of opportunity types on stay dates. The mean value of rates is calculated from all the available opportunity types which are observed on each stay date.

Table 2. The total number of rooms located in each prefecture of Japan. 961,974 rooms in 16,650 accommodations are included in the data. This data was collected on 20 January 2012.

ISO-3166:JP	prefecture	# hotels	# rooms
JP-01	Hokkaido	987	72,327
JP-02	Aomori	178	12,632
JP-03	Iwate	222	12,128
JP-04	Miyagi	227	18,547
JP-05	Akita	138	8,375
JP-06	Yamagata	266	10,830
JP-07	Fukushima	363	17,034
JP-08	Ibaraki	510	18,587
JP-09	Tochigi	562	17,878
JP-10	Gunma	205	11,774
JP-11	Saitama	146	9,466
JP-12	Chiba	458	32,102
JP-13	Tokyo	798	116,542
JP-14	Kanagawa	559	35,283
JP-15	Niigata	449	12,158
JP-16	Toyama	1,425	37,559
JP-17	Ishikawa	555	23,993
JP-18	Fukui	140	9,170
JP-19	Yamanashi	261	14,740
JP-20	Nagano	225	7,366
JP-21	Gifu	1,267	41,689
JP-22	Shizuoka	389	13,650
JP-23	Aichi	407	35,814
JP-24	Mie	397	17,385
JP-25	Siga	148	8,716
JP-26	Kyoto	479	24,114
JP-27	Osaka	351	57,082
JP-28	Hyogo	551	25,511
JP-29	Nara	118	4,254
JP-30	Wakayama	247	9,843
JP-31	Tottori	169	7,389
JP-32	Shimane	167	7,400
JP-33	Okayama	193	12,327
JP-34	Hiroshima	213	17,063
JP-35	Yamaguchi	160	10,733
JP-36	Tokushima	89	4,392
JP-37	Kagawa	136	8,901
JP-38	Ehime	155	9,734
JP-39	Kochi	124	6,991
JP-40	Fukuoka	285	36,330
JP-41	Saga	122	5,622
JP-42	Nagasaki	198	13,128
JP-43	Kumamoto	404	16,949
JP-44	Oita	441	14,034
JP-45	Miyazaki	106	9,640
JP-46	Kagoshima	231	15,027
JP-47	Okinawa	444	29,765

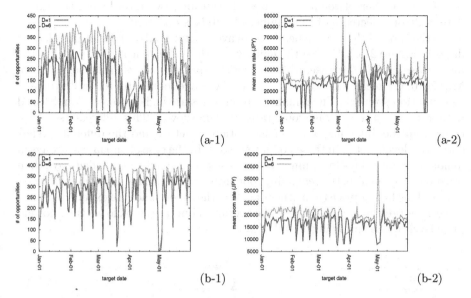

Fig. 3. The number of available room opportunity types for (a-1) Higashiyama and Gion in Kyoto for the period from 1 January to June 1 2010 and (b-1) Yuda and Oufu in Yamaguchi. The average room rates for (a-1) Higashiyama and Gion in Kyoto for the period from 1 January to June 1 2010 and (b-2) Yuda and Oufu in Yamaguchi. D represents the difference between the sample date and target date.

Figure 1 (bottom) shows a probability density of room rates on 15 April 2010 ($D = 6$). The highest density is found at 9,000 JPY. Several peaks exist for more than 20,000 JPY. These peaks are related to rounding effect of room pricing. This shows that hotel managers prefer some specific prices when they determine room prices. Throughout the investigation we regard the number of available opportunity types as a proxy variable of the remaining opportunities of rooms.

3   Overview of Data

The total number of opportunity types at which two adults have possibility to stay was counted up from the data throughout the whole sample period. Figure 2 (a) shows the total number of opportunity types per day. There exists weekly seasonality for the total number of opportunity types. In the case of Japan, Saturday nights are the highest demanded date in a week. The total number of opportunity types is low on Saturday. There is a strong dependence of the remaining number of rooms on the Japanese calendar. Namely, holidays influence reservation activities of costumers. For example, during the golden week holidays (from 1st to 5th May) and public holidays in the spring season (around 20th March) the total number of the remaining rooms shows a rapid drop. The summer vacation season (from the end of July to the beginning of September) the total number of opportunity types decrease. As D decreases, the remaining number of rooms decreases. Specifically, two days before the target date, a drastic decrease of the number of hotels is observed.

Furthermore, I show the dependence of average rates all over Japan on calendar dates in Figure 2 (b). During the holidays in March 2010, it is observed that the mean rates rapidly decrease, while during the spring holidays in the beginning of May 2010, the mean rates rapidly increase. Average room rates tell us customers' consumption strength. The remaining room rates are higher (lower) than booked room rates, then the average room rates goes up (down). This difference seems to come from the difference of the consumption structure and the different preference price levels between these holiday seasons. In the case of Japan, Saturdays in holidays or Saturdays in summer vacation season are more demanded days than other days. Therefore, the remaining room rates on Saturdays in summer vacation season show higher price than other days. On the other hand, the expensive rooms sometimes preferred. For example, the average room rates took lower than normal price levels on Saturday in the end of March.

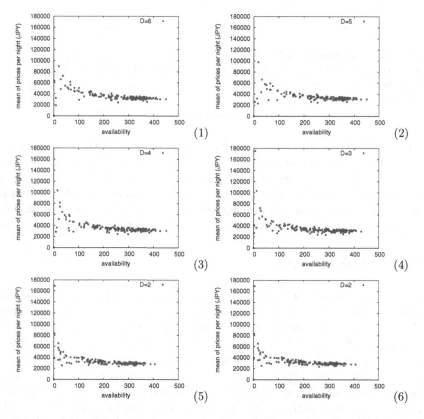

Fig. 4. The relationship between averaged rates and the number of available opportunity types at Higashiyama and Gion in Kyoto for the period from 1 January to 1 June 2010. (1) $D = 6$, (2) $D = 5$, (3) $D = 4$, (4) $D = 3$, (5) $D = 2$, and (6) $D = 1$. D represents the difference between the sample date and target date.

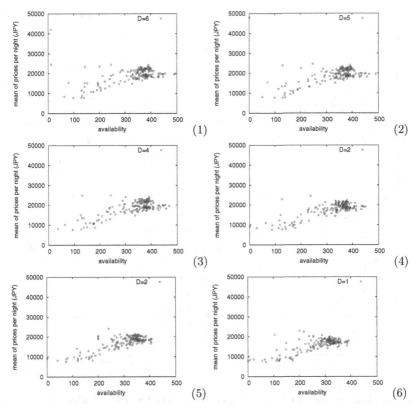

Fig. 5. The relationship between averaged rates and the number of available opportunity types at Yuda and Oufu in Yamaguchi for the period from 1 January to 1 June 2010. (1) $D = 6$, (2) $D = 5$, (3) $D = 4$, (4) $D = 3$, (5) $D = 2$, and (6) $D = 1$. D represents the difference between the sample date and target date.

4   Empirical Analysis

In this section, I show the result of an empirical analysis of the demand-supply situation. First of all, we study the number of available opportunity types for different regions. Figure 3 shows the number of available room opportunity types for two regions (Higashiyama and Gion in Kyoto and Yuda and Oufu in Yamaguchi) for the period from 1st January to 1st June 2010. It is found that there are regional dependencies of the temporal development. In the case of Gion in Kyoto, in the spring season there is high demand. In the case of Yuda and Oufu in Yamaguchi, there is higher demand in the summer season than in the spring season.

A demand-supply situation determines the price direction. Namely, the excess demand (supply) increases (decreases) prices of goods or services. In order to understand such effects on room prices, we computed relationship between the number of available room opportunity types and averaged prices for each region.

Figures 4 and 5 show the relationship between availability and the average rates for two sub areas. On the one hand, Figure 4 shows that even in the case

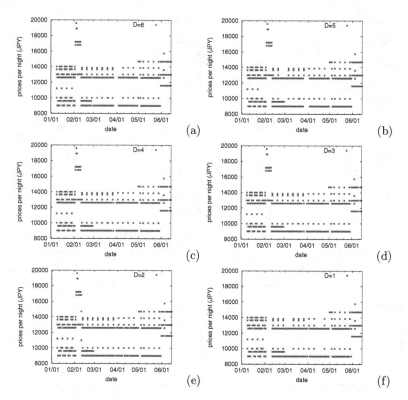

Fig. 6. The temporal dependence of prices per night for a hotel which serves a few kinds of rooms (a) 6 days before the stay, (b) 5 days before, (c) 4 days before, (d) 3 days before, (e) 2 days before, and (f) 1 day before. The x-axis represents target date, and the y-axis a room prices. These graphs depict temporal development of room prices.

of a small number of room opportunity types there still exist opportunity types to book expensive rooms. In this case, we may assume that consumers prefer to book cheaper rooms with available room opportunity types in a sub area. On the other hand, Figure 5 shows that opportunity types cheaper than the mean price for the large number of available opportunity types remain in the case of the small number of available opportunity types. In this case we may assume that consumers demand mainly distributes at the higher side than prices of which they serve.

Figures 6 and 7 show the temporal dependence of prices per night for two hotels during the period from January 1 to June 1, 2010. It is found that hotel managers offered prices depending on calendar dates. Specifically it is confirmed that relatively higher prices were offered on new year holidays and spring holidays. Furthermore in these cases, rates on the spring season are higher than ones on the winter season.

The hotel shown in Figure 6 provides different kinds of room opportunity types. The prices ranged from 9,000 JPY to 20,000 JPY. The price suddenly

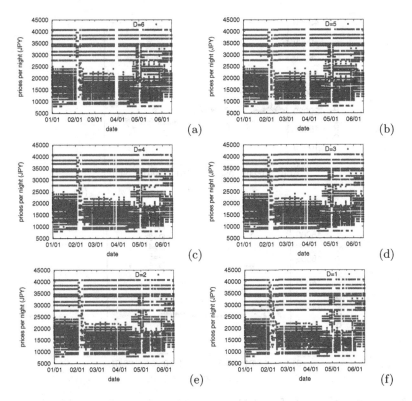

Fig. 7. Temporal dependence of prices per night for a hotel which provides various kinds of rooms for the period from 1 January to June 1 2010. (a) 6 days before the stay, (b) 5 days before, (c) 4 days before, (d) 3 days before, (e) 2 days before, and (f) 1 day before. The x-axis represents target date, and the y-axis a room prices. These graphs depict the temporal development of room prices.

jumped up to a range from 12,000 JPY to 20,000 JPY from 3 to 10 February 2010. After June 2010 the price range shifted up. The Figures (a) to (f) confirm that room opportunity types were eventually booked. The cheap opportunity types ranging from 8,000 JPY to 12,000 JPY were booked 6 days before the stay. The expensive opportunity types ranging from 12,000 JPY to 20,000 JPY were booked from 6 days before in the case of the beginning of February.

The hotel shown in Figure 7 provides many different kinds of room opportunity types. The prices ranged from 9,000 JPY to 40,000 JPY. The price suddenly jumped up to a range from 20,000 JPY to 40,000 JPY from 3 to 10 February 2010. After June 2010 the price range also shifted up. The Figures (a) to (f) also confirm that room opportunity types were eventually booked. The expensive opportunity types were booked 6 days before the stay. The cheaper opportunity types ranging from 12,000 JPY to 20,000 JPY were booked from 6 days before in the case of the Golden week (the beginning of May).

Figure 8 shows the average values of available room rates on two types of hotels. It is found that the average room rates on Saturday is higher than weekdays.

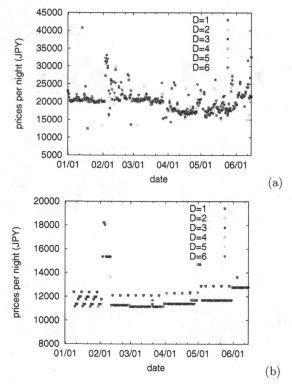

Fig. 8. The average values of available room rates on (a) medium and (b) large hotels for the period from 1 January to 1 June 2010

The average room rates during the high demand days (4-6 February, 1-4 May) are higher than ordinary days. The hotel managers adjust room rates according to expected demand in advance.

This is a kind of regime switching of hotel prices. The hotel managers have to catch up demand and supply balance of rooms. The balance normally changes in time. The room rate distribution is a kind of finger print of the demand and supply balance.

5   Conclusion

We analyzed data of a Japanese hotel booking site. There is a strong dependence of the number of available opportunity types on the calendar date.

We investigated the relationship between the mean room price and the excess supply at each target date. The relationship between average room price and the number of available opportunity types is often a decreasing function in terms of the availability of rooms. However, at several districts, the number of available rooms shows an increasing function in terms of the availability. The difference seems to come from the imbalance between the quality of rooms and demand of customers. In such districts it is often observed that customers preferred to book expensive rooms than cheaper rooms during high demand season. Consequently, the average room fair sometimes increases in terms of availability of rooms.

The hotel booking data is sensitive to macroeconomic conditions significantly since this reflects temporal migration for both the business and tourism purposes. It is found that this large-scale data on hotel opportunity types provides us with insights into properties of human activities related to travels and tourism in Japan.

Acknowledgement. The author acknowledges the financial support from the European Community Seventh Framework Programme (FP7/2007-2013) under Socio-economic Sciences and Humanities, grant agreement no. 255987 (FOC-II). The author expresses his sincere gratitude to Prof. Dr. Hideaki Aoyama, Prof. Dr. Frank Schweitzer, Dr. Stefano Battiston and Prof. Dr. Dirk Helbing for his constructive comments.

References

1. Samuelson, P.A.: Reply in Critical Essays on Piero Sraffa's Legacy in Economics. In: Kurz, H.D. (ed.) Critical Essays on Piero Sraffa's Legacy in Economics. Cambridge University Press, Cambridge (2000)
2. Japan Tourism Agency,
   http://www.mlit.go.jp/kankocho/siryou/toukei/shukuhakutoukei.html
3. Cha, S., Mccleary, K.W., Uysal, M.: Travel Motivations of Japanese Overseas Travelers: A Factor-Cluster Segmentation Approach. Journal of Travel Research 34, 33–39 (1995)
4. Dann, G.M.S.: Anomie, ego-enhancement and tourism. Annals of Tourism Research 4, 184–194 (1977)
5. Freeman, R.E.: Strategic Management: A Stakeholder Approach. Pitman, Boston (1984)
6. Tkaczynski, A., Rundle-Thiele, S., Beaumont, N.: Destination Segmentation: A Recommended Two-Step Approach. Journal of Travel Research 49, 139–152 (2010)
7. Cuccia, T., Rizzo, I.: Tourism seasonality in cultural destinations: Empirical evidence from Sicily. Tourism Management 32, 589–595 (2011)
8. Liu, P.H., Smith, S., Orkin, E.B., Carey, G.: Estimating unconstrained hotel demand based on censored booking data. Journal of Revenue and Pricing Management 1, 121–138 (2002)
9. Sato, A.-H.: Patterns of regional travel behavior: An analysis of Japanese hotel reservation data. International Review of Financial Analysis 23, 55–65 (2012)
10. Mantegna, R.N., Palagyl, Z., Stanley, H.E.: Applications of Statistical Mechanics to Finance. Physica A 274, 216–221 (1999)
11. Sornette, D.: Why Stock Markets Crash: Critical Events in Complex Financial Systems. Princeton University Press, Princeton (2003)
12. Takayasu, H. (ed.): Practical Fruits of Econophysics: Proceedings of The Third Nikkei Econophysics Symposium. Springer, Tokyo (2006)
13. Plerou, V., Gopikrishnan, P., Amaral, L.A.N., Meyer, M., Stanley, H.E.: Scaling of the distribution of fluctuations of financial market indices. Physical Review E 60, 5305–5316 (1999)
14. Anderson, J.: A Theoretical Foundation for the Gravity Equation. The American Economic Review 69, 106–116 (1979)
15. Gheorghiu, A., Spanulescu, I.: An Econophysics Model for the Migration Phenomena. Hyperion International for Econophysics and New Economy 4, 272–284 (2011)

Identification of Chordless Cycles
in Ecological Networks

Nayla Sokhn[1,2], Richard Baltensperger[2], Louis-Félix Bersier[1],
Jean Hennebert[1,2], and Ulrich Ultes-Nitsche[1]

[1] University of Fribourg, CH 1700 Fribourg, Switzerland
[2] University of Applied Sciences of Western Switzerland, CH 1700 Fribourg,
Switzerland

Abstract. In the last few years the studies on complex networks have
gained extensive research interests. Significant impacts are made by these
studies on a wide range of different areas including social networks, technology networks, biological networks and others. Motivated by understanding the structure of ecological networks we introduce in this paper
a new algorithm for enumerating all chordless cycles. The proposed algorithm is a recursive one based on the depth-first search.

Keywords: ecological networks, community structure, food webs, niche-
overlap graphs, chordless cycles.

1 Introduction

Food webs are well known networks in ecology. They depict feeding connections
between species in natural communities. They are represented as directed graphs
where each vertex corresponds to a kind of organism and each directed link
corresponds to a flow of energy or biomass, see Fig. 1. From this directed graph
it is possible to construct a new undirected graph called the niche-overlap graph
that represents the competition structure between predators. In other terms if
two predators have at least one common prey they will be connected in the niche-
overlap graph. According to Fig. 1 v_1 and v_2 (predators) have v_6 as a common prey
therefore in the niche-overlap graph they are linked by an edge. Fig. 2 illustrates
the corresponding niche-overlap graph of the food-web graph shown in Fig. 1. Let
$[v_1, v_2, ..., v_n]$ be a sequence of n distinct vertices. By definition a cycle of length
$k > 3$ is chordless if there is only one link from a vertex v_i to v_{i+1} (for all $i = 1, ..., k$)
and there is no other link between any two of these vertices. It has been suggested
in [1,2,3] that real systems almost completely lack chordless cycles. This indicates
that species can be arranged along a single hierarchy (e.g., body size). Previous
analyses in [4,5] have shown that recent and high-quality food webs possess many
chordless cycles. This implies that species can no more be ordered along a single
hierarchy. Therefore identifying those cycles is important in order to better understand the structure of ecological networks.

The literature contains several algorithms able to find cycles and elementary
cycles[1] in graphs. Some of them are based on vector search space and others on

---

[1] In this type of cycles, vertices are not allowed to be repeated.

K. Glass et al. (Eds.): COMPLEX 2012, LNICST 126, pp. 316–324, 2013.

backtracking algorithm [7,8,9,10,11,12]. On the other hand only few were seeking the enumeration of all chordless cycles [13,14,15,16]. In [13], Spinrad presents an algorithm that determines whether an undirected graph has chordless cycles of size at least K in $\mathcal{O}(n^{K-3} \cdot M)$, where n is the number of vertices and M is the time required to multiply two n by n matrices. In [14], an algorithm that detects one chordless cycle in undirected graphs is described. In [16], an algorithm which enumerates all chordless cycles in directed graphs is introduced. It is based on the use of the asymmetry therefore applying it directly to undirected graphs is worthless. In this paper, we propose an algorithm for enumerating all chordless cycles in undirected graphs. The algorithm is a combination of proper steps found in [12,14,16].

The rest of the paper is organized as follows. Section 2 introduces some fundamentals about graphs that are important for the rest of this paper. Section 3 describes the algorithm and illustrates its flowchart. Section 4 presents the results. Section 5 concludes and exhibits some future work.

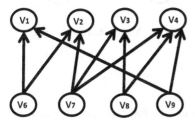
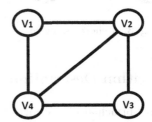

Fig. 1. A food web graph   **Fig. 2.** A niche-overlap graph

2   Fundamentals about Graphs

A graph G consists of two finite sets: a set $V(G)$ of **vertices** and a set $E(G)$ of **edges** where each edge is associated with a set consisting of two vertices called its endpoints. The order (number of vertices) and the size (number of edges) of the graph are denoted by n and m respectively. A graph is **undirected** if the edges have no orientation and it is **directed** if they have orientation. By definition a **walk** is a finite alternating sequence of adjacent vertices and edges. It has the form $v_0 e_1 v_1 e_2 ... v_{k-1} e_k v_k$ where the v's represent vertices, the e's represent edges. A **path** is a walk of the form $v = v_0 e_1 v_1 e_2 ... v_{k-1} e_k v_k$ where all the e_i are distinct [17]. A path $v_0 e_1 v_1 e_2 ... v_{k-1}$ is **chordless** if $v_i v_j \notin E(G)$ for any two non-adjacent vertices v_i, v_j in the path [14]. A **cycle** (a closed path) $[v_0, v_1, ..., v_k]$ is **chordless** if no edge $v_i v_j$ exists in $E(G)$ such that $|i - j| \neq 1$ $mod\ k$ [14]. Fig. 3 and Fig. 4 illustrates a chordless cycle of order 4 and a non-chordless one of order 4 respectively.

A graph G can be represented by an **adjacency matrix** or **adjacency lists**. The adjacency matrix of order n is a $n \times n$ binary matrix A with entries given by

$$a_{ij} = \begin{cases} 0, \text{ if } v_i v_j \notin E(G), \\ 1, \text{ if } v_i v_j \in E(G). \end{cases}$$

It is filled with a 1 in position (v_i, v_j) if v_i and v_j are adjacent and with a 0 otherwise. An adjacency list for a vertex v_i is a list containing all vertices adjacent to v_i. These vertices are named **neighborhood** and denoted by $N(v_i)$. In this paper, we treat undirected graphs with no loops (an edge with just one endpoint) and no multiple edges (two or more edges connecting the same two vertices).

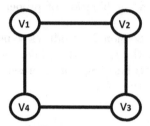

Fig. 3. A chordless cycle

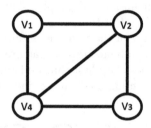

Fig. 4. A non-chordless cycle

3   Algorithm Description

The general principal of the algorithm is to create for each vertex expanding paths (using the depth-first search startegy) that respect the conditions of a chordless path and a chordless cycle. To limit the search space, several optimizations are proposed. The process of the algorithm is described below and its flowchart is illustrated in Fig. 5.

1. Create the adjacency matrix and the adjacency lists for the given graph. Label the vertices from 1 to n $[v_1, v_2, ..., v_n]$.

2. Select the first vertex v_{start} (until all $[v_1, v_2, ..., v_n]$ are handled) and add it to the path P initially empty. Now P contains v_{start}.

3. Select the first adjacent vertex v_j (if it exists) of the last vertex v_{end} in the path P. Two conditions are imposed on v_j, it should not exist in the given path P and it must be bigger than v_{start}. If v_j does not exist, delete the last vertex of P and go back to this step. When all the adjacent vertices v_j of v_{start} are handled, go to step 2 and select a new vertex $v_{start+1}$.

4. If the size of P (denoted by $|P|$) = 2, go to step 3.

5. If $|P| = 3$, check if v_{end} and v_{start} are connected. If they are not connected, go to step 3 to potentially expand the path looking for chordless cycles. If they are connected, a complete cycle K_3 is detected, therefore delete the last

vertex in P and go to step 3 to look for other potential paths leading to chordless cycles.

6. If $|P| > 3$, check if any two non-adjacent vertices are connected ignoring v_{start} (the first vertex in P). If any, delete the last vertex of P and go to step 3. On the contrary, see if there is an edge between v_{end} and v_{start}. If it exists, a chordless cycle of order $|P|$ is then detected, therefore delete the last vertex in P and go to step 3. Otherwise go to step 3.

7. When the list $[v_1, v_2, ..., v_n]$ is handled, the algorithm is finished and all the chordless cycles are found.

3.1   Clarification of Some Steps in the Algorithm

In step 1, it is important to create the adjacency matrix because it is then possible to detect the presence or absence of a specific edge in constant time. The use of the adjacency lists is important too since the selection of an adjacent vertex occurs in constant time. In step 3, the first condition (v_j should not be in the path) avoids to have a path where vertices are repeated. The verification if a vertex is already in the path is performed in the following way : a vector of size n is initialized as 'False'. Each time an adjacent vertex v_j is added to the path, the status of this vertex is changed to 'True'. Accordingly v_j is added to a path only if its status is set to 'False'. The second condition ($v_j > v_{start}$) presents two important advantages described hereafter :

The first one is the possibility of running concurrently the algorithm. In others terms, detecting chordless cycles may be performed in parallel for different vertices. The way of dividing the vertices is important to balance the computation load. To simplify the task, we consider two identical computers. Running the first set $[v_1, v_2, ..., v_k]$ ($k = \frac{n}{2}$, $k \in \mathbb{N}$) on the first one and the second set $[v_{k+1}, ..., v_n]$ on the other one will lead to unbalanced loads since most of the chordless cycles appears in the first part of the set due to the following condition: adjacent vertices $v_j > v_{start}$. Nevertheless creating two sets by interlacing the vertices is a more suitable solution. In this case, the first set starts with vertex v_1 (odd numbers are selected) $[v_1, v_3, v_5, ..., v_n]$ and the second set starts with vertex v_2 (even numbers are selected) $[v_2, v_4, v_6, ..., v_{n-1}]$. In that way, vertices that contain most of the chordless cycles are separated. Results confirming this useful separation are presented in section 4.

The second advantage is that duplicates chordless cycles are avoided. Suppose we found the chordless cycle $[v_1, v_2, v_3, v_4]$. Removing this condition on the adjacent vertex v_j will provide same chordless cycles ($[v_2, v_3, v_4, v_1]$, $[v_3, v_4, v_1, v_2]$, $[v_4, v_1, v_2, v_3]$). However in this algorithm each chordless cycle is found twice. By symmetry $[v_1, v_4, v_3, v_2]$ is another copy of $[v_1, v_2, v_3, v_4]$. In order to keep only one cycle a condition is imposed : the second element of the cycle (v_2) should be always smaller than the last one (v_{end}). $v_2 > v_{end}$ implies that the cycle is a symmetry of a previous detected one. In step 5, the size of the current path is 3. Checking if v_{start} and v_{end} are connected is required. If no edge exists there is then the guarantee that the path is chordless. But if an edge exists between v_{start} and v_{end}, a complete cycle K_3 is then detected. Therefore proceeding with the current path and choosing an adjacent vertex v_j is useless since the new path is no more a chordless one. For that reason we delete the last vertex in the path. In step 6, the size of P ($|P|$) is bigger than 3. Verifying whether all non-adjacent vertices are not connected ensures that the enumerated path is chordless. Accordingly whenever v_{start} and v_{end} are connected the cycle is then a chordless one.

Studying the structure of the graph before applying this algorithm could also lead to optimization in terms of running time computation. For example if the graphs can be separated in several biconnected components, the algorithm could be applied separately on each one for a faster detection of chordless cycles.

3.2   Space and Time Complexity

The proposed algorithm explores the graph using the depth-first search strategy with an additional condition : the selected path must be chordless. The space complexity required by this algorithm is determined by the storage of set V of vertices $\mathcal{O}(n)$, the current path P $\mathcal{O}(k)$ where k is the length of the current path, the adjacency lists $\mathcal{O}(n+m)$ and the adjacency matrix $\mathcal{O}(n^2)$. Consequently the algorithm requires $\mathcal{O}(n^2)$ space.

Estimating the time complexity is more difficult. In the current stage of our studies, we believe that only an empirical estimation of the complexity is possible. There is indeed a relation between the number and the length of chordless paths that exist in the graph and the running time of the algorithm. In fact, adding adjacent vertices to the current path P is performed as long as the path is chordless. The more a graph has chordless paths the longer the running time will be. The complexity is then increasing with the length and number of chordless paths which is actually not known in advance. A worst case estimation of the complexity could potentially be expressed but is not treated in this paper.

4   Results

The running time T (in seconds) for the niche-overlap graphs is given in Table 1. The implementation is executed using C++ (Microsoft Visual Studio 2010) on a 2.93 GHz processors and a 4 GB memory running on windows 7. In table 2 the number and the order of chordless cycles are shown. Note that C_k (for $k = \{4, 5, ..., 8\}$) represents a chordless cycle of order k.

As it was mentioned in section 3 the way of dividing the vertices for concurrent computation is important. To confirm this we choose two matrices : "LRL South Summer" and "Floridabay". We apply two different separations for "LRL South Summer". The first one is simply splitting the vertex set in two equal parts $[v_1, v_2, v_3, ..., v_{59}]$ (the first half) and $[v_{60}, v_{61}, v_{62}, ..., v_{119}]$ (the second half). This separation is not useful because even though the duration to detect some of the chordless cycles is **10 seconds** for the second set, it takes **7226 seconds** (see

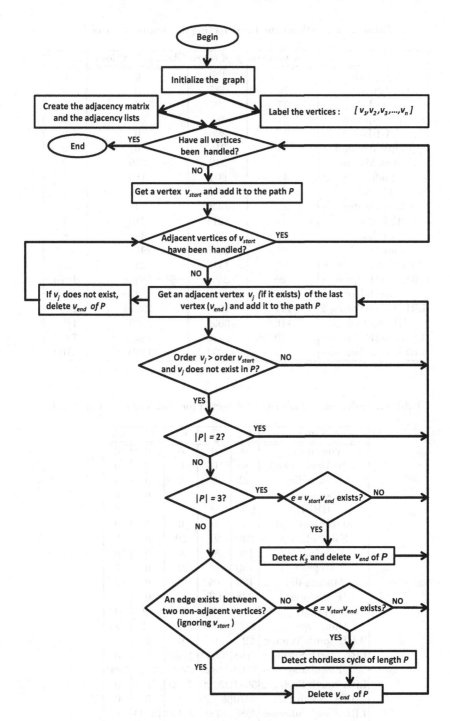

Fig. 5. Flowchart of the algorithm

Table 1. Execution time for detecting all chordless cycles C

Name	# of vertices	# of edges	Chordless cycles C	Time T
Volcan dry	19	170	0	0
Quebrada wet	19	137	0	0
Coachella	27	297	42	0
Chesapeake	27	95	0	0
HBBL	27	306	0	0
SkipWithPond	34	318	0	0
Saint Martin	38	312	356	2
Aguafria dry	45	904	180	2
Cypwet	53	854	130	6
Macara dry	55	1346	249	2
Everglades	58	1214	710	7
Ythan	84	1306	391	8
Mangrovedry	86	2315	29178	359
LRL South Winter	83	1418	224	23
LRL North Spring 1	105	2594	18032	1070
Floridabay	107	3249	85976	4569
LRL North Spring 2	111	2520	25824	2297
LRL North Fall	116	3095	32695	1850
LRL South Summer	119	2420	48921	7409
LRL North Summer	121	3064	16904	3700

Table 2. Order and number of C detected for each niche-overlap graph

Name	C_4	C_5	C_6	C_7	C_8
Volcan dry	0	0	0	0	0
Quebrada wet	0	0	0	0	0
Coachella	42	0	0	0	0
Chesapeake	0	0	0	0	0
HBBL	1768	0	0	0	0
SkipWithPond	0	0	0	0	0
Saint Martin	230	91	29	6	0
Aguafria dry	126	54	0	0	0
Cypwet	130	0	0	0	0
Macara dry	184	65	0	0	0
Everglades	568	130	12	0	0
Ythan	213	156	10	12	0
Mangrovedry	7329	8506	8259	4448	636
LRL South Winter	224	0	0	0	0
LRL North Spring 1	5168	10944	1920	0	0
Floridabay	5769	15825	35824	21158	7400
LRL North Spring 2	5664	13120	7040	0	0
LRL North Fall	8970	23725	0	0	0
LRL South Summer	3689	8788	23340	13104	0
LRL North Summer	6956	9948	0	0	0

Table 3. Experimental results on "LRL South Summer" and "Floridabay"

LRL South Summer	First set	Second set	First set	Second set
Vertices	$[v_1, v_2, ..., v_{59}]$	$[v_{60}, v_{61}, ..., v_{119}]$	$[v_1, v_3, v_5, ..., v_{119}]$	$[v_2, v_4, v_6, v_8, ..., v_{118}]$
Number of C	48903	18	22580	26341
Time in seconds	7226	10	3594	3960

Floridabay	First set	Second set	First set	Second set
Vertices	$[v_1, v_2, ..., v_{53}]$	$[v_{54}, v_{55}, ..., v_{107}]$	$[v_1, v_3, v_5, ..., v_{107}]$	$[v_2, v_4, v_6, ..., v_{106}]$
Number of C	85247	729	51437	34539
Time in seconds	4557	12	2459	2247

Table 3) for the first set. Note that **7226 seconds** is close to the total running time (**7409 seconds**) see Table 1.

This observed result is actually the consequence of the condition " adjacent vertices $v_j > v_{start}$ " explained in section 3.1

The second one is the even/odd version : $[v_2, v_4, v_6..., v_{118}]$ and $[v_1, v_3, v_5.., v_{119}]$. This separation leads to more balanced loads and takes respectively **3594 seconds** and **3960 seconds** to find all the chordless cycles, see Table 3 . Time is reduced by **3449 seconds** (7409 − 3960) which is more or less half of the total running time (**7409 seconds**).

Similarly results are observed the even/odd division for "Floridabay". It takes **2459 seconds** for the first set and **2247 seconds** for the second one, (see Table 3). In that case too, time is reduced by almost half (**2110 seconds**). Note that a splitting the vertices in more than two sets will parallelize further the computation for larger graphs.

5 Conclusion

In this paper, we presented an algorithm for enumerating all chordless cycles in a given undirected graph. We also experimented the algorithm on real niche-overlap graphs used in ecological studies, meaning the number of detected chordless cycles and the corresponding execution time. The algorithm is based on several optimizations a depth-first search strategy including constraints of chordless path and chordless cycles. One important advantage of the algorithm is the possibility to run the search concurrently on different nodes with a fair balancing of sets. It is also possible to change in a simple way the algorithm to find particular chordless cycles, for example of specific length.

In future work, we will apply this algorithm on quantitative niche-overlap graphs (including information on the weights of interactions) and we will analyze the distribution of the weighted chordless cycles. Moreover we will try to clarify the reasons why some ecological networks have many chordless cycles and others do not.

Acknowledgment. The authors would like to acknowledge R. Bisdorff who provided some initial ideas that contributed to this work.

References

1. Cohen, J.: Food Webs and Niche Space. Princeton University Press, Princeton (1978)
2. Sugihara, G.: Niche Hierarchy: Structure Assembly and Organization in Natural Communities. PhD thesis, Princeton University, Princeton (1982)
3. Cohen, J., Briand, F., Newman, C.: Community Food Webs, Data and Theory. Springer (1990)
4. Bersier, L.F., Baltensperger, R., Gabriel, J.P.: Why are cordless cycles so common in niche overlap graphs? In: Ecological Society of America, Annual Meeting, p. 76 (2002)
5. Huxham, M., Beaney, S., Raffaelli, D.: Do parasites reduce the changes of triangulation in a real food web? Oikos 76, 284–300 (1996)
6. Golumbic, M.: Algorithmic graph theory and perfect graphs, 2nd edn. Elsevier (2004)
7. Mateti, P., Deo, N.: On algorithms for enumerating all circuits of a graph. SIAM J. Comput. 5, 90–99 (1976)
8. Tiernan, J.: An efficient search algorithm to find the elementary circuits of a graph. Communications of the ACM 13, 722–726 (1970)
9. Tarjan, R.: Enumeration of the elementary circuits of a directed graph. SIAM J. Comput. 2, 211–216 (1973)
10. Liu, H., Wang, J.: A new way to enumerate cycles in graph. In: AICT/ICIW, pp. 57–59 (2006)
11. Sankar, K., Sarad, A.: A time and memory efficient way to enumerate cycles in a graph. In: ICIAS, pp. 498–500 (2007)
12. Johnson, D.: Find all the elementary circuits of a directed graph. SIAM J. Comput. 4, 77–84 (1977)
13. Spinrad, J.: Finding large holes. Inform. Process. Lett. 39, 227–229 (1991)
14. Nikolopoulos, S., Palios, L.: Hole and antihole detection in graphs. In: Proc. 15th ACM-SIAM Sympos. Discrete Algorithms, pp. 843–852 (2004)
15. Hayward, R.: Weakly triangulated graphs. J. Combinatorial Theory Series B 39, 200–208 (1985)
16. Bisdorff, R.: On enumerating chordless circuits in directed graphs, http://charles-sanders-peirce.uni.lu/bisdorff/documents/chordlessCircuits.pdf
17. Epp, S.: Discrete mathematics with applications, 2nd edn. Brooks/Cole Publishing Company (1995)

In Search of Prudential Leverage Regulation Regimes

Caihong Sun, Wenying Ding, and Ruxin Han

School of Information, Renmin University of China, Beijing, 100872, P.R. China
chsun@ruc.edu.cn, dwy1988@163.com, hanruxin888@sohu.com

Abstract. In year 2009, Thurner, Farmer and Geanakoplos construct an agent-based model of leverage asset purchases with margin calls. The interesting research shows that leverage could cause fat tails and clustered volatility. In this paper, we study the effects of leverage regulation regimes on financial markets based on their model, by introducing two types of leverage regulation policy: risk-based policy and incentive-based policy. Besides examining fat tails and clustered volatility stylized facts, we analyze macroeconomic indicators such as bankruptcy ratio, total social wealth and the efficiency of banking system for identifying prudential leverage regulation regimes.

Keywords: leverage regulation, risk-based policy, incentive-based policy.

1 Introduction

In recent years, the financial crisis has engendered lots of debates on the prudential regulation of bank leverage, since the 2007 financial crisis was blamed in part on excessive leverage. Michael Simkovic(2009)[1]explains that hidden leverage is the root of financial crises in one of the oldest and most fundamental problems of commercial law. Leverage is used usually in investments or corporate finance, which means using debt to finance an activity. It is a general term of risk evaluation is measured as the ratio of total assets owned to the wealth of the borrowers. Is it necessarily to regulate leverage? How to regulate leverage effectively to benefit financial market? In this paper, we examine leverage regulation effects on financial market from the view of banks' local control strategies.

The existing literature on leverage can be divided into three strands.

1. Impact analysis. These papers mainly discuss the consequence of using leverage. Fostel and Geanakoplos(2008)[2]show that leverage cycles can cause contagion and financial crisis in an anxious economy through providing a pricing theory for emerging asset classes. Thurner, Farmer and Geanakoplos (2009) [3]construct an agent-based model of leverage asset purchases with margin calls. The research shows that leverage causes fat tails and clustered volatility and causes financial crisis under special conditions. Feldman(2010)[4]reveals that the portfolio managers taking on excessive leverage when they become risk averse

K. Glass et al. (Eds.): COMPLEX 2012, LNICST 126, pp. 325–338, 2013.

could create harder hit global crisis. Above researches support the viewpoint that it is imperative to regulate the leverage.

2. Leverage computation. Friedman and Abraham(2009)[5]compute leverage in response to the payoff gradient and study the equilibrium and dynamics. Peters(2009)[6]analyzes the optimal leverage for self-financing portfolios by considering time-irreversibility and non-ergodicity. These methods try to explore a reasonable or optimal leverage from the view of borrowers for maximizing their return, not from the side of lenders (such as banks).

3. Leverage regulation. It is important for financial institution and government since excessive leverage can cause crisis of a country or even the whole world. Leverage regulation involves two questions: when and how to regulate. Hodas, Tagliabue, Schmidt and Barofsky(2009)[7] build a model on the work of Thurner, Farmer and Geanakoplos, they present an economy consisting of a banking sector and an equity market,with traders transferring money between the two. Their research mainly analyzes the banks behaviours and set the leverage on the basis of banks balance sheet. Feldman(2011)[8]uses an agent based model to find that regulating leverage by using margin calls could lead to less but harder financial crisis hits. Feldman compares four regulatory regimes: no regulation, fixed leverage, the amount of risky asset limiting and constrained based on detrended price. None of them considers funds' performance. Christensen, Meh and Moran(2011) [9], Raberto, Teglio and Cincotti(2012)[10] focus on stduying banking regulation. The former paper finds that countercyclical bank leverage reguation is likely to stabilize the economy and there exists strong interations between monetary policy and bank regulation policy. The latter shows that the dynamic regulation of capital requirements is more effective than fixed tight one.

While, the purpose of this paper is to focus on bank's leverage strategies. We introduce a framework for examining the effects of banks' leverage regulation policies by using an agent-based financial market model constructed by Thurner, Farmer and Geanakoplos (2009)[3] in which leverage is allowed. We explore a variety of banks' leverage regulation regimes that adjust each fund's leverage level dynamically based on their performance or market volatilities, and try to search prudential leverage regulation regimes which could reduce excessive volatility, the damage of defaults and stabilize the financial market. In the finanial makert model of Thurner et al., there are two types of traders in the model: noise trader and hedge funds. Hedge funds are value investing and can borrow from a bank under leverage. The loan is a collateralized one in which the debt is guaranteed by an asset. The asset price is determined through market clearing mechanism, i.e. the equilibrium between the market demand and supply. A rational investor will buy assets at a low price and sell at a at a high price. But a fund with collateralized loan may be forced to sell as the value of the collateralized asset falls. When a group of fundsselling occurs together,it may cause defaults or crashes.When a fund invest with borrowed money, it can potentially earn more due to larger scale of investment, but also can lose more because of default. That is to say, leverage can magnify the expected revenue, but also cause bankruptcy of hedge funds. Therefore the using of leverage is followed with risk. Especially, this risk can be much more complex as that the financial market is unstable and the financial derivatives are expanding rapidly. In addition, the banks may face expanding risk for the bad debt that the funds bankrupcies bring about. To maintain safety itself and keep profit, a bank has to limit the borrowing, i.e. setting a reasonable leverage limit. The banks leverage policy has great influence on financial market. If the leverage is over high, then there are more bankruptcies. If leverage is too low, the market is less flexible. Intuitively, a bank sets leverage on the basis of funds' performance and market volatility. Banks evaluate funds' performance by their revenue, and assess risk based on the fluctuations in prices. Thurner et al.(2009) consider banks' leverage regulation according to market volatility. But for all the funds, they use the same leverage level. We argue that homogeneous leverage can make the high-risk funds to borrow excessively, even cause defaults.

In this paper, we explore leverage regulation regimes of banks in three main perspectives. The first one is Homogenous or heterogeneous policy. All the funds have the same leverage under the homogenous policy, but each fund has its own individual leverage level under heterogenous policy. Heterogenous policies are more realistic and could perform well. The second perspective considers long term and short term policy. The difference between these two policies is the length of time that a policy considers. The third perspective is incentive based and risk based policy, in the former, leverage level is constrained on each funds' performance, and in the latter, leverage is constrained mainly on market volatility.

This paper is organized as follows. Section 1 is introduction. In Section 2 we describe the model and different leverage policies. Section 3 shows the computer experiments and results, we use comparative analysis method to study the consequences that different policies cause. Section 4 concludes the paper.

2   Model and Policies

2.1   Leverage Asset Purchases Model

We build our model on the basis of a leverage asset purchases model proposed by Thurner, Farmer, Geanakoplos (2009). In this model, agents consist of two types of traders including noise traders and hedge funds and commercial banks, investors. Hedge funds can borrow from the commercial banks to buy. Commercial banks compute a leverage according to the funds transaction data to limit their borrow amount. When a funds wealth goes below a threshold, it has to get out of the market for waiting. The banks have no capital limit and will not default. Investors value the funds' performance to determine investing or withdrawing money at every timestep.

In their model, Noise traders buy and sell assets randomly. The hedge funds are value investors, their demands are depended on a mispricing signal

$$m(t) = V - p(t) \tag{1}$$

In the equation (1), V is the perceived fundamental value, which is held constant as 1. $p(t)$ is the asset price. Hedge fund i computes its demand $D_i(t)$ based on the mispricing at time t. As the mispricing increases the dollar value of the fund's position increases linearly until it reaches the maximum leverage, at which point it is capped. The hedge funds' demand can be written as following :

$$
\begin{array}{ll}
m < 0 & : \quad D_i = 0 \\
0 < m < m_{crit} : & D_ip = \beta_i mW_i \\
m \geq m_{crit} & : D_ip = \lambda_i^{Max} W_i
\end{array}
\tag{2}
$$

In the equation (2), β_i is the aggressiveness of the hedge fund i. m_{crit} is λ_i^{Max}/β_i, this is the critical mispricing that can limit the leverage. λ_i^{Max} is the leverage ratio that the banks based on to provide loans. If the price decreases, the fund may have to sell assets even though the mispricing is high.

And we compute the risky assets ratio of hedge fund i, $\lambda_i(t-1)$ in the same way in [3]:

$$
\lambda_i(t) = \frac{D_i(t) \cdot p(t)}{W_i(t)} = \frac{D_i(t) \cdot p(t)}{D_i(t) \cdot p(t) + C_i(t)}
\tag{3}
$$

The risky assets ratio can not exceed the leverage ratio λ_h^{Max}, Otherwise, funds have to sell assets for repayment.

This model is a simplified framework for the real financial market, it ignores many elements, such as the banks' economic behavior, banks capital limit, investors profit model. But it can be used to focus on studying banks' local strategies without other influencing factors.

2.2   Leverage Regulatory Regimes

A fund can be valued from its profit and risk. Accordingly, a bank can set a leverage for a fund from these two aspects. In this paper, we explore different types of leverage policies as below.

Risk-Based Policies. A risk-based policy assesses a hedge fund through its market risk. In this paper, we consider three different Risk-based policies.

Regime I: This regime monitors the volatility of asset price to set the leverage which is negative correlation with the volatility. Asset price volatility can be described as variance of the price within a time window. As the price determined in the market is same to all the hedge funds, the leverage is homogeneous to all the funds. Regime I has two parameters k and τ which we will discuss in following analysis. In the equation (4), σ_τ^2 reveals the market risk, it is the price volatility within the observation period of time steps τ . λ is a variable ranges from 1 to 16. Regime I is a baseline model which we quote from the model of Thurner, Farmer, Geanakoplos (2009).

$$
\lambda_i^{Max}(t) = \max[1, \frac{\lambda}{1 + k\sigma_\tau^2}]
\tag{4}
$$

Regime II: This regime considers variation of market risk, the risky asset ratio and the leverage at one previous time step. RegimeII is heterogeneous as the risk assets ratios of different funds are different. It is also a short term policy, in another word, it is a point to point policy.

$$\lambda_i^{Max}(t) = \max\{1, \lambda_i^{Max}(t-1) + \theta_{t-1} * \max\{0, [\lambda_i^{Max}(t-1) - \lambda_i(t-1)]\}\} \quad (5)$$

In equation (5), the variation of market risk can be computed as the gradient of σ_τ^2 , which is

$$\theta_{t-1} = \frac{-\sigma_{t-1}^2 + \sigma_{t-2}^2}{\sigma_{t-2}^2} \quad (6)$$

Regime III: This policy is similar to Regime II, the difference is that it computes a arithmetic average of the risk gradient. So this policy is a long term policy, and also a heterogeneous one.

$$\lambda_i^{Max}(t) = \max\{1, \lambda_i^{Max}(t-1) + \overline{\theta}_{t-1} * \max\{0, [\lambda_i^{Max}(t-1) - \lambda_i(t-1)]\}\} \quad (7)$$

In the equation(7), the risk parameter is

$$\overline{\theta}_{t-1} = average(\theta_1, \theta_2 \ldots \theta_{t-1}) \quad (8)$$

Incentive-Based Policy. An incentive-based policy assesses a hedge fund based on its ability of making money. In this paper, we consider two different incentive-based policies: long term and short term. Since each hedge fund has different profit situations, incentive-based policies are all heterogeneous policies.

Regime IV: This policy determine the current leverage according to the risky asset ratio, the leverage level and rate of return at the previous time step. Regime IV is a short term policy.

$$\lambda_i^{Max}(t) = \max\{1, \lambda_i^{Max}(t-1) + r_i(t-1) * \max\{0, [\lambda_i^{Max}(t-1) - \lambda_i(t-1)]\}\} \quad (9)$$

In the equation(9), the yield rate is computed as

$$r_i(t) = \frac{D_i(t-1)(p(t) - p(t-1))}{W_i(t-1)} \quad (10)$$

Regime V: We get Regime V on the basis of Regime IV. The difference is that leverage adjustment is based on a fund's yield curve over a long period. In our model, we assign the time window as $t-1$ at the time step t, so this policy is a long term one.

$$\lambda_i^{Max}(t) = \max\{1, \lambda_i^{Max}(t-1) + \overline{r_i(t-1)} * \max\{0, [\lambda_i^{Max}(t-1) - \lambda_i(t-1)]\}\} \quad (11)$$

In the equation (11), $\bar{r}_i^{NAV}$ is the geometric average of the yield rate over the time window. It is

$$\bar{r}_i(t) = \sqrt[n]{\prod_{t=1}^{n} (1 + r_i(t))} - 1 \tag{12}$$

3 Simulation Experiments and Result Analysis

In our simulations, there are 10 hedge funds, each one with an initial wealth W_0. A fund has to get out of the market when its wealth goes down to a level, then after a while (100 time periods), a new fund with initial wealth will enter the market. For hedge fund i, its aggressiveness is $5i$. We set the parameters as follows, the other parameters that we not list here are the same with the model in [3].

Parameters of market:

- The amount of assets: $N = 1000$;
- Perceived fundamental value: $V = 1$;

Parameters of hedge fund i:

- Initial wealth(Cash) of i: $W_i(0) = C_i(0) = 2$;
- Initial demand of i: $D_i(0) = 0$;
- Aggressiveness of i: $\beta_i = 5 \cdot i$;
- Bankruptcy level of i: $W_i(t)/10$;
- The waiting time that need to return to the market: $T_wait = 100$;

Parameters that we vary to discuss:

- Initial leverage: range from 1 to 16 for each regime;
- Volatility monitoring parameter k of Regime I: 0, 1, 10, 100;
- Volatility monitoring parameter τ of Regime I:10,100;

As mentioned before, our analytical framework of leverage policies in three aspects: homogenous and heterogeneous, long term and short term, incentive-based and risk-based. Under each policy, we simulate the model 5 times for each initial leverage limit ranging from 1 to 16, then get the average as the results. For the whole market, we analyze the total wealth, bankruptcy rate, fat tails and clustered volatility. Furthermore, we examine the effects of bank's local leverage regulation policies from banks' perspective, including the mortgage amount and bad debts of banks.

3.1 Default

In Regime I, there are two volatility monitoring parameters. We firstly discuss how the two parameters affect the funds' default. Fig. 1 illustrates correlation between these parameters and funds' default rate. In Fig. 1(a), $k = 0$ corresponds to constant maximum leverage. The default rate under Regime I has no great differences over different k. When the initial leverage exceeds 8, the default rate amplifies along with the increasing of k. In our paper, we set $k = 100$, under which the default rate is less with initial leverage below 8 and greater with initial leverage above 8. Fig. 1(b) measures the default rate over initial leverage under different τ. It shows that the default rate decreases as τ increases. We conjecture that this phenomenon shows that long term policy can reduce the default ratio. τ is the time window of computing the variance of asset price, when τ increases, computational cost increases. We assign τ to a neutral value 10.

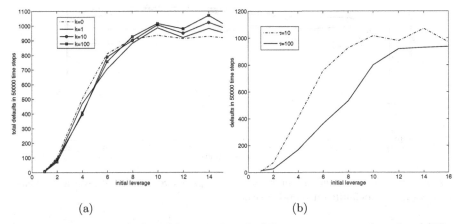

(a)   (b)

Fig. 1. Discuss the defaults of Regime I with different parameters k and τ.(a)The effects of k with different initial leverage limit(in 50000 time steps).(b)The effects of τ.

Fig. 2 shows the defaults of the 5 regimes and the situation with no leverage regulation. From this figure, the policy of no leverage (dotted line) has the most defaults in most instances. Regime I (solid line) takes the second place, i.e. the homogeneous leverage policy faces more bankruptcies than heterogeneous ones, and the odds are very large from the figure. Comparing the risk-based policies with the incentive-based policies(with the same time window), risk-based policies have less defaults. Moreover, the long term policy has less defaults than the short term policy in both risk-based policies and the incentive-based policies.

This result shows that leverage regulation is necessary and the funds' performance should be involved when determining leverage. And the long term policies and risk-based policies are more effective to reduce defaults. But the long term policies may amplify the computational cost and regulation cost, policy makers should consider the costs in the real market when they implement regulation regimes.

3.2   Wealth in the Market

The purpose of trading is to gain wealth. A fund borrows money to maintain its long position. Wealth is a symbol of market activity. We compute the total wealth of all funds within 50000 time steps. Fig. 3 shows the total wealth of funds under 5 regimes. The homogeneous policy (Regime I) which has the most

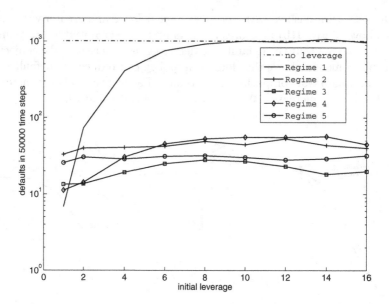

Fig. 2. Comparisons of defaults in 50000 time steps between different leverage policies. We measure the defaults on semi-log scale.

default rate has the least total wealth. Under incentive-based policies(Regime IV and Regime V) and risk-based policies (Regime II and Regime III), the former have higher amount of total wealth. By comparison of two risk-based policies, the long term one has more total wealth,so are the incentive-based policies.

3.3   Fat Tails

In this section , we draw the probability distribution of logarithmic price returns $p(r|m > 0, r(t) = logp(t) - logp(t - 1))$, and compare it with the price return of noise traders that can be treated as normal distribution. We only consider the situation of $m > 0$, as when $m < 0$, funds sell all the assets and be not active. Fig. 4(a) illustrates the probability distribution under risk-based policies, Regime I has obvious fat tails on the negative side, Regime II and Regime III have inconspicuous fat tails. Fig. 4(b)show the probability distribution of incentive-based polices, the policies have fat tails with respect to the situation that there are only noise traders.

To compare our 5 regimes, we plot the cumulative distribution $P(r > R|m > 0)$ of r. The fat tail is more obvious as the P is larger. In Fig. 4(c), the incentive-based policies have more fat tails than the risk-based ones. And the long term policies have no much difference with the short term ones. Regime I has the most fat tail.

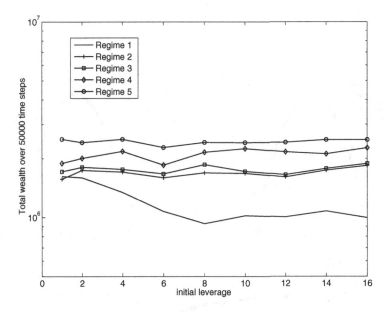

Fig. 3. Market wealth of 5 regimes over different initial leverage. Vertical axis is the total wealth of 10 funds within 50000 time steps. The total wealth use semi-log scale.

3.4 Clustered Volatility

Volatility means the dispersion degree in time series, it reflects the uncertainty of asset price, and can be used to measure risk and yield rate. Volatility has great impact on the financial market and macro-economy, it is significant to measure the volatility accurately. Mandlbrot(1963)[11]described volatility clustering as "large changes tend to be followed by large changes, of either sign, and small changes tend to be followed by small changes". This fact has a quantitative manifestation: while returns themselves are uncorrelated, absolute returns or their squares display a positive, significant and slowly decaying autocorrelation function.

Fig. 5 compares the conditional standard deviations between Regime V and the situation that only have noise traders. The distribution of noise traders is nearly a straight line while the Regime V has obvious fluctuation of the standard deviations. We draw a conclusion that the price under Regime V has clustered volatility.

Fig. 6 displays the autocorrelation coefficient of return under our 5 leverage regulation regimes. We can conclude from the figure that: (a).The incentive-based policies have larger autocorrelation coefficient, i.e.has more obvious clustered volatility. (b).The clustered volatility of short term policies is obvious. (c)Regime I which is homogeneous has the most obvious clustered volatility.

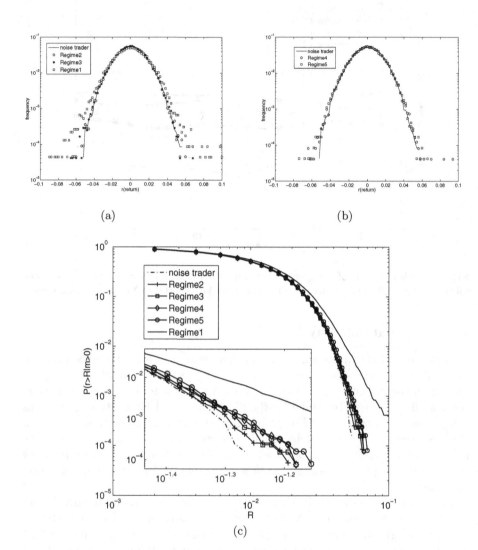

Fig. 4. The distribution of logarithmic price returns. (a)plots the probability distribution of risk-based policies.(b)plots the probability distribution of incentive-based policies. In(a)and(b), we can see fat tails at the negtive side, the vertical axis is $p(r|m > 0)$, uses semi-log scale. (c)illustrate the cumulative distribution of r under of regimes and noise trader, the vertical axis is $P(r > R|m > 0)$, (c)is log-log scale.

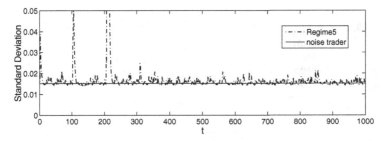

Fig. 5. The Conditional Standard Deviations of noise trader and Regime V

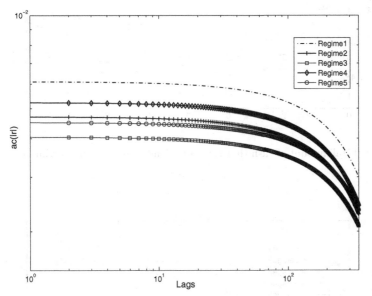

Fig. 6. Autocorrelation coefficient of the absolute values of log-returns of 5 regimes,the lag varies from 0 to 1000

3.5 Bank's Behaviour

In our paper, the function of bank is to provide cash for hedge funds,the banks have no profits. When a fund default, it causes bad debt of relevant bank. In the real market, commercial banks are profits seekers, so a rational bank will certainly consider the default risk when it provides lending and sets funds' leverage limit.

Fig. 7 illustrates the lending amount of banks under 5 different regimes. Regime I has the most loan amount, and banks with incentive-based leverage policies provide more lending than the ones with risk-based polices. Under the incentive-based policies, the long term one has larger lending amount. But under the risk-based policies, short term policy brings larger lending amount.

In Fig. 8, we show unit loss of banks over the initial leverage. The homogeneous policy(Regime I) has the largest rate of bad accounts. While other regimes have little bad accounts under our simulations.

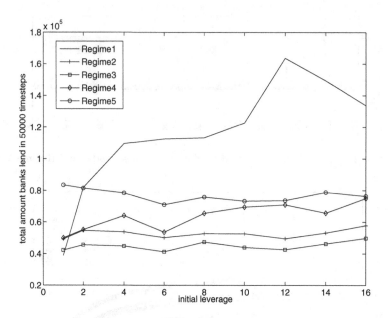

Fig. 7. The total amount of cash that the banks lend to hedge funds within 50000 time steps

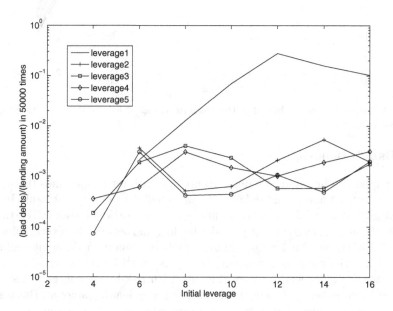

Fig. 8. The Unit lost of cash that the banks lend to hedge funds within 50000 time steps

4 Conclusions

Our paper proposed a framework to examine the effects of banks' leverage regulation regimes based on the model in [3] . We discuss 5 different leverage regulation policies. Through computer simulations, some findings are as follows.

Firstly, leverage regulation is necessary. As in our experiments, no regulation causes multiple more defaults than the risk-based polices and incentive-based polices. In fact, the financial crisis in recent years are induced by the excessive leverage.

Secondly, when banks implementing leverage regulation, it is useful to set the leverage based on the funds performance, as the homogeneous policy could cause worse market situations including more defaults, larger fat tails and more obvious clustered volatility, less market flexibility, i.e., less wealth and less lending amount, and the larger union loss of the banks meanwhile.

Thirdly, the incentive-based policy in this paper is pro-cyclical which brings about more defaults, obvious fat tails and obvious clustered volatility that reduces the stability of the market. But at the same time, the market has larger wealth and lending amount which means higher market flexibility. The two types of polices cut both ways, it may need to consider the purpose of regulations when set leverage in real markets.

Long term polices may have less defaults, larger wealth, smaller fat tails and less obvious clustered volatility that means the long term polices can have stable market and high market flexibility. But Long term polices can also bring about larger cost including computational costs and data acquisition costs. In real life, computational cost matters little with the advanced technologies, but data acquisition matters much. Therefore leverage policies involve many factors including the market conditions, capital of banks, monetary policy, costs and so on.

The main limitations of this paper is the lack of banks behavior, the activities of banks in the model is providing money for funds, and setting leverage level. And there is no capital limit and monetary policy, the funds do not pay interest on the loan. In future work, we extend our framework to simulate banksbehaviors and features and monetary policy. The banks will be a financial entity, and it can provide loan, make profits and have interbank leading.

Acknowledgements. The work is supported by the Ministry of Education of China, Humanities and Social Sciences Fund Project No. 11YJCZH148.

References

1. Simkovic, M.: Secret Liens and the Financial Crisis of 2008. American Bankruptcy Law Journal 83, 253 (2009)
2. Fostel, A., Geanakoplos, J.: Leverage Cycles and the Anxious Economy. American Economic Review 2008 98(4), 1211–1244 (2008)
3. Thurner, S., Farmer, J.D., Geanakoplos, J.: Leverage Causes Fat Tails and Clustered Volatility. SFI Working Paper 09-08-031 (2009)

4. Feldman, T.: Portfolio Manager Behavior and Global Financial Crises. Journal of Economic Behavior & Organization 75(2), 192–202 (2010)
5. Friedman, D., Abraham, R.: Bubbles and crashes: Gradient dynamics in financial-markets. Journal of Economic Dynamics and Control 33(4), 922–937 (2009)
6. Peters, O.: Optimal Leverage from non-Ergodicity. SFI Working Paper, 09-02-004 (2009)
7. Hodas, N., Tagliabue, J., Schmidt, M., Barofsky, J.: The Effect of Leverage on Financial Markets. Market Simulation Working Paper (2009)
8. Feldman, T.: Leverage regulation: An agent-based simulation. Journal of Economics and Business 63, 431–440 (2011)
9. Christensen, I., Meh, C., Moran, K.: Bank Leverage Regulation and Macroeconomic Dynamics. Bank of Canada Working Paper 2011-32 (2011)
10. Raberto, M., Teglio, A., Cincotti, S.: Macroprudential policies in an agent-based artificial economy. In: Workhshop on New Advances in Agent-based Modeling, Paris, France, June 19-20 (2012)
11. Mandelbrot, B.: The variation of certain speculative prices. Journal of Business 36, 394–419 (1963)

Author Index